Darwin's Mentor

JOHN STEVENS HENSLOW, 1796–1861

John Stevens Henslow is known for his formative influence on Charles Darwin, who described their meeting as the one circumstance 'which influenced my career more than any other'. As Professor of Botany at Cambridge University, Henslow was Darwin's teacher and eventual life-long friend, but what of the man himself? In this new biography, much previously unpublished material has been carefully sifted and selected to produce a rounded picture of a remarkable and unusually likable academic. The time in 1829–31 when Darwin 'walked with Henslow' in and around Cambridge was followed directly by Darwin's voyage around the world. The gradually changing relationship between teacher and pupil over the course of time is revealed through their correspondence, illuminating a remarkable friendship which persisted, in spite of Darwin's eventual atheism and Henslow's never-failing liberal Christian belief, to the end of Henslow's life.

John Stevens Henslow: The young Professor of Botany. (Artist unknown.)

Darwin's Mentor

JOHN STEVENS HENSLOW
1796–1861

S.M. WALTERS & E.A. STOW

CAMBRIDGE
UNIVERSITY PRESS

CAMBRIDGE UNIVERSITY PRESS
Cambridge, New York, Melbourne, Madrid, Cape Town, Singapore, São Paulo, Delhi

Cambridge University Press
The Edinburgh Building, Cambridge CB2 8RU, UK

Published in the United States of America by Cambridge University Press, New York

www.cambridge.org
Information on this title: www.cambridge.org/9780521117999

First published 2001
This digitally printed version 2009

A catalogue record for this publication is available from the British Library

Library of Congress Cataloguing in Publication data

Walters, Max.
  Darwin's mentor ; John Stevens Henslow / S.M. Walters & E.A. Stow.
    p.   cm.
  Includes bibliographical references (p.).
  ISBN 0 521 59146 5
    1. Henslow, J.S. (John Stevens), 1796–1861.   2. Botanists–England–Biography.   3.
  Church of England–Clergy–Biography.   4. Clergy–England–Biography.   5. Hitcham
  (England)–Biography.   I. Stow, Anne, 1931–   II. Title.

QK31.H35 W35   2001   580´.92
[B]

ISBN 978-0-521-59146-1 hardback
ISBN 978-0-521-11799-9 paperback

Additional resources for this publication at www.cambridge.org/9780521117999

With respect to a biography of Henslow, I cannot help feeling rather doubtful, on the principle that a biography could not do him justice. His letters were generally written in a hurry & I fear he did not keep any Journal or Diary. If there were any vivid materials to describe his life as Parish-priest, & manner of managing the poor it would be very good.

I am never very sanguine on literary projects. I cannot help fearing his Life might turn out flat. There can hardly be marked incidents to describe. — I sincerely hope that I take a wrong & gloomy view; but I cannot help fearing. I would rather see no life than one that would interest very few. It will be a pleasure & duty in me to consider what I recollect; but at present I can think of scarcely anything. The equability & perfection of Henslows whole character, I shd think would make it very difficult for anyone to pourtray him.

Charles Darwin to Joseph Hooker, 24 May 1861

To Lorna

Contents

Illustrations

*These plates are available in colour as a download from www.cambridge.org/9780521117999

Acknowledgements

In addition to thanking the Syndics of Cambridge University Library, the Board of Trustees of the Royal Botanic Gardens, Kew, the Master and Fellows of Trinity College, Cambridge, Suffolk Record Office, Bath Royal Literary and Scientific Institution, and Ipswich Borough Council Museums and Galleries for permission to quote from manuscript and archival material in their possession, we want to thank the following owners and copyright holders for permission to reproduce their material.

R. Gray: Fig. 1; K. Harrison: Fig. 2; P. Akerheim: Figs. 3, 6 and 41; University of Cambridge, Department of Earth Sciences: Fig. 4; Cambridge Philosophical Society: Plate 4, Figs. 5 and 22; M. Henslow: Plates 1, 2, 7, 8, 9 and 10, Fig. 9; English Heritage: Plate 6; University of Cambridge, Department of Plant Sciences: Plate 13, Figs. 10, 12, 15, 21 and 40; Bath Royal Literary and Scientific Institution: Fig. 13; Syndics of the Cambridge University Library: Plate 6, Figs. 14, 18, 20, 24, 25, 26, 27, 31 and 46; Board of Trustees of the Royal Botanic Gardens, Kew: Figs 28, 38 and 47; Cambridge University Botanic Garden: Plate 14, Figs. 29 and 49; M.G. Walters: Fig. 30; S.J. Roles: Fig. 32; Oliver Rackham: Plate 12, Fig. 33; Colchester & Ipswich Museum Service: Frontispiece, Fig. 34; Suffolk Record Office, Ipswich: Fig. 35, and Bury St Edmunds: Figs. 42 and 44; Hitcham Parochial Church Council: Fig 36; J.G. Duckett: Fig. 37; A. Barnard: Fig. 39; U.S. National Archives, Group 373: Fig. 45; Whipple Science Museum: Plate 3.

Foreword
by Professor Patrick Bateson,
Provost of King's College, Cambridge

John Henslow had a remarkable effect on British scientific life. He was opposed to the ineffectual traditional methods of teaching Natural Science. An anti-traditionalist, he established a tradition of rational enquiry which persists to this day and permeates the teaching and research of his old University here at Cambridge. Henslow was an inspirational teacher and formed strong bonds with his pupils. One of his favourites, Charles Darwin, became a friend for life. Darwin repeatedly referred to the strong effect that Henslow had on him. Meeting Henslow, said Darwin, was the one circumstance 'which influenced my career more than any other'. That was quite a tribute from a man who did so much through his writings to influence the careers of a host of biologists ever since.

As part of his campaign to vitalise the scientific life of Cambridge, Henslow was co-founder of the Cambridge Philosophical Society. The Society continues to this day to hold regular meetings. These are not about philosophy in the modern sense but about all aspects of natural science. Given its own origins, it was only right that the Society should have given Max Walters and Anne Stow a grant to help them carry out research for this book about Henslow.

Although John Henslow is regarded as a botanist, just as Charles Darwin is usually regarded as a zoologist, they both had a grand view of natural science which was lost in the growth of specialised approaches that occurred in the following 150 years. Information has been gathered at accelerating speed about genes, cells, organisms and communities. The language spoken in any one of these many fiefdoms is simply not understood by those working in the others. Fortunately, the walls erected between these territories by the sheer pace of discovery are starting to crumble. This is because unifying principles of biology have started to emerge and they apply to both plants and animals. Molecular biology has startled everyone by showing how alike are organisms of all kinds in their biochemical and genetic machinery: animal genes can be transferred to plants and vice-versa. We have the astonishing picture of tobacco plants glowing in the dark when they are desiccated as a result of the introduction of a gene from a firefly. While some people find these developments frightening, I find them exhilarating. It has become sensible to refer to myself simply as a biologist. Something of both Henslow's and Dawin's big picture has returned and it is a testament to their long-lasting influence that it has done so.

Another monument to Henslow's influence on those around him is the wonderful Botanic Garden in Cambridge. The old Botanic Garden at Cambridge was close to the centre and was little more than a 'physic' garden, like the one at Chelsea, although Henslow and Darwin paced around it together many times so it cannot have been all that small. Henslow persuaded Cambridge University to buy a much more extensive tract of land about a mile from the centre of the City so that a botanic garden of real service to teaching and the scientific community could be laid out. Henslow's conception is now a mature and remarkable collection of varied botanical habitats.

When my wife and I lived nearby and were just starting a family, I used to walk round and round the Cambridge Botanic Garden on Sundays in an attempt to get my elder daughter to fall asleep in her push chair. But she was as fascinated as I was by the sheer richness and variety of the place. We never tired of it. Max Walters lived in the middle of the garden at the time – one of the perks of being Director of the Botanic Garden. He and his wife Lorna kept a donkey which was a zoological deviation for my daughter and myself in this extraordinary collection of botanical treasures. Max presided over this wonderful garden for many years and, just because it owes so much to Henslow, he has repaid the debt by writing this excellent book.

Patrick Bateson

Preface by S.M.W.

Almost exactly twenty years ago, whilst I was Director of the University Botanic Garden in Cambridge, much of my spare time in between teaching and administration was going into preparing a little book entitled *The Shaping of Cambridge Botany*, with an explanatory sub-title 'A short history of whole-plant botany in Cambridge from the time of Ray into the present century'. Although I greatly enjoyed the challenge of putting on record how much I appreciated what great botanists of earlier generations had done, I also discovered the limitations of my own tastes and abilities. In particular, I came to understand how important the rôles of librarian and archivist are, and indeed how the work I was attempting was only possible because such excellent people existed. Nothing in my own scientific training had supplied any expertise of this sort. Indeed, until the age of forty I found all history dull and irrelevant – an attitude I now find quite incomprehensible!

A particular part of the story presented in *The Shaping of Cambridge Botany* concerned the vision of the fourth of our Professors of Botany, John Stevens Henslow, as to the kind of Botanic Garden he thought worthy of a great scientific University, and how, with some difficulty, this vision was translated into reality in a series of developments which began with Henslow persuading the University in 1831 to buy the land on which our excellent Botanic Garden stands. The book was published as part of the celebration of the sesquicentenary of that important step, which we held in the summer of 1981. Writing the chapter on Henslow and his 'New Botanic Garden', I soon realised that a new biography of this remarkable man was much needed, and I naïvely thought that someone else would undertake this task, in which I would be happy to play perhaps some minor rôle.

Early retirement in 1983 brought opportunities to study and write, and interest in Charles Darwin, Henslow's most famous pupil, grew more and more strongly in Cambridge through the 80s, but no potential Henslow biographer appeared. Rehabilitating after a very successful heart by-pass operation in the early 1990s I began to think the unthinkable: perhaps I should write the biography myself. So it was that one day some six years ago I found myself in the office of the Cambridge Philosophical Society enquiring about some trivial matter. Judith Winton-Thomas, the Society's

Secretary, was there, and there was Anne Stow also. From their conversation I learned that Anne had just retired from her post as Librarian of the University's Science Periodicals Library. I asked Anne what she was going to turn her talents to in retirement and received the entirely reasonable reply that she hadn't decided. I then said, half-jokingly, 'Would you like to help me to write a biography of Henslow?' She said Yes, she thought she would. This book is a witness to the success of our cooperation.

Preface by E.A.S.

When Max Walters suggested that I could help him with a new biography of Henslow I was familiar with the well-known Maguire portrait hanging in the Henslow Room at the Scientific Periodicals Library. My stock explanation to visitors 'He was one of the three founders of the Cambridge Philosophical Society, responsible for the move of the Botanic Garden from the New Museums Site to Trumpington Road, and encouraged Darwin to accept the post of naturalist on the Beagle' had no depth of knowledge. But I had always been intrigued by the kindly, intelligent face in the print. This collaboration has enabled me to 'walk' with Henslow from his childhood home in Rochester, through Cambridge to my native county of Suffolk. Although no collection of his papers exists I have been able to read many of his letters in other collections and these more than anything have revealed the man, his devotion to his family, his humour, his enthusiasm for new interests, his attention to the minutest detail and his inevitable frustration at the lack of time to encompass all he wanted to do. It has been a fascinating journey.

Joint Preface and Acknowledgements

It is often said that, just as 'good wine needs no bush', introductory explanations and apologies from the author(s) can be dispensed with, and a
book must speak for itself. We have felt unmoved by such considerations, if
only because it seemed important to put on record the nature of our collaboration, which has been satisfying to both authors and will, we hope, prove
so to the reader.

Foremost among the list of individuals and organisations to which we are
indebted must come the Cambridge Philosophical Society. As S.M.W.
records in his Preface, the idea of our collaboration actually arose in the
Society's offices, and both of us had already our own areas of interest in
Henslow: E.A.S. as Librarian of a prestigious and historic Library that was
in origin that of the Philosophical Society of which Henslow was a founder-
member, and S.M.W. as Director of 'Henslow's Botanic Garden'. Our
application to the Society for a grant to help in the preparation of the book
was generously accepted at a meeting on 16 May 1994 at which our cause
was strongly supported by the then President, Professor Patrick Bateson,
who has kindly supplied the Foreword to our book.

Armed with this support, we approached Cambridge University Press as
potential publishers, and were rewarded with a contract in 1996. Initially
only black and white illustrations were envisaged, but with the further help
of a grant from the Botanic Garden, it has been possible to include some
coloured plates. We are most grateful to the present Director, Professor
John Parker, himself a Henslow enthusiast, for making this extra grant.

When we began we had little idea of the size of the relevant 'new' or
unpublished material, though, of course, some of it, especially the Darwin
correspondence, was already more or less familiar and accessible. It gradually became apparent that much more material was available than we at
first thought. A chance conversation with Neil and Anna Hitchin in 1996
after Sunday morning service in Grantchester Church introduced us via
the Hitchins to the American Henslow(e) descendants, in particular Paul
Akerheim and Peter Henslowe, from whom we have received valuable
information and the use of family portraits. From Mrs Maria Henslow,
widow of Major John Henslow, who was present with her husband at the

1981 Botanic Garden celebrations, we have also had access to unpublished manuscript notebooks and diaries belonging to the family. Through the good offices of Professor Michael Akam, Director of the Museum of Zoology, we were introduced to Canon Anthony Barnard, of Lichfield, who has provided us with much information about the links between the Henslows and the Barnards. Similarly we owe much to Mr Roger Vaughan, who has made available to us his great knowledge of Henslow's brother-in-law and biographer Leonard Jenyns, and to Mr Roger Jenyns of Bottisham Hall, who has kindly given us access to material in the Jenyns house.

Many other people have helped us with particular aspects of our work. In the alphabetically arranged list that follows we have briefly indicated the kind of help provided.

David Allen has provided much helpful information from his expert knowledge of the history of natural history.

David Briggs has read and commented helpfully on the drafts of some chapters.

Janet Browne has helped with questions arising from her book entitled *Charles Darwin: Voyaging*, and kindly made available material from the unpublished second volume of her Darwin biography.

Alec Bull has helped us with the Henslow lists of Hitcham plants.

Betty Bury has freely given us information on Henslow from her work on editing Romilly's Diaries.

Nigel Cooper has helped with the theological background to Henslow's life.

Gigi Crompton, who is preparing a *Historical Flora of Cambridgeshire*, has exchanged with us information on Henslow's records in Cambridgeshire.

Brian Dolan has answered questions arising from his PhD thesis concerning E.D. Clarke.

Katherine Edgar has discussed her study of E.D. Clarke.

Rosemary and David Gardiner introduced us to Thomas Underwood and the relevant correspondence containing Henslow references.

Keith Goodway, whose early interest in the history of Cambridge botany enabled us to find 'lost' documents in the Library of the Department of Plant Sciences in Cambridge.

Scilla Hall has given us archival information from the Libraries of Corpus Christi and Peterhouse Colleges in Cambridge.

Rachel and Robin Hamilton, who live by Hitcham Church, introduced us to Hitcham residents interested in Henslow and parish
history.

Michael Hickey, botanical artist, whose career began in 'Henslow's
Garden', has contributed the illustrations beginning the three
Parts of our book, and also told us of Youmans' books.

David Hunt has provided expert advice on the nomenclature of
'Henslow's Cactus'.

Anne James, who typed very promptly and efficiently the first drafts
of much of the text.

David Kohn, who is assessing new manuscript evidences of
Darwin's early interest in botany, has discussed his work with
us.

Elizabeth Leedham-Green has been very helpful on matters of
Cambridge University history.

Duncan McKie has supplied information on Clarke's Oriental Plane
tree in Jesus College.

Clodomiro Marticorena has answered questions about Darwin's
seeds from Chile grown by Henslow.

Edward Martin has helped with Henslow's archaeological interests.

Gina Murrell has helped with many questions about Henslow specimens in the Cambridge University Herbarium.

John Pickles, co-editor with Betty Bury, has helped with Romilly
Diaries references.

Duncan Porter, Director of the Darwin Correspondence Project in
the Cambridge University Library, has freely helped with Darwin
references.

Richard Preece has helped with Henslow's studies of Snails and
other Mollusca.

Chris Preston has helped with Henslow's bryophyte records.

Oliver Rackham, expert on woodland ecology, shared with us his
knowledge on a field trip to the site of Hitcham Great Wood.

Fritz Rehbock, author of *The Philosophical Naturalists*, discussed
with us Henslow and his contemporaries.

Michael Roberts has helped us with the relationship between
science and theology in the thought of Henslow and his contemporaries.

Jim Secord has shared with us his knowledge of Lyell in particular.

Richard Stow has transferred e-mail images into usable portraits.

Hugh Torrens was most helpful in preventing our ignorance of Sedgwick and the early geological controversies from being transferred to the book.

David Turner, Churchwarden of Hitcham and local historian, has given us much valuable help with Henslow's parish history.

Richard West, 12th Professor of Botany in Cambridge (retired), has provided the information we needed on 'celts' and the Hoxne remains studied by Henslow in his later years.

We are also grateful to the staff of the following institutions, who have almost without exception been helpful in our enquiries:

Archives of the Royal Botanic Gardens, Kew: especially Lesley Price, then Archivist.

Ipswich Museum, especially Stephen Plunkett and Martin Sanford.

The following Record Offices: Cambridgeshire, at Cambridge and Huntingdon; Kent, at Rochester and Maidstone; and Suffolk at Bury St Edmunds and Ipswich.

The Natural History Museum and the National Portrait Gallery, London.

The Departmental Libraries in the Science Departments of Cambridge University, especially the Botanic Garden, Plant Sciences, Zoology, Archaeology and Geology.

We owe a special debt to the staff at the Cambridge University Library, especially Archives, Manuscripts, Map Room, Peridodicals, Photography and the Scientific Periodicals Library.

Two people deserve our very special thanks. Caroline Burkitt who has taken the often messy, handwritten text produced by S.M.W. and turned it, through many drafts, into an impeccable typescript; and S.M.W.'s wife Lorna, who has provided with cheerful equanimity innumerable helpings of tea and home-made cakes, and will have compiled when our book is published the sort of index that all good books need.

It is inevitable that we shall find too late that we have omitted to thank some people who have significantly helped in preparing the book. We can only ask to be forgiven, and assure them that we have appreciated their help.

A final word of thanks to the staff at Cambridge University Press, in particular Maria Murphy, Stephanie Thelwell and Jane Bulleid, who have been tolerant and positive even through delays and postponements. We hope they finally feel it was worth the effort.

E.A. Stow

S.M. Walters

30.9.2000

Part I
Origins

1 Family background
growing up in Kent and London

This is a book about an admirable man whose qualities have been over-shadowed, even distorted, by the reputation of his most famous pupil, Charles Darwin. Both Henslow and Darwin came from richly-endowed families whose local wealth and esteem had their roots in the rise of industrial England and the increasing power of the British Empire. There are, however, significant differences between the two stories. John Stevens Henslow was the eldest of a large family of eleven children, born at the very end of the eighteenth century into a stable, happy home in Rochester, Kent. Charles Darwin, thirteen years younger, was the fifth child of Robert Darwin, a rich Shrewsbury physician, whose wife Susanna was a daughter of Josiah Wedgwood, founder of the famous Wedgwood chinaware business. The picture we have of Darwin's child-hood is of a somewhat withdrawn little boy, apparently of a placid temper-ament, with little contact with his ailing mother, who died when Charles was only eight years old, and rather afraid of his large, stern and frugal father.[1] No such complications are present in our account of John Henslow's childhood, which seems to have been uniformly smooth and happy, as we shall see.

The Henslow family probably originated in Devon, but moved to Burhunt, now Boarhunt, a village in Hampshire north of Portsmouth, in the six-teenth century. Another branch of the family, which moved to Sussex, included Philip Henslowe, an Elizabethan theatrical manager who held offices at Court, and left a manuscript diary (1591–1604) which is a rich storehouse of facts about contemporary productions of Shakespearean and other plays. The spelling of the surname, with or without a final 'e' throughout the sixteenth and seventeenth centuries, seems to have remained uncertain until well into the eighteenth.[2] Ralph Henslowe of Burhunt was granted a coat of arms which was confirmed by his son Thomas in 1591,[3] and another Thomas helped to conceal Charles II in September 1651. From then on the line can be traced from contemporary wills, with the name Thomas persisting through the generations until we reach a Thomas Henslow who married in 1686, and had two sons, the elder Thomas and the younger John. It was this John Henslow who became a master carpenter in the dockyard at Woolwich, marrying Mary David in 1719, and receiving from his father-in-law the family Bible which passed down in the family.[4] The master carpenter had seven children, of whom the seventh, John (1730–1815) became Sir John Henslow(e), the grandfather of John Stevens Henslow (Plate 1).

Sir John was Chief Surveyor to the Navy, a post he held jointly or solely for 23 years from 1784 to 1806. He was responsible for the design of 156 naval vessels in a period when British sea power was increasingly asserting itself. A clever draughtsman whose talent developed during his apprenticeship at the age of 15 to Sir Thomas Slade in the naval dockyard, Sir John carved out his career as naval architect, with promotion to Master Boat Builder at Woolwich in 1762, then to posts in Chatham and Portsmouth naval dock-yards, leading to his appointment as Assistant Surveyor of the Navy in 1771. After the death of his first wife, Frances Hooper, niece of Sir Thomas Slade, in 1764, he married Ann Prentis, daughter of Edward and Damaris Prentis of Maidstone in 1766. Ann had been christened in 1739 in Maidstone Presbyterian Church, which can be taken as evidence that her parents were Presbyterians. Ann bore him two daughters before their first son, John Prentis, was born in 1770. Sir John was knighted in 1793, widowed in 1803, and retired in 1806 to Sittingbourne in Kent, where he died in 1815. His name is perpetuated in the Solomon Islands where there is to this day a Cape Henslow on Guadalcanal. We should not, however, assume from this that he ever undertook extensive sea voyages: it was the tradition in the Navy to name significant geographical features after eminent naval men, whether or not they were present on the occasion of the first charting of 'unknown' seas.[5]

An obituary declares Sir John to have been 'scrupulously just, active, per-severing . . . a good husband and father and warm and constant friend'.[6] Since John Stevens Henslow was born in 1796, and therefore was his eldest grandchild, it seems certain that as a child and a youth he would have known his eminent grandfather, and more than likely that he would have been influenced by him. It has not, however, been possible to trace any direct evidence of such a relationship, and we can only speculate; but we might note that Jenyns himself (Henslow's brother-in-law and author of the *Memoir*, 1862) felt that some of Sir John's 'ingenuity & skill in design-ing' had obviously passed to the grandson, in 'whom they were equally conspicuous'.[7]

John Prentis Henslow, after what must have been a very comfortable childhood shared with his two elder sisters, Ann and Frances, and his younger brother, Edward, was early launched into a career as a solicitor in Enfield 'in consequence of the death of his Principal by a fall from his horse'.[8] He was soon, however, beguiled by an attractive offer to enter into partnership with his uncle Walter Prentis, a wine-merchant in Rochester. There he was fortunate in meeting Frances, the eldest daughter of a rich brewer, Thomas Stevens, and married her in 1795 when she was just 20. His father-in-law took him into partnership in the brewery business, and John and Frances set up their home in Rochester, where their eldest child, John Stevens, was duly born on 6 February 1796.[9]

There seems to be little doubt that this, their firstborn child, was especially dear to his parents. We learn that he was 'a beautiful boy, with brown curling hair, a fine straight nose, brilliant complexion, soft eyes, and a smile that reached everybody's heart'.[10] A regular succession of brothers and sisters followed him, in the way of Georgian and Victorian families, until 1814 when the youngest child, baptised Alexina Frederica, was born and lived for only three months. In all, John Stevens was the eldest of eleven children, three of whom died in infancy; he outlived all his four brothers, but four sisters outlived him, Louisa (Kirkpatrick) living until 1903.

It was obviously a relatively prosperous and happy childhood, and both physically and mentally the eldest child was apparently specially favoured. Jenyns paints a picture from which we can learn something of the early influences from each parent:

> He may have inherited some part of his taste for natural history from his mother, an accomplished woman, who, though she never studied it as a science, was . . . a great admirer as well as a collector of natural and artificial curiosities. His father, too, was extremely fond of birds and other animals, as well as of his garden. At one period of his life his father had an extensive aviary, comprising a great variety of species, some of which are not often seen in cages in private houses. His library also contained a good many books on Natural History. This was quite enough to create an interest for such things in the child, while the taste thus excited was, as might be supposed, duly encouraged by his parents.[11]

Since both his parents indulged in various aspects of collecting natural objects, it is not surprising that their bright, active little son should follow in their footsteps. Jenyns tells us that:

The passion for collecting was first exemplified by his bringing home the different natural objects he met with on his walks. In one instance, while yet a child in a frock, he dragged all the way home from a field a considerable distance off a large fungus, which when exhibited to the family was said to be almost as big as himself. This fungus being dried was hung up in the hall of his father's house, and often pointed out to strangers as an indication of the future botanist.[12]

Young John's schooling began when he was seven, first at a small private school in Rochester kept by Mr and Mrs Dillon who were French émigrés, and then at Rochester Free School, at which Mr Hawkins was headmaster. When he was nine years old, in March 1805, he was sent as a boarder to a school in Camberwell run by the Revd W. Jephson, where he remained until the time for entering on his University career.

The Camberwell schooling had a profound influence on the development of young Henslow's already keen interest in natural history. By a fortunate chance, the drawing-master at Jephson's school was George Samuel, who was a keen entomologist. Finding that his pupil was enthusiastic, Samuel encouraged the collection and study of insects. As Jenyns puts it: 'Young Henslow was often seen running about an orchard adjoining the school with his green gauze butterfly net; and occasionally the drawing-master and he went out to longer distances insect-hunting together.' Samuel also taught his young pupil the art of 'setting' and mounting the captured butterflies and other insects, and these collections were proudly displayed to his sister Charlotte and doting parents when he came home for the holidays.[13] It seems very likely that it was through Samuel that he was introduced to two established naturalists in London who greatly influenced his further development. The first of these was William Elford Leach, six years his senior, who was already a widely-respected zoologist and was appointed Assistant Keeper at the British Museum in 1813 when he was 23. The other was James Francis Stephens, a celebrated amateur entomologist reputed to have the best collection of British insects in the country.

We can let Jenyns tell us of the happy influence these two eminent naturalists had on their young friend:

> Naturalists of this stamp and standing were not likely to let the young collector's zeal evaporate. They at once fixed him down to the pursuits, which had been hitherto taken up as a mere boyish amusement, but which were henceforth to be made regular studies. The woods of Kent were now well

searched for insects, while the Medway was explored for shells. He was always active and busy. He had, indeed, before this, shown a desire to become acquainted with the inhabitants of the water as well as the inhabitants of the air. His father was in the habit of making yearly picnic excursions with his family up the Medway, and the boy was delighted with the opportunity thus afforded him of fishing for all he could get, while the family were enjoying themselves in a very different way. He has been heard to say that these were the happiest days of his life, though he generally got well scolded by his parents for spoiling his clothes. The fruits of his industry, however, now acquired more value from his superior knowledge of the science. Among the acquisitions made to his collections were many interesting and little-known species, – crustacea and shells from the Medway and the adjoining salt-marshes, – lepidoptera, of which few specimens previously existed in cabinets. Some of these being shown and given to his patrons Dr. Leach and Mr. Stephens, the habitats were recorded by those gentlemen on the authority of the captor, and the specimens found a place in their respective drawers in the British Museum.[14]

Another talent developed early by Henslow was the ability to draw. No doubt in the Henslow household in Rochester drawing and painting were encouraged, though perhaps more amongst his sisters as a lady-like accomplishment; but we have evidence that he was already enjoying exercising his skill in drawing when a schoolboy (Plate 2). We shall see how important this skill became in the development of his botanical teaching career and indeed in the wider sphere of teaching science in primary schools.

In 1810, Henslow was awarded as a school prize a book which greatly fired his enthusiasm to become an explorer-zoologist in the African continent. This was *Travels in Africa*, by Levaillant, a traveller's tale of the marvellous wild animals of that 'Dark Continent' which, as the nineteenth century proceeded, yielded up its secret interior to European colonisation. Apparently, in his desire to become an African explorer and zoologist, the young Henslow was strongly backed by William Leach himself, and other friends who saw how talented and enthusiastic the boy was, but his parents were utterly opposed to such a dangerously outlandish profession. Jenyns tells us that

> his mother had many anxious moments, from the pertinacious way in
> which he clung to the idea of going out. He himself, too, often came home

depressed and out of spirits. And even long after he had given in to the wishes of his relations, he still continued to think much upon the subject, read with the greatest interest many other African travels that were published from time to time, and the volumes, procured to gratify the taste, continued to occupy a place on the shelves of his library to the day of his death.[15]

In these ways the future career of the young Henslow was shaped. In spite of parental refusal to countenance any career in African exploration for their beloved eldest child, it was a generally fortunate set of circumstances and influences that prepared the schoolboy for further education and academic success in Cambridge. He duly arrived to begin his undergraduate studies at St John's College in October 1814, in his eighteenth year.

Part II

Cambridge

2   The young Henslow at Cambridge

Henslow entered St John's College in October 1814. The reasons why St John's was chosen are not immediately clear: neither his father nor grandfather had preceded him there, nor is there any other obvious family link. The likeliest reason is that William Jephson, Headmaster of Wilson's Grammar School, Camberwell, where Henslow was a boarder from 1805 to 1814, recommended St John's to his bright, hardworking young pupil because he himself had been educated there from 1792 to 1796. Moreover, William's father, Thomas, from whom he had taken over the position of Headmaster of the school, had himself also been educated there. By an odd coincidence, another Jephson, Thomas, who was the younger brother of William and a Fellow of St John's during Henslow's time there, re-appears in our story as a contender for the Chair of Mineralogy.

That St John's was a good choice for an intelligent and enthusiastic student seems, however, to be quite evident from Henslow's academic progress in his undergraduate years. In several ways, the College had advantages. It was large and relatively wealthy, second only to Trinity in size and endowments – indeed by the end of the eighteenth century Trinity and St John's between them accounted for half of all Cambridge undergraduates[1] – and it was therefore able to employ tutors and assistant tutors to ensure that the students were taught properly, so that they were regularly placed in the competitive Senate House examinations. As Garland, referring to the first decades of the nineteenth century, explains:

> At St John's & Trinity . . . there were large numbers of excellent students and capable fellows. The diversity of talents among the senior members made flexible, specialised teaching arrangements possible. Their huge endowments meant that the colleges could afford to improve their educational arrangements, paying tutors and fellows enough to guarantee their presence within college walls. A history of success in the University degree competition gave Trinity and St John's a sense of confidence, so that they felt able to experiment with their teaching methods without fear of destroying their reputations for excellence.[2]

Indeed, in one respect St John's was a pioneer in the field of University reform. As early as 1765, the College had instituted a regular system of twice-yearly public examinations, on the basis of which the progress of all its undergraduates could be assessed and College prizes awarded – a practice eventually followed by Trinity (though not until 1790) and, later, other colleges.[3]

Henslow had rooms in First Court,[4] and made such good progress in his first year that his name heads the list of freshmen prize-winners in the year 1815. These prizes are stated to be awarded to 'those who are in the first class at two general examinations, one about the 16th of December, the other about the 6th of June', so that we can safely assume that Henslow's first year as 'Pensioner' at St John's was academically in the first class. His progress in the second year was equally meritorious, and he was awarded a scholarship for his diligence; and in his third and final year, his name appears second on the list of seven prizemen from among the 'Senior Sophs.' (final-year undergraduates) awarded in June 1817.[5]

Jenyns, in his biography, comments on the academic ability of the young Johnian as follows:

> Though devoted to Natural History, he did not allow himself to be drawn away from mathematical studies, which at that day were the prevailing studies of the place (especially at the college to which he joined himself), to the exclusion of many subjects which now [1862] form part of the academical course. His powers of reasoning & clear faculties well suited him for studies which necessitated much application of the mind and close thought. Nor do mathematics, which are needed for the successful cultivation of some of the sciences in which he took interest, appear to have been at all distasteful to him.[6]

A feature of Cambridge education in Henslow's undergraduate days was the rise in popularity and importance of the various lecture courses given by some of the science Professors. Although none of these courses provided material on which the Senate House examinations were based, they were available and relatively inexpensive and, at least in St John's and Trinity, students with an interest in the natural sciences were not discouraged from attending such courses. Henslow, in fact, seems to have benefited a good deal from two sets of lectures, those of Cumming, Professor of Chemistry, and Clarke, Professor of Mineralogy.

James Cumming's chemistry lectures were presumably attended by Henslow in his first or second year, and Jenyns tells us that when he himself attended the same lecture course in 1820, Henslow was acting as Cumming's assistant in the lecture-room.[7] The lectures given by Edward Daniel Clarke, Professor of Mineralogy, which Henslow also attended, must however have had a much greater impact, not because of any practical knowledge of the subject the young Henslow might have gained, but from the nature of the unique performances that Clarke provided. The first Professor of Mineralogy, whom Henslow was to succeed in that chair, was a very remarkable, colourful character who had already become nationally and even internationally famous before Henslow's time. He deserves closer inspection.

Edward Daniel Clarke (1769–1822) was descended from a line of literary ancestors: his great-grandfather was William Wotton, an accomplished linguist and the author of many antiquarian and classical studies in the early part of the eighteenth century, and said to be the first Englishman to have preached a sermon in Welsh! His grandfather, William Clarke (1696–1771), became a fellow of St John's College, Cambridge, at the age of 20, and went on to an exemplary career in the Church, during which he made a considerable reputation as an antiquarian scholar. His father Edward (1730–86), who inherited much of his own father's literary skill and enthusiasms, followed him to a fellowship in St John's, and made some contribution to the rapidly-developing market for travel books after a period as Chaplain to the British ambassador in Madrid when he published *Letters concerning the Spanish Nation* in 1762.

With such a pedigree, the career of Edward Daniel Clarke, Edward's second son, is not too difficult to understand. We are told he was 'a most amusing and attractive child who showed early enthusiasm for collecting "stones, weeds and other natural productions" of the countryside & open enjoyment of adventure', as when he was lost for a day and found eventually 'in a rocky & remote valley above a mile from his father's house [in Buxted, Sussex], surrounded by a group of gypsies, and deeply intent upon a story which one of them was relating'. Edward did not do well with his formal schooling, neglecting his own studies and at times irritating his schoolfellows with 'ingenious & good-natured tricks'. It is not, therefore, too surprising that he failed to follow his elder brother, father and grandfather at St John's, but nevertheless entered Jesus College through the good offices of the Master, Dr Beadon, as Chapel Clerk at the age of 16. When his father died, later in the same year (1786), he was appointed to a Rustat Scholarship, reserved for the orphans of clergy.

His undergraduate career was not distinguished, and he just scraped through the Senate House examinations in 1790, being placed third in the list of Junior Optimes. As his friend and biographer William Otter tells us:

> In this irregular & careless manner, undistinguished as an academic in his own College, and altogether unknown as such to the University at large, was formed and educated almost to the age of twenty-one, a man who in his maturer years was numbered both at home and abroad amongst the most celebrated of its members.[8]

What was Edward Clarke to do, a young graduate with a profound distaste for the Church as his obvious profession, and a total unwillingness to undertake even the minimum study required of him to enter 'holy orders'? The answer was provided for him: within a few months of his graduation, Dr Beadon, now Bishop of Gloucester, recommended him to the Duke of Dorset as a fit person to tutor his nephew, the Honourable Henry Tufton. So began a career into which Clarke grew and expanded his extraordinary and diverse talents, culminating in a three-year Grand Tour with another rich young pupil, John Marten Cripps; they began in Scandinavia, then came south through Russia, taking in Jerusalem, Egypt and Greece, and returned to England via Paris in 1802 during the brief lull in hostilities after the Peace of Amiens.

Clarke was a tough but sensitive observer of diverse human societies, with considerable practical ability and an insatiable drive to collect and record everything, from fossils and rocks, through plants and animals to classical statues and archaeological remains. His young pupil soon turned out to be even tougher and more devoted to 'collector's pieces': apparently Clarke himself sent back to Cambridge 76 crates of 'plunder', but was easily beaten by Cripps, who provided another 107 crates! Since Cripps was paying for the whole expedition, including Clarke's collections, his greater zeal may, of course, simply reflect his greater financial resources. Tutor and pupil returned to England amid general acclaim, and in The Times of 8 December 1802 we read:

> The rare manuscripts and other valuable Books brought to this country by Mr Cripps, of Sussex, and the Rev Mr Clarke, are principally intended to

enrich the library of Jesus College, Cambridge. – The whole collection made by these gentlemen is contained in 183 cases illustrating the Natural and Moral History of the various people they visited, in a journey from the 69th degree of North latitude to the territories of Circassia and the shores of the Nile. The Botanic part contains the Herbary of the celebrated Pallas, enriched by the contributions of Linnaeus and his numerous literary friends. – With the minerals are several new substances, and the rarest productions of the Siberian mines. Among the Antiquities are various Inscriptions and Bas Reliefs, relative to observations made in the plain of Troy . . . The manuscripts are in Hebrew, Coptic, Arabic, Abyssinian, Persian, Turkish, and the language of Thibet Tartary; and in the Greek and Latin languages are several manuscripts of the Classics, of the Gospels, and the writings of the earliest Fathers of the Church. In addition to these, the collection contains Greek vases, Gems, Sculptures, and many remarkable Egyptian monuments from the ruins of the city of Sais discovered by these travellers in the Delta, after the evacuation of Egypt by the French. Also numerous original drawings, Maps, Charts, Plans, Models and Seeds of many rare and useful Plants, the Habits, Utensils, Idols of the inhabitants of the Aleutian Isles, brought by Billings to Russia, after his expedition to the countries lying between Kamschatka, and the N W Coast of America, with many geological observations; the publication of which was so long withheld by order of the Russian Government.

By 1807, seven years before Henslow came up to Cambridge, Clarke was established as a remarkable University figure with a considerable reputation throughout England and abroad, and had persuaded the University authorities to allow him to give an official course of lectures on Mineralogy, for which he prepared and had printed a Syllabus.[9]

Clarke's own account of the success of the first lecture held on 17 February 1807, taken from a letter to Cripps, written on the following day, speaks for itself:

> I have only time to say, I never came off with such flying colours in my life. I quitted my papers and spoke extempore. There was not room for them all to sit. Above two hundred persons were in the room. I worked myself into a passion with the subject, and so all my terror vanished. I wish you could have seen the table covered with beautiful models for the Lecture.[10]

By the end of the following year, having delivered a second course of lectures, Clarke (with powerful friends having the ear of the Vice-Chancellor and Heads of Houses) was the recipient of a newly-created Chair of Mineralogy.

This Lecture Course, begun with such éclat in 1807, and becoming the annual Lent Term course of the new Professor of Mineralogy, took place in Professor Thomas Martyn's lecture room in the Botanical Museum in the Walkerian Botanic Garden (see Chapter 5). By good fortune we have a sketch of the room made by one of Clarke's students about the year 1813. We can reasonably assume that the layout was much the same when Henslow took the lecture course. The walls of the room were hung with Clarke's own oil-paintings and the wide front bench behind which the lecturer stood would be covered with appropriate demonstration material. A former student remembers Clarke's performances as follows (he is comparing them with lectures on machinery given by Professor William Farish):[11]

> Far more amusing, but much less beneficial, were the lectures upon Mineralogy of the great traveller, Professor Clarke. His delivery was a master-piece of didactic eloquence. Even the very commands to his attendants were, some how or other, squeezed into the sentence so as to produce no interruption to the flow of his discourse. His manner was ease and even elegance, although occasionally he would burst forth into the most energetic pathos and sublimity ... Comedy as well as Tragedy was in requisition, and he would continually relate some anecdote of persons or things he had met with both at home and abroad. From every stone, as he handled it and described its qualities – from the diamond, through a world of crystals, quartz, lime-stones, granites, &c. down to the common pebble which the boys would pelt with in the streets, would spring some pieces of pleasantry.[12]

Although we have no direct evidence, there can be little doubt that Henslow would have appreciated Clarke's inspired use of his own oil-paintings with which he enriched the thrilling accounts of his travels – for example, the eruption of Vesuvius, for which he had both wall-paintings and a working model![13] He would have remembered how valuable these were, so that an integral part of his own botanical lecture course was the use of large coloured wall-charts prepared by the lecturer himself. Moreover, the idea

FIGURE I Clarke's *Platanus orientalis* in Jesus College Fellows' Garden.

of the demonstration, involving as appropriate both observation and experiment, though not unique to Clarke amongst the Science Professors, was developed *par excellence* by the showman–lecturer on Mineralogy. Again, this was an important part of Henslow's later teaching. On the other hand, Henslow had no wild, extrovert streak in his nature, and it seems very likely that he found succeeding Clarke as Professor of Mineralogy a very difficult act to follow (see Chapter 5).

Before we leave Clarke and his talents, we should note that one remarkable survivor of his botanical collections can still be admired in his College: in the Fellows' Garden of Jesus College is one of the finest specimens in Cambridge of the Oriental Plane, *Platanus orientalis*, raised from seed collected by Clarke in 1802 on the classical site of Thermopylae. The existence of this beautiful tree, now approaching its bicentenary, reminds us that, in the days before Henslow's New Botanic Garden was laid out from 1846, College gardens were the obvious place for Cambridge men to find a home for any special tree seedlings or saplings they may have acquired[14] (Figure 1).

Clarke died in 1822, not before he had played a significant part in the formation of the Cambridge Philosophical Society (see Chapter 3). The last two years of his life were beset with illness: his wife and children were very seriously ill with typhus fever in autumn 1821, and his efforts to bring them through, though successful, probably contributed to his own decline and death.

Towards the end of his undergraduate days, Henslow was proposed as a Fellow of the Linnean Society of London by his entomologist friends Leach and Stephens, and duly elected on 9 February 1818. The original proposal, dated 30 August 1817, signed by Alexander Macleay, entomologist and Secretary of the Society, and by Leach and Stephens, reads as follows:

> John Francis [sic] Henslow Esq., St. John's College Cambridge, a gentleman well versed in Entomology and a zealous student of other branches of Zoology, being desirous to become a fellow of the Linnean Society of London. We whose names are hereunto subscribed, do from our personal knowledge recommend him as highly deserving of that honour being perfectly satisfied that he will prove an useful and valuable member.

It is odd that there should be such a mistake in Henslow's second name in what is the official filed document recording the election. It was, presumably, corrected in later Linnean membership lists. No mention is made of any interest in botany – further evidence, if it were needed, that the subject that was to become his specialist interest occupied very little of his time as a student in Cambridge.[15]

The Senate House examination, which Henslow sat in January 1818, was a formidable endurance test that lasted several days. We do not have Henslow's own account of his ordeal, but from surviving documents and other contemporary accounts it is possible to construct the picture. A preliminary stage, which can be seen as a vestigial survival of the medieval disputations, consisted of a formal verbal contest between pairs of disputants, the pairs being arranged by the proctors and moderators (examiners), taking as a basis the College tutors' reports on the ability of each candidate. The three propositions to be defended or attacked were chosen from mathematics (Newton's *Principia* or some branch of Newtonian physics) or moral philosophy, and all the candidates were divided on the basis of their performance at their disputation into a number of provisional classes. After this preliminary test by disputation, which survived in modified form until finally discontinued in 1839, the whole body of students proceeded to the Senate House, where they were arranged and seated in order of their provisional class. The moderators then read out questions in mathematics and moral philosophy, to be answered in writing by the students. By the time Henslow took his degree, this annual Senate House examination took several days to complete. The weaker candidates, the so-called 'poll men' who were aiming for only an 'ordinary degree', were eliminated first and released, and the examiners finally produced a list of honours graduates in continuous order of merit and divided into three classes: Wranglers, Senior Optimes and Junior Optimes. Printed question papers were only introduced in the 1820s: Henslow must have received the earlier examination in its full rigour, whilst Charles Darwin would experience a written examination more nearly resembling any modern test.[16]

Such an examination must have been a test of qualities of stamina and endurance: the Senate House, like most of the main University buildings, was unheated, and the ordeal took place in January! Henslow, we are led to believe by Jenyns, was unusually tough and hardy as a young man, and these qualities, together with an ability to learn and an enthusiasm for all intellectual pursuits, must explain his success in appearing in the Wrangler list. The Senior Wrangler of the year was, as so often happened, an outstandingly able Trinity man, J.G. Lefevre, but the second place, as Junior Wrangler, went to John Hind, a fellow-Johnian. Out of a list of 28 Wranglers for 1818, no fewer than nine are Trinity men, whilst St John's achieved only three, namely Hind, Henslow (at 16th place), and Beech (at 20th). Not a vintage Johnian year, in spite of the fact that one of the two moderators for the examination was a Johnian, Fearon Fallows, College Lecturer in Rhetoric![17]

Amongst Henslow's Johnian friends when he was an undergraduate was William Porter, son of Thomas Porter of Rockbeare House, an estate near Exeter. According to family records Henslow and William's sister, Christina, became deeply attached but 'inequality of means . . . prevented the union, and for a time at least there was much unhappiness on either side'.[18] There is independent evidence of Henslow visiting Devon in 1817 in the form of 'Henslow's Swimming Crab', *Polybius henslowi*, the type specimen of which was sent by Henslow to Leach from the north coast of Devon. (For detail of this, see Appendix 4, p. 285, and Figure 2). Christina went on

FIGURE 2 Henslow's crab,
Polybius henslowi.

to marry Thomas Barnard. The story was not to end there. Years later, when her son, Robert Cary Barnard, decided to study at Cambridge, she encouraged him to meet Henslow. He not only studied under him, and took the Natural Sciences Tripos, but met and married Henslow's daughter Anne.

Henslow did not obtain a Fellowship at St John's, or indeed at any other College although, as we shall see, his later association with Downing was a very close one.[19] By 1820 he was living outside the College in a set of rooms in All Saints' Passage as described by Jenyns, who tells us how his friendship with his future brother-in-law began. Writing in 1862, he says:

> It was, I think, in the year 1820 that I had the happiness of making my first acquaintance with Henslow. He, at that time, occupied lodgings in All-Saints Passage, since pulled down to make room for the new court at Trinity, erected by the present master, Dr Whewell [Whewell's Court] . . . Henslow was busily engaged at that period in arranging his cabinets of British insects and shells, both which collections he presented to the Cambridge Philosophical Society, the year after it was instituted, & for which he received the thanks of the Society on the 13th November, 1820.[20]

Jenyns was four years younger than Henslow: he had entered St John's College in October 1818, and it was in his second year that he discovered that the young Henslow was such a keen collector of insects and shells. They began to cooperate and to exchange specimens, and Henslow must soon have received and accepted an invitation to Bottisham Hall, only a few miles from Cambridge, to meet his family. Here he was a frequent visitor, and a friendship developed between this young academic with a promising career ahead (Figure 3), and Jenyns's sister Harriet, an accomplished young lady with a good education and interests in music, literature and painting. We follow the story of Henslow's marriage to Harriet in Chapter 4.

FIGURE 3 Henslow as a young man.

3 Henslow:
men who influenced him at Cambridge

History has decided for us who was Charles Darwin's mentor, as the title of
our book makes clear, but it is by no means so easy to decide who played the
key roles in shaping the character and career of the young Henslow
himself. This difficulty arises in part from the fact that, unlike Darwin,
nowhere does Henslow tell us how he, in later life, assessed the influence
of his teachers and colleagues. There is nothing surprising in this: for most
people it seems to be true that we see no one teacher, colleague or friend
who played the principal role in shaping our lives. We can, however, point
with some confidence to two men, both of whom achieved some eminence
in Cambridge University, whose long friendship with Henslow would
qualify them for consideration as shaping his career and his interests.

The senior of these men was Adam Sedgwick, who outlived all the others
including Henslow himself.[1] Sedgwick, born in 1785 and elected to a
fellowship in Trinity College in 1810, became Woodwardian Professor of
Geology in 1818 and held the post for fifty-five years. He was a
Yorkshireman, born in the village of Dent, in a beautiful part of the North
Country near the Westmorland border. His father was a well-loved parish
priest who had himself received a Cambridge education, and throughout
his long life Adam Sedgwick retained a great affection for Dentdale and his
northern roots. Henslow was introduced to Sedgwick by J.G. Lefevre, a
Trinity man who had graduated together with Henslow in January 1818.[2]
From this fortunate introduction blossomed a strong friendship which
was to last until Henslow's death.[3]

Adam Sedgwick gave his first course of lectures as Woodwardian
Professor of Geology in the Lent Term of 1819, a course he delivered
almost without interruption for just over half a century. Although the conditions of Woodward's will establishing the Chair required that he should
deliver only four lectures a year, he was already informing the Vice-
Chancellor in 1820 that he had given 22 public lectures in that year – and
asking that his stipend might be increased as a result! In this request he
was successful, and the stipend was increased to £200 on condition that his
annual course should contain an extra 15 lectures in addition to the statutory four.

For the next half-century

> Sedgwick's geological lectures were one of the features of Cambridge life,
> and were attended by large and enthusiastic audiences. His inspirational
> qualities, summed up by a statement made at one of his lectures 'I cannot
> promise to teach you all geology, I can only fire your imaginations' . . . must,
> if the many comments and accounts of his skill are to be accepted, make
> Sedgwick one of the greatest scientific lecturers and teachers of his day.[4]

There is evidence that Henslow attended Sedgwick's first lecture course in
the Lent Term of 1819, and we can assume was greatly impressed by it, as he
had been by Clarke's mineralogy lectures in the preceding years. It must
have been true that Sedgwick's eloquence and enthusiasm, combined with
his undoubted ability to present a logical story of geological method,
quickly confirmed Henslow's already keen interest in the subject, linked as
it was with mineralogy. This enthusiasm which lecturer and pupil shared
was greatly extended when Sedgwick invited Henslow to join his geological
tour of the Isle of Wight in the Easter Vacation immediately following the
lecture course, an offer gratefully accepted by Henslow.

A keen horseman, Sedgwick became famous in both University and Town
circles for his equestrian field excursions to sites of geological interest
around Cambridge, as attested by the following news item from the
Cambridge Chronicle of 10 April 1835:

> Geological Excursion.
> On Tuesday last, a field lecture on Geology, attended by circumstances of
> unusual animation and novelty was given by the Woodwardian Professor, to
> a class of 60 or 70 academic horsemen.

Before we continue the story of the Sedgwick–Henslow friendship, we
might note that Sedgwick's lectures were truly 'public' in that, not only
were they attended by the young gentlemen of the University, but also by
townspeople of both sexes and a wide age-range. The admission of women
to his lectures was a conscious policy on Sedgwick's part. He took a delight
in feminine company: as his biographer tells us: 'Genial as he was to all-
comers, [Sedgwick's] special pleasure was to entertain ladies and children,
whom he amused in all manner of quaint ways, and sent home with a store
of memories that never faded from their minds.'[5] Some of these young
girls were destined to be more or less willing recipients of a regular and
constant series of quite intimate letters. For example: ' . . . His particular

friend [amongst the children of Sir John Malcolm of Hyde Hall, near Sawbridgeworth, Essex] was the third daughter Kate. He won her affection in the first instance by carrying her on his back, and as she grew up he established a correspondence with her, which was carried on regularly until his death'.[6] Here is one extract from this voluminous correspondence, of particular interest, not least because it was one of many practical questions on which Henslow had sought advice from the elder Hooker before he began his regular botanical lectures in Cambridge: namely, should ladies be admitted to the lectures? Sedgwick was in no doubt, as we see in the following extract from his letter to Kate Malcolm sent from Cambridge on 7 February 1848:

> I began my lectures in the latter part of October [1847], immediately after my return from Dent. It was my 30th course, and I had a very large class. Do you know that the Cambridge daughters of Eve are like their mother, and love to pluck fruit from the tree of knowledge? They believe, in their hearts, that geologists have dealings with the spirits of the lower world; yet spite of this they came, and resolved to learn from me a little of my 'black art'. And, do you know, it is now no easy matter to find room for ladies, so monstrously do they puff themselves, out of all nature, in the mounting of their lower garments, so that they put my poor lecture-room quite in a *bustle*. Lest they should dazzle my young men, I placed them with their backs to the light, on one side of my room. And what do you think was the consequence? All my regular academic class learnt to squint, long before my course was over. If you can't understand this, come and see for yourself; and I will promise you that when you set your foot in my lecture-room, and sit down with your back to the light, you will make them all squint ten times worse than ever.[7]

How should we assess the relationship between Sedgwick and Henslow? Clark & Hughes tell us that, in their opinion, 'the characters of the two men were very similar . . . in religion, love of truth, and hatred of wrong, they were in exact agreement'.[8] Whilst we can readily accept this list of areas of agreement as their friendship developed, it is surely misleading to suggest they were very similar in character without noting some obvious and important differences. Sedgwick, who remained a bachelor,[9] comes across both through his own writings and those of his contemporaries as a gifted, extrovert Northerner who was increasingly fond of hearing his own voice, even to the extent of being thought to enjoy polemical controversy for its own sake. Both had the inestimable advantage of a happy childhood on which they could look back with affection; but in Henslow's case there is very little evidence that he felt any strong, life-time allegiance to the county of Kent and the countryside in which his early years were spent, whereas the proud, no-nonsense Northerner never lost his affection for his native land. It would be easy to illustrate this side of Sedgwick's character – Clark & Hughes provide many rich examples – but one must suffice here.

In 1837, when he was 52, he paid one of his regular return visits to his beloved Dent, and was involved in an extraordinary dedication of a foundation-stone for a 'little chapel in the wild part of my native valley' at which 'for the first time in my life [I] turned field-preacher, as I addressed about eight hundred wild people for more than an hour, having a large rock of mountain-limestone for my pulpit, & the vault of heaven for my sounding-board'.[10]

This picture of a bluff, self-satisfied Northerner is not, however, an adequate one. In Sedgwick's life, illness, real or imaginary, increasingly seems to play an important part, and the dark side of his character emerges from many of his letters to friends. Clark & Hughes say: 'It must be admitted that he allowed his ailments to occupy too large a space in his conversation and his letters; but in judging him – and in taking stock of what he accomplished during his long-life – it must be recollected that from 1813 to his death he could never count upon robust health for even a single day'.[11] This could hardly be more different from Henslow's life. Jenyns tells us that Henslow 'had naturally a very strong constitution, and in his younger days was capable of enduring great fatigue'.[12] Until 1853, in his 57th year, Henslow seems never to have suffered from anything other than minor indispositions, and certainly showed little interest in the state of his own health.

In spite of these temperamental differences between Henslow and Sedgwick, the friendship, originally of the older for the younger man, was quickly converted on the Isle of Wight field trip into a firm, life-time bond. This seems to have come about because, to a degree, Henslow and Sedgwick were both exploring and learning together. It was certainly the view of Sedgwick's biographers that neither Sedgwick himself, nor his rival for the post of Woodwardian Professor of Geology, George Gorham, knew much about the subject. They write:

... the worthlessness of [Gorham's] geological knowledge has been too
hastily assumed from Sedgwick's celebrated account of himself and his
opponent, which is still remembered in the University: 'I had but one rival,
Gorham of Queens', and he had not the slightest chance against me, for I
knew absolutely nothing of geology, whereas he knew a great deal – but it
was all wrong!' This remark, however, was not made seriously, and it would
be unjust to Gorham's memory to quote it as a deliberate judgment,
without making a large allowance for that departure from literal truth
which is permitted to a brilliant antithesis.[13]

Clark & Hughes go on to say: 'Precedents are not wanting at Cambridge for
the election of a man of ability to a Professorship in a subject of which he
knew nothing' – and of course Henslow himself illustrates this in his own
career (see Chapter 6).

Looking back from today, the reader is inclined to say that some false
modesty must have been involved in such disclaimers, but we should
remember that the group of Chairs in what was then broadly encompassed
in 'Natural Philosophy' (and which today would be the Natural Sciences)
showed little or no coordination or clear definition. Indeed, even the terms
and conditions of the Chairs were often the subject of quite vicious argument. Nor was a prior knowledge of the subject thought to be a particularly
important criterion for the selectors to weigh against other qualities such
as social standing or religious belief.

Following the initiatory field trip with Sedgwick, the young Henslow was
confident enough of his ability to learn from the rocks to plan and carry out
himself a visit to the Isle of Man in the Summer Vacation in 1819, taking
with him a few Cambridge students who were his pupils. On this visit, few
details of which unfortunately seem to have survived, he 'attended to the
tuition' of his students, and devoted his leisure time to studying the
geology of the island. The geological material he collected was used in
what was his first published paper, which appeared in 1821.[14] By means of
this paper, which appeared soon after he had been elected a Member of the
Geological Society of London,[15] he had begun to make his own reputation
as a geologist rather than a mineralogist (Figure 4). His love of zoology
continued, however, as he succeeded in collecting a number of animal
specimens some of which he sent to his old teacher Dr Leach at the British
Museum. Botanical collecting came a poor third: Jenyns tells us he collected 'a few of [the] rarer plants' of the island, and two of his Isle of Man

FIGURE 4 Fossil ammonite collected by Henslow in the Isle of Man, 1819 (Sedgwick Museum, E3900).

finds – a pondweed (*Potamogeton polygonifolius*) and an eyebright (*Euphrasia micrantha*), both constituting first records for the island, are in the Cambridge University Herbarium.[16]

During their field trip to the Isle of Wight, Sedgwick and Henslow had discussed the state of scientific studies at Cambridge, and in particular found that they shared the feeling that there was an obvious need of a place where people of the same scientific interests could meet and exchange ideas. They returned to Cambridge determined to canvass opinion and support. Henslow volunteered to speak to his contacts in London, and these must have included James Stephens and William Leach, the entomologists who had exerted such influence on him while still a schoolboy. Certainly William Swainson was to recall in a letter to Henslow in 1826: 'I recall the zeal with which you once in company with our poor friend Leach spoke upon this subject.'[17] Among Cambridge men E.D. Clarke, the Professor of Mineralogy, was an enthusiatic supporter and, indeed, was always spoken of by Sedgwick as the founder of the Society.[18]

On 30 October 1819, at Clarke's instigation, a public meeting was called by 33 signatories, including five heads of houses and six professors, and at the subsequent meeting on 2 November it was agreed to set up the Society. After a few meetings in the Botanic Garden the Society set itself up in rooms above Bulstrode's shop where Whewell's Court now stands, and regular Monday evening meetings began. In 1820 it was agreed to publish some of the papers read at meetings in a journal to be called *Transactions* and to establish a reading room which took newspapers and general journals as well as scientific ones.

A detailed account of the Society is given by John Willis Clark (1891); here we are concerned with Henslow's own contribution to the Society's early years. Henslow was only newly graduated, which is probably why he did not immediately serve as an officer or a member of council. Nevertheless he was active behind the scenes. In 1820 he was donating books and maps to the newly formed reading room and on 1 May 1820 he was thanked for his donation of his collection of British Shells and Insects. Council agreed to order cabinets to house the gift, which formed the basis of a rapidly expanding collection of geological, mineralogical and ornithological as well as entomological specimens. Leach, who became an Honorary Fellow, and Stephens also made gifts showing practical support for the project so dear to their protégé's heart. On 29 June 1821 the minutes recorded Henslow's first attendance at Council and from then on his contributions to the affairs of the Society became more obvious. In November of the same year he was elected one of the Secretaries, a position he held until he moved to Hitcham in 1839. His co-secretary was George Peacock, who was succeeded by Whewell.

Henslow's hand can regularly be detected in both General and Council minute books. In 1821 he was refereeing papers for the first volume of *Transactions* and contributed his first important paper, *The Geological Description of Anglesea*, to that journal. In 1822 Henslow and Whewell were requested to draw up regulations for the Reading Room, and they became involved in the building up of informal contacts with other societies at home and overseas which developed into the formal exchange of publications. The general meetings of the Society were held on Monday nights and for many in Cambridge that night became the 'Philosophical'. Joseph Romilly noted in his diary 'Mon 18 April 1831 . . . Philosophical where experiments of spectra wheels . . .' The General Meetings minute book regularly records contributions from Henslow, both as formal papers and also at the end of the meeting as less formal demonstrations or 'savouries'. The Minute for 18 April 1831 reads: 'April 18, 1831 After the meeting Professor Henslow exhibited a series of appearances produced by two wheels revolving one behind the other.' Romilly fails to mention the more serious papers by Whewell and Airy which preceded Henslow's demonstration. The Anniversary dinner was also an important date in the calendar and evidently a date Romilly was reluctant to miss: 'Tu 9 May 1826 Philosophical dinner at wch Cumming presided; Sedgwick &c spoke & it was the best Philosophical dinner I was at.' (See Chapter 4, p. 42, and end-note 4.12.)

In 1832 the Society moved to a purpose-built house in All Saints' Passage on land leased from St John's.[19] A detailed building specification for this house exists. It principally comprised a general lecture theatre, a reading room, a museum and a council room and domestic quarters for a custodian. It is difficult to believe that Henslow would not fully involve himself in this project, especially in the museum to which he had continued to donate specimens from the beginning. In the same year the Society received its Charter, a distinction which had been actively canvassed by Henslow.

Whilst Henslow's party were on the Isle of Man, the discovery there of the bones of a giant Irish Elk caused much local excitement, no doubt reported to Sedgwick by Henslow on his return. Sedgwick, anxious to obtain the skeleton for his Woodwardian Museum, seems to have persuaded Henslow to return in March 1820 to buy the bones from the local blacksmith on behalf of the Cambridge Museum. This proved to be a physically demanding excursion: gales in the Irish Sea gave Henslow a 'stormy passage of 30 hours from Liverpool', at the end of which, in Ramsay, on the following day he 'borrowed a horse of one man, and a saddle of another' and rode to Bishop's Court, where his friend, Dr Murray, the Bishop of Sodor and Man, resided. Alas, he failed to persuade the blacksmith to sell the Irish Elk to Cambridge, and it went eventually to the Museum at Edinburgh.[20]

Armed with the knowledge and experience gained from his two geological field trips, Henslow felt confident enough to undertake a Long Vacation trip, unaccompanied by any other geologist, to Anglesey, an island with a complicated geology which, in Jenyns' words, 'had never been satisfactorily worked out by any previous observer'. He took with him, in 1820, as in the Isle of Man the previous summer, a group of his pupils. These, it should be remembered, were not students of geology or any other branch of

Natural Science, but undergraduates, presumably 'poll men' or candidates for ordinary degrees (as opposed to the 'honours men' who were eventually classed in order of merit as Wranglers, Senior Optimes and Junior Optimes). How far his pupils were used by him as labourers in the geological field is not recorded: it would be surprising if Henslow's obvious enthusiasm did not impart itself to some extent to the group and, indeed, helping with mapping the island and acquiring a set of rocks and fossils would provide the healthy exercise that such student tours were supposed to supply. There is indirect evidence to suggest that Calvert of Jesus was on the party: on the title page of a copy of the Anglesey paper is the following inscription in Henslow's hand: 'To F. Calvert Esq. from his Fellow-Traveller & obliged Friend the Author.'

Henslow's first published paper, containing observations on Isle of Man geology, came out in 1821, and his more important paper on the work in Anglesey appeared later in the same year. This paper, nearly 100 pages long and illustrated with maps and sketches by Henslow himself (Figure 5), established a reputation for the young Cambridge scientist in a wider national scientific community, as witnessed by the fact that the volume containing the paper 'was speedily out of print, mainly it is believed, from the great demand for the memoir in question'.[21]

Henslow had prepared for the Anglesey study with characteristic thoroughness. He was evidently in touch with Thomas Richard Underwood,[22] an English painter resident in Paris, who had become, earlier in his life, much interested in field geological studies in the Isle of Anglesey. There seems little doubt that Sedgwick must have effected this introduction, for the Sedgwick archive contains several letters from Underwood in which they are discussing geological matters, and Henslow is mentioned in some of them. One letter, written in haste from Paris on 18 March 1822, apparently in reply to a letter to Underwood from Henslow, implies strongly that Henslow paid a visit to Paris about this time.[23]

Underwood (1772–1835) had travelled to the Continent with his friend Thomas Wedgwood during the Peace of Amiens in March 1803, but was trapped in Paris when war was declared, and interned there. His friend escaped and returned to England, but Underwood spent the rest of his life in France, and after the battle of Waterloo in 1815 travelled freely back to England on visits. He was a friend of the geologists Brongniart, father and son, and seems to have become very enthusiastic and knowledgeable himself about geology. He greatly facilitated visits to Paris by many English scientists, and introduced them in Parisian scientific circles.

Both Whewell and the Oxford Professor of Geology Buckland were guests of Underwood in Paris in the years 1822–23, as evidenced by a long letter to Sedgwick dated 20 January 1823, which begins: 'I was on the point of writing to you when Mr. Whewell brought me your letter.' Part of this letter concerns Professor Cordier's examination of rock samples collected by Sedgwick in a variety of English localities; Henslow used the same Parisian expert to give an opinion on a number of his Anglesey rock samples and acknowledges his help in the published paper. We further learn from the same letter to Sedgwick that Underwood knows and values Henslow's Anglesey paper: he writes: 'I beg you will do me the favour of examining the specimens from No 405 to 421 of Professor [sic] Henslow's Anglesea paper and ascertain whether there is a trilobite among them if there is let me know the locality.' Underwood was obviously exchanging geological specimens and information with both Henslow and Sedgwick, and Henslow's numbered rock specimens from Anglesey are still preserved today in the University Museum in Cambridge which now bears Sedgwick's name. Sedgwick recorded, in his Report dated 1 May 1822 to the auditors of the Woodwardian Museum that he now curated:

> 'The Museum has also received a very valuable accession in a collection presented by Mr Henslow, which consists of nearly 1000 specimens carefully selected during a geological survey of the Isle of Anglesea, and illustrated by a memoir and sections which will be published in the next number of the Cambridge Transactions. Mr Henslow has undertaken the arrangement of this collection, which occupies twenty-four drawers.'[24]

There is another very important area in which Sedgwick's ideas must have greatly influenced Henslow, namely the early geological controversy between the Huttonian and Wernerian schools, evident in Edinburgh at the turn of the century and later to be labelled 'catastrophism' versus 'uniformitarianism'. What the young Henslow's views were on the problem of reconciling the geological and biblical evidence via 'catastrophist' interpretations of sedimentary rocks is difficult to assess, but, as we shall see in Chapter 9, his paper *On the Deluge*, published in 1823, is by no means a literal, miraculous speculation on Noah's Flood. It seems that both Sedgwick and Henslow gradually accommodated their views to a long

Transactions of the Cambridge Phil. Soc. Vol. 1 Pl. 16.

FIGURE 5 Geological illustrations by Henslow from the paper on Anglesey (*Trans. Camb. Phil. Soc.* vol. 1.).

EXPLANATION OF THE PLATES.

———◆———

PLATE XV

Contortions in the strata of the quartz rock at Holyhead mountain. Sketched from the South Stack .... p. 363.

PLATE XVI.

Fig. 1. Cleavages exhibited by the strata of the quartz rock ... p. 364.

Fig. 2. Vertical section of a mass of breccia (*a*), and a quartzose vein (*b*) connected with it, which rises through the chlorite schist, near its junction with the quartz rock .... p. 366.

Fig. 3. Junction of the quartz rock (*a*), and chlorite schist (*b*) to the West of Rhoscolyn .... p. 366.

Fig. 4. Section of the stratified chlorite schist .... p. 371.

Fig. 5. Serpentine (*a*) rising abruptly through the chlorite schist (*b*), which dips in various directions .... p. 376.

Fig. 6. Massive serpentine (*a*) gradually assuming a schistose character (*b*) .... p. 376.

Fig. 7. Appearance presented by the greywacké slate on the shore near Monachdy .... p. 383.

> (*a*) Hard, green, and unlaminated portion, passing gradually on one side to a schistose black slate (*b*), and terminated abruptly against a similar rock on the other.

Fig. 8. Arrangement of particles in the stratified grit at Bodorgan, p. 395.

FIGURE 5 (*cont.*)

geological time-scale, though neither of them were prepared to accept the arguments of Charles Lyell, the first volume of whose *Principles of Geology* was published in 1830. Lyell's book had the most profound effect upon the young Darwin, and could with complete justification be said to have laid the foundation for Darwin's career and eventual fame.

Charles Lyell was an almost exact contemporary of Henslow, born in November 1797 at Kinnordy House, in Forfarshire, the son of a rich Scottish landowner who was a cryptogamic botanist and a Dante scholar. The family moved to Hampshire when Charles was a baby, and the boy was reared 'in a cultured environment typical of the moderate Tory gentry'. His later education was at Exeter College, Oxford, where he attended the geological lectures given by William Buckland, an experience curiously parallel to that of Henslow in Cambridge, for Clarke and Sedgwick in Cambridge and Buckland in Oxford were extrovert, powerful lecturers with a great influence upon susceptible young pupils. Buckland held what were increasingly seen to be extreme 'catastrophist' views which made all geology subservient to a more or less literal interpretation of Genesis: in his *Reliquiae Diluvianae* (1823) he maintained that all superficial geological deposits were the result of Noah's Flood. Lyell was gradually able to free himself from his mentor's 'catastrophist' views, but continued to acknowledge his debt, for example in dedicating the second edition of the first volume of the *Principles* in 1832 to Buckland: 'who had first instructed me in the elements of geology, and by whose energy and talents the cultivation of science in the country has been so eminently promoted'.[25]

When Henslow recommended Charles Darwin to read Lyell's 'new book' – the first volume of the *Principles* – it was for the practical, factual information it contained, and Henslow explicitly warns his pupil that Lyell's book is 'altogether wild as far as theory goes'. This theory was the uniformitarian view of the history of the earth, which stated that no 'forces of nature' need be assumed to have operated over geological time which scientists could not see and measure operating at the present day. Such a view opposed all 'catastrophist' interpretations, whether of the naïve Genesis-based 'Noah's Flood' type, or of more modified and attenuated versions such as Sedgwick moved on to hold in later life.

Lyell and Henslow remained on good terms throughout life, although there is no evidence that they kept in touch to anything like the extent that 'the Cambridge group' did. This must have been at least in part because Henslow's taste was never for theoretical and philosophical questions, nor

had he developed the wide cultural and artistic interests that were second
nature to Lyell. This difference is admirably and sympathetically assessed
by their mutual friend Sir Charles Bunbury, in a letter to his brother:

> *February 2nd, 1857. Mildenhall.* We have had Charles and Mary Lyell and
> Joanna Horner with us. Also a visit from Professor and Mrs. Henslow and
> their youngest daughter. I had not seen Henslow (I think) since the summer
> of 1854. He is a little older in looks, but as lively, cheerful, and active as ever.
> His activity and versatility of mind are indeed wonderful; and living as he
> does, in an obscure out-of-the-way country parish, his attention is awake
> to everything that is going on. He has a powerful as well as extremely active
> mind, (deficient however in the imaginative or poetical element), and if
> he had devoted it in earnest to science, he might no doubt have made great
> advances; yet I think on the whole his talents are more essentially practical
> than scientific. Certainly he has now devoted his time and his talents so
> absolutely to the improvement of his parish, and to teaching there and at
> Ipswich, that the pursuit of pure science has fallen quite into the back-
> ground with him. He is one of those men of great ability (and there are not
> a few such in our time) who devote themselves rather to the spread than the
> advancement of science.
>
> Lyell, on the other hand, though zealous in the cause of education and of
> general improvement, is pre-eminently a man of science, and to my think-
> ing, a true philosopher. It is delightful to see his eagerness about every new
> scientific discovery and every subject of scientific research; and this not
> merely in relation to his own especial pursuit of geology, but to all branches
> of natural history. I had many interesting and scientific talks with him
> during this visit, and I have put down in another book some notes of what
> I have learned from him in this way; but no mere notes of facts or opinions
> can do justice to the amount of scientific and philosophical instruction that
> I gain (or ought to gain) from his conversation.[26]

The relationship between Henslow and William Whewell, that other
influential member of the small group of reforming scientists[27] in the university, is much more difficult to assess than the Henslow–Sedgwick relationship. Whewell, like Sedgwick, was a Trinity man of Northern English
origin; but he was only two years older than Henslow, and one feels that, in
their relationship, there was not the warmth that existed between
Sedgwick and his former pupil. Whewell was a polymath who rose rapidly
in the stimulating academic atmosphere of Trinity College to become

Master of the College in 1841. He differed from Sedgwick in many ways, not least that his output of published books and papers was unbelievably large and, as Becher explains: 'Whewell [in 1840] was far better known to the world outside Cambridge than any other Fellow of Trinity, with the possible exception of Adam Sedgwick, and as he had already published treatises upon mechanics, architecture, mineralogy, astronomy and a *History of the Inductive Sciences* in three thick octavo volumes, he was supposed to be omniscient or, at least, by way of becoming so.'[28]

Whewell in fact became Professor of Mineralogy in 1828 when, after the protracted Richardson enquiry into the Cambridge Professorships, Henslow formally resigned the Chair – though he continued to lecture in Mineralogy for two years after he obtained the Chair of Botany in 1825. In spite of this link, there is little evidence that Henslow shared with Whewell any of the practical pursuits in mineralogy or geology that he obviously enjoyed with Sedgwick. Whewell's interest in mineralogy seems to have been largely mathematical and theoretical. His first paper on the subject was read to the newly-formed Cambridge Philosophical Society in 1821, and he followed this with two papers to the same society in 1823 on aspects of coordinate geometry, and then a more general paper applicable to crystal structure, based on his own studies and those of his student John Lubbock, to the Royal Society in London. This paper, which laid out a mathematical foundation for crystallography, was praised by Humphrey Davy, then President of the Royal Society, as 'an admirable application of mathematical to physical science'. Further, Whewell studied at Berlin and Freiburg, and 'converted to the continental natural classification method from the artificial system then in vogue [in mineralogy] in England'.[29]

Henslow was, of course, quite competent in mathematics, but throughout his life excelled as a practical man interested in detail rather than a philosopher interested in the underlying ideas of his subject. In view of this deep difference between Henslow and Whewell, we can reasonably conclude that there was no very close bond between them, although, as we shall see, their friendship and correspondence continued throughout Henslow's life. It is, however, significant that Jenyns, in his Memoir, makes no mention of Whewell, and apparently did not consult him when he was eliciting suitable appreciations of Henslow's worth from colleagues – a list of contacts that included both Darwin and Sedgwick.

There is, however, another way in which both Sedgwick and Whewell must have been important influences on the young Henslow, although we get little direct acknowledgment of this in Henslow's surviving writings: their strong liberal Christian faith which they both expressed on many occasions and in many publications with which Henslow was obviously familiar. We try to assess this kind of influence in Chapter 9.

4 Harriet

Harriet was the youngest daughter of the Revd George Leonard Jenyns (1763–1848) and his wife Mary (1763–1832). George Jenyns had inherited the property of Bottisham Hall, some five miles north-east of Cambridge, in 1787; he was a landowner of some standing in the County, a Magistrate and a Prebendary of Ely Cathedral, Chairman of the Bedford Level Corporation, and Chairman of the Board of Agriculture in London. The present Bottisham Hall, still today in the hands of the Jenyns family, was built in 1797, the year Harriet was born. An earlier house, the foundations of which are near the present one, was pulled down. The surrounding park covers about 140 acres and includes an avenue and plantations.[1] The Jenyns, like many landed gentry of the period, had a London house, in Connaught Place, from which Harriet was married to Henslow in St James' Church, Paddington, on 16 December 1823.

Harriet's mother, Mary Jenyns, née Heberden, was the daughter of William Heberden senior and his second wife Mary Wollaston, who was herself a grand-daughter of William Wollaston (1660–1724), daughter of Francis Wollaston and sister to the Revd Francis Wollaston (1731–1815), a Fellow of the Royal Society with a great interest in astronomy. The latter's sons, cousins to Mary Heberden, included Francis John Hyde Wollaston, Jacksonian Professor from 1792 to 1813, and William Hyde Wollaston, 'more especially known by his researches in Chemistry & Optics, & a great friend of Sir Humphrey Davy'.[2]

On both sides, therefore, of her mother's family, Harriet had impressively able, well-educated men and women; we learn, for example, from her brother Leonard that William Hyde Wollaston 'was often at my father's house' and must have been very familiar to Harriet as to Leonard and the other children. The younger William Heberden (1767–1845), Harriet's uncle, must also have been a regular visitor. Both Heberdens were, coincidentally, Fellows of St John's College, and the senior gave, for some ten years from 1749, a series of lectures on Materia Medica which were important in the history of the teaching of medicine and botany in Cambridge.[3]

There seems little doubt that all the Jenyns sisters, including Harriet, received a good education in the arts and languages in a general intellectual atmosphere that was encouraged by their mother, whose contacts with the world of London aristocracy via her father and brother, the latter physician to the King and Queen, were clearly very close. Not long before her marriage to George Jenyns, Mary Heberden had her portrait painted by Gainsborough.[4] As the youngest daughter in a family of eight, Harriet would have been free from the primary responsibility to help with the household that fell upon the unmarried eldest daughter, Mary, born in 1790. Two of the eight children died young, Soame (the eldest child) and Emily; the other sister, Charlotte, died in her twenty-first year. Harriet grew up with an elder sister and three brothers, George, Charles and Leonard, in that order of seniority. It was, of course, through Leonard that Harriet met her future husband.

We have, as is so often the case with the wives of men who achieve some fame, remarkably little direct evidence about their own character and talents. Leonard Jenyns himself is not prepared to tell us anything about Harriet in the Memoir, deeming it inappropriate to say more than the laconic opening sentence of Chapter 3: 'On the 16th of December, 1823, the year after he was appointed to the Professorship of Mineralogy, Professor Henslow laid down his bachelor life, and married Harriet, second daughter of the Rev. George Leonard Jenyns, of Bottisham Hall, in the County of Cambridge, and sister of the writer of this memoir.' Throughout the whole of the 266-page Memoir, Jenyns makes no other reference to his sister, and indeed only by an oblique reference in the final chapter on his brother-in-law's death does he even tell us that Harriet had died in the autumn of 1857: 'He was buried on Wednesday the 22nd of May in Hitcham Churchyard, in the same vault in which his wife had been interred in the autumn of 1857.'[5]

We can, however, glean a little more from other published sources. Thus, Jenyns himself tells that, with regard to: 'French – I had the advantage; having been taught by my sisters at home'. So far as painting was concerned – an accomplishment generally considered to be particularly suitable for the female members of a gentleman's household – Jenyns informs us that all his brothers and sisters were artists, Mary in particular being 'a most successful miniature painter; while both my brothers were quite first-rate artists . . .'[6]

Given the dearth of information on Harriet from Jenyns himself, it is fortunate that we can fill in the picture somewhat from surviving letters. A letter from Henslow to Jenyns, dated '8 April 1823' and evidently from

Cambridge, contains, in addition to *minutiae* of zoological interest, the following.

> Calvert & I intend to visit Mr Brocklebank [Rector of Teversham] early on
> Thursday & then continue our [route?] to Swaffham as he wishes to call on
> Mrs Jermyn. I dare say that I shall find time to call at Bottisham before one
> in my way back: . . . prepare a magnificent suite of all the desiderata . . .
> Execute the following commissions for me: tell Mrs Jenyns . . . [t]hat I have
> examined the 3-ton lid of the Sarcophagus wch is decidedly good & (wch
> she will be delighted to hear) that it contains very large crystals of felspar.
> Tell Miss J. that I can learn nothing concerning Mons. Alexandré. Tell Miss
> Harriet – That Schiller's works are in 12 volumes & that I have sent 3 & that
> if this is not enough I will forward an additional supply – & also say – That
> Miss Benger's Queen of Scots is <u>ordered</u> but has not yet arrived.

This passage suggests that the young Henslow, now well established in the
University as the new Professor of Mineralogy, is already 'paying court' at
Bottisham Hall to his future wife, by providing for his future mother-in-
law up-to-date information on a local event – the arrival in the Fitzwilliam
Museum of the lid of an Egyptian sarcophagus – by showing willing with
regard to Mary Jenyns' interest in a visiting stage performer, a M.
Alexandré, and most impressively by providing Harriet with three volumes
of Schiller's works, which must have been in the original German, and
confirm that Harriet's linguistic accomplishments included a knowledge
of German as well as French.[7]

A further quality in Harriet's mother mentioned specifically by Jenyns
is that, while not being 'particularly talented or studious', she brought up
her children with 'high moral and religious principles'.[8] Harriet certainly
occupied her mind at times with theological questions, as we learn from a
letter from Henslow to William Hooker on 27 April 1836, where he tells
Hooker they have moved into a new house [Park Terrace] and continues

> . . . as we have sometimes [?talked] upon higher subjects than botany will
> you allow me to introduce to your reading a work entitled [Knox's remains]
> I have not yet read it myself but Mrs Henslow is so delighted with it, & tells
> me that she is so fully persuaded that it will meet my own views on religion
> & I hear it so highly spoken of by others that I intend to read it at the first
> leisure I have when my lectures are done – she has pointed out several
> passages to me in which I fully agree . . .[9]

Harriet's small literary career developed after the Henslows moved to
Hitcham in 1839; we know from a letter from Henslow himself to John
Lindley that her commentary on the Pentateuch or 'Five Books of Moses'
was already prepared by early 1840. Writing on 4 June 1840 Henslow
says:

> Mrs Henslow has been at some pains in writing a comment of the 5 Books
> of Moses – showing the connections between the Old and New Testament –
> The work will make a decent octavo, if published. Mr Maund spoke to
> Rivingtons but they will only undertake it on her account & I don't want to
> be at personal expense. They have not seen the M.S. She would put her
> name to it – Now do you think it likely that any one would publish it – giving
> her some kind of honorarium – I should state that it is not intended for
> learned divines – but for the rising generation from 12 to 18 – tho I suspect
> some older persons might profit by it – I think it would be very useful to
> Mothers who pay attention to the discipline of their daughters.[10]

We do not have Lindley's reply, but can only assume it was not helpful,
because the book was not published until eight years later. In the meantime
Harriet had written a small 'improving' work of fiction entitled *John Borton;
or a word in season*, a 30-page moral tale which appeared in paper-back in
1843, and sold for four pence. Her main work, entitled *A Practical Application
of the Five Books of Moses, adapted to young persons*, was a substantial book of
185 pages, with a four-page Preface in which the authoress explains her
intentions, no doubt derived from her activities in the Hitcham Sunday
School, where Harriet and her daughter took a regular part in the teaching.
Here is how Harriet's Preface begins:

> In the present age, when so many religious books are issuing from the
> press, I should scarcely have ventured to add another to the list, had it not
> been for the persuasion that if we would 'train up our children in the way
> they should go', we must not only impress them with early habits of piety,
> but must be careful that the great truths of God's Holy Word be laid before
> them in such a way that they may not rest contented with the outward form
> of religion only, but be enabled to grasp the substance also.[11]

It is clear that both Henslow and his wife look upon the duties of the
mother, with regard to religious belief and practice, as being primarily to
bring up their *daughters* in the appropriate way, since (unlike the sons) their
education is largely at home. Harriet seems to have been the model

Rector's wife – though increasing illness must greatly have limited her ability to fill the role.

Joseph Romilly (1791–1864), a bachelor friend of the Henslows, whose Diaries[12] provide us with entertaining accounts of the professional and private lives of many Cambridge figures, records that he 'dined at Henslow's to meet his wife and Mrs Jenyns [Harriet's mother]' on Monday 9 February 1824. There are further entries about Henslow's hospitality in later years, including the following:

> 'Wed. 6 May [1829] in evng to Mrs Henslow's where marchesa &c rather dull & very hot
> Tues 1 Dec [1829] Mrs Henslow's: no women there except Skrines who played guitar & Miss Jennings – staid but little . . .'

After these two rather trying parties, Romilly records that on 11 February 1830 he 'shirked Henslow's party'. But things improve, after 'tea with Marchesa where met Henslows' on Sunday 9 May. We can imagine Harriet asking Romilly why *her* parties are not acceptable, and persuading him to come to her *salon* on Thursday 23, for which he records: 'Mrs Henslow's where Mrs Airy sang well: shook hands 2ce with the hostess.' Harriet's parties are back in favour with Romilly who records 'tea with Mrs Henslow . . . on Saturday 6 November & an occasion at "the observatory" on the following Wednesday at which "Mrs Airey sang the Marseillaise, Mrs Henslow Beethoven, Miss Smith a singularly coarse milkmaid (haydin)"' [*sic*].

These short diary entries by Romilly provide us with the clearest evidence that Harriet kept up a circle of art-loving friends who regularly met to play and sing and discuss the theatre, literature and music. These soirées or tea-parties were quite separate from the more famous Friday soirées which so impressed Darwin and other students, although there was obviously some overlap in the clientèle, as we shall see in Chapter 6 when we record Miles Berkeley's comments on the Henslow hospitality. From 1832, when Romilly became the University Registrary, the diary entries are much longer and more informative: the following indicate Harriet's continued social activity in the Henslow home. On Wednesday 24 February 1836 Romilly records 'Mrs Henslow's ball – all the world there . . .', followed by Tuesday 15 March 'at 10 oclock went to a ball at the Henslows; staid 10 minutes & then home to bed'. Why did Harriet run two balls within a month of each other? Frances, her eldest daughter, was still only 11, and hardly old enough to be 'coming out'. By the autumn there was more social activity at the Henslows, and for Tuesday 13 December the entry reads:

> Drank tea with Mrs Henslow to meet Marchesa &c.: here also I met Mr Darwin (g.s. of the Botc Garden) who is just returned from his travels round the world: he declares that in 'terra del fuego' whenever a scarcity occurs (wch is every 5 or 6 years) they kill the old women as the most useless living creatures: in conseq when a famine begins the old women run away into the woods & many of them perish miserably there.–[13]

New Year celebrations were held on 3 January 1838, as Romilly records:

> . . . in the evening to Henslow's where I met the Marchesa, &c & – heard that at Lord Northamptons's the other day one of the charades they acted was in honor of Whewell: they began by representing 'hewing' a tree, then a well: in hunting for truth they find a book (Hist Of Ind. Sciences), on wch fame seizes & afterwards delivers to immortality: the party then crowned Whewell with Laurel – At this Henslowian party were Miss Trevor, Mrs & Miss Keown (Irish people) &c: MrsH. sung.

And on Twelfth Night, yet another Henslow party, this time for the children:

> Sat 6 Jan Took the boys to dine in Hall with me: Prof Henslow was also my guest – Bought a Twelf cake for the lads to eat at home, they declining the invitation to the children's party at the Henslows. – I went to the Henslow party: the Cummings, Dorias, & Hughes there: little Louisa Hughes' dancing quite beautiful. – Mrs Henslow drew in pen and ink some clever characters with appropriate poetry. I took off one called 'the Phrenologist'. Among others was Mrs Airy repres as a lady in a swing.

The 'Marchesa' who appears in a number of references to Henslow's parties was the second wife of the teacher of Italian (and French) in the University, the Marchese (Marquis) di Spineto. She was born Elizabeth Campbell, 'a Scotch lady of good family', and she and her husband must have been very popular in the University. (See also Chapter 6, where Miles Berkeley refers to the 'Marquesa Spinosa'.)[14]

So far as Charles Darwin was concerned, Mrs Henslow seems to have been a somewhat intimidating figure. Darwin's acquaintance with Harriet dates from 1828, when the Henslows had been married for more than four years, and the famous soirées had begun. His letters to his cousin W.D. Fox in 1830, when Darwin was in his final year at Christ's, contain some interesting passages. Written on 'Friday evening', 8 October, from his 'most snug & comfortable rooms' in Christ's, Darwin, who had just come up for the Michaelmas Term, reveals to his friend his problems with the Professor's wife: 'I have not seen Prof. Henslow, but am going to a Party there to night; you have not told me half enough what you think about Mrs Henslow. She is a devilish odd woman. I am always frightened whenever I speak to her, & yet I cannot help liking her' (Figure 6).

In an earlier letter to Fox, written from Barmouth on 25 August, Darwin says 'I suppose you are now living with Prof. Henslowe [sic]. I shall enjoy hearing from you some gossip about him & his wife.' Fox lived with the Henslows during the summer of 1830 as a paying guest receiving instruction from Henslow before his ordination. It would be fascinating to know what 'gossip' about his host and hostess Fox related to Darwin, but the letters have not been traced.[15]

Was it perhaps some quality of facial expression and speaking voice in Harriet that terrified the young Darwin? It could be. According to Darwin's wife, Emma, Harriet had 'a good, loud, sharp voice'. In her description of a dinner party given by the Darwins on 1 April 1839 – an occasion about which apparently, according to Emma, 'Charles is much more alarmed at the thoughts of them [i.e. the guests] than I am', Emma much approved of Mrs Henslow's voice, and indeed of the Henslow pair, and found them much more acceptable than the other guests. Writing to her sister Elizabeth the day after the party, she says:

> I must tell you how our learned party went off yesterday. Mr. & Mrs. Henslow came at four o'clock and she, like a discreet woman, went up to her room till dinner. The rest of the company consisted of Mr. & Mrs. Lyell and Leonora Horner [daughter of Leonard Horner, geologist], Dr. Fitton [geologist] & Mr. Robert Brown. We had some time to wait before dinner for Dr. Fitton, which is always awful, and, in my opinion, Mr. Lyell is enough to flatten a party, as he never speaks above his breath, so that everybody keeps lowering their tone to his. Mr. Brown, whom Humboldt calls 'the glory of Great Britain', looks so shy, as if he longed to shrink into himself and disappear entirely; however, notwithstanding those two dead weights viz., the greatest botanist and the greatest geologist in Europe, we did very well and had no pauses. Mrs. Henslow has a good, loud, sharp voice which was a great comfort, and Mrs. Lyell has a very constant supply

FIGURE 6 Harriet Henslow.

of talk. Mr. H. was very glad to meet Mr. Brown as the two great botanists had a great deal to say to each other. Charles was dreadfully exhausted when it was over, and is only as well as can be expected today. There never were easier guests than the Henslows, as he has taken himself off all day, and she is gone out in a fly to pay calls, and Charles and I have been walking in the garden. He is rather ashamed of himself for finding his dear friends such a burden. Mr. Henslow is so very nice and comfortable and it is a pleasure to look at him. It is said of him that he never wishes to eat, but always eats everything offered to him. The dinner was very good . . .[16]

During the Henslows' last year in Cambridge Charles Darwin had married Emma Wedgwood, and Charles had obviously written to announce his engagement. Harriet's reply, combining correct etiquette – 'My dear Mr Darwin' – with a light, almost playful style, is fortunately preserved for us.

Here it is:

Cambridge

Nov. 22 – [1838]

My dear Mr Darwin,

I do believe there are few pieces of intelligence that could have delighted me more than that which your letter of day contains, and you may be assured of the warmest congratulations of both Mr Henslow and myself – That your fit of *insanity* may long continue is my sincere wish, but pray don't let me hear a word of the *calamities* of domestic life, all such matters ought now to appear to you en couleur de rose, and if it is your Intention in future to resign *all such* cares to your fair Lady, it strikes me you do not intend to trouble *yourself* about them henceforward, *under such circumstances* I do not doubt you will prove a most *dutiful* husband, joking apart, I cannot help observing, what you are not to take as a compliment, because they tell me I never paid but two in my life, that I think Miss Wedgewood a very fortunate being; and that she has every reason to look forward to as great a share of matrimonial happiness as falls to the lot of most people. I am glad to hear it is your intention to reside in town, I feel that we have not quite lost sight of you, had you told me you were going to take up your abode in Shropshire, why then – I wonder if I should have read your letter with the disinterested satisfaction I ought to have done – Now I am on the subject of *Selfishness*, I must just say, that I do hope this event will not put a stop to the visit Mr Henslow has held out hopes of, of your coming to us at Christmas. I am sure you must require a great deal of information on household matters, so pray come, and talk it all over –

Believe me | Very Sincerely Yrs | H Henslow

Mr Henslow returned from Hitcham yesterday –[17]

Henslow's own reply, written nearly a month later and rather apologetic for the delay, refers in its opening lines to his own marriage exactly fifteen years earlier, and then develops into a small homily:

Cambridge

16 Decr. 1838

My dear Darwin,

This day 15 yrs ago I entered on that state which, it rejoices my pericardium to think that, you are about to enter – I have been remiss in not telling you so sooner, but I am sure you will not think me unmindful of your happiness from having added one more specimen of my carelessness to the many you have witnessed before – All I can wish you is, that you may experience as great content in the marriage state as I have done myself – & all the advice, which I need not give you is, to remember that as you take your wife for better for worse, be careful to value the better & care nothing for the worse – Of course it is impossible for a lover to suppose for an instant that there can be any worse in the matter, but it is the prudent part of a husband, to provide that there shall be none – It is the neglect of this little particular which makes the marriage state of so many men worse than their single blessedness – if there is such a thing – for it is now so long since I have enjoyed my double blessedness that I cannot fancy myself in my Bachelor days . . .[18]

It seems that Charles, beset with self-doubt and his customary ill-health in the winter months preceding his marriage on 29 January 1839, was unable to accept the Henslows' invitation to visit them over Christmas, so Harriet was deprived of the opportunity of imparting 'a great deal of information on household matters' to the reluctant Darwin! Nor were the Henslows present at the wedding, which took place at Maer Church, Emma's Shropshire home, very quietly indeed, with only a few relatives.[19] A letter from Henslow to Darwin reads as mildly complaining that they have been kept in the dark about the wedding:

Cambridge
7 Feb. 1839

My dear Brother Benedict,

Whewell assures me that you are certainly twin, though I have seen no account of the important event in the Gazette or elswhere – I must therefore take it for granted that the Queens Speech or something else has so much occupied the thoughts of all the Printers that they have forgotten you – I need not tell you how heartily I wish you all health & happiness in your new estate – nor I trust assure you that I know no one beyond the circle of my own family in whom I feel a more hearty interest. Should it please God that we & our better halves are to see many days, I hope our friendship will go on increasing & our intercourse be more & more frequent – Mrs H. hopes with myself that you will accept a trifling wedding present for Mrs C. Darwin's table. It is only a little silver Candlestick for a wax taper; but the use of it will occasionally remind you of us, if we should never meet again – I don't know how to send it to you, but as I shall be in town in 2 or 3 weeks I can at all events bring it up with me . . .[20]

The friendship between the Henslows and the Darwins did stand the test of time, as we shall see in the later chapters. Although Harriet's early death in 1857, at the age of 59, belongs to this later period, we feel it appropriate to report here how Joseph Hooker, Henslow's son-in-law, informed Darwin of the events. Writing on 2 December 1857, Hooker says:

> Dear Darwin
>
> I arrived last Monday, having left my wife at Hitcham[21] all were well & cheerful there – when I left – Old Henslow was very much cut up, though being wondrous stolid he did not show it much & he, in common with all the family, occupied themselves throughout industriously. Poor Mrs. H went off as peacefully as possible, – without a struggle, having had no pain for the last 12 hours. – She was a very gentle lady, who I had become very deeply attached to of late years. She inherited the calm self-possession of the Jenyns to the full, had long known that any severe attack would certainly carry her off, & led that kind of life which regarded death as going out of one room into another – happy soul![22]

5 The young Professor

As we saw in Chapter 2, Henslow had attended as an undergraduate the lectures of both the Professor of Chemistry, James Cumming, and the Professor of Mineralogy, Edward Daniel Clarke, and continued after taking his degree in 1819 to assist both Professors in preparing and using demonstration materials in their lectures. This practical experience he was to turn to very good effect in his own lecture courses in later years. Not only this, but his friendship with Clarke and Sedgwick, cemented by their joint work in the formation of the Philosophical Society, obviously placed him in an ideal position to apply for the Chair of Mineralogy on the death of Clarke on 9 March 1822, and he announced his attention of becoming a candidate if the Chair should be advertised. This duly took place on 15 May in the form of a Grace passed by the Senate.

Henslow's appointment as Professor of Mineralogy on 28 May 1822 was accompanied by a very complex, tedious legal controversy in which, inevitably, Sedgwick played a leading rôle. By a curious coincidence, Thomas Jephson, fellow of St John's and younger brother of William Jephson, Henslow's teacher at his boarding-school, is involved in this intricate story as a rival for the Chair of Mineralogy. The details we have judged to be of little interest to the modern reader, who might well react that it seems a pity that the participants, from the Vice-Chancellor downwards, had nothing more important to do with their time. The complex legalities, which affected elections to all three of the Chairs of Anatomy, Botany and Mineralogy, were not clarified for more than five years, being finally settled by the 'determination' of Sir John Richardson dated 1 December 1827.[1] This delay, however, had no practical effect on Henslow's activities as Professor of Mineralogy: the new Professor lost no time in preparing a printed *Syllabus* for his first course of lectures, which he delivered in the Lent Term of 1823.

The Preface to his *Syllabus* warns his readers and audience that the new Professor is offering a very different lecture course from that of his illustrious predecessor. The very first sentence makes this clear: 'Some explanation seems necessary for having deviated from the plan pursued by my predecessor in forming his Syllabus . . .' Henslow's plan involves an explanation of the relevance of Dalton's Atomic Theory to the study of mineralogy: in other words he attempts to elevate mineralogy from the level of mere description and anecdote to a science based on chemistry in which patterns and generalisations become evident. At the same time, he warns that since chemistry is obviously a rapidly-developing science, his particular presentation must not be taken as 'correct':

> Many of the theoretic results, as well as some notices, are not to be considered as conclusions upon which any reliance can be fixed, or as exhibiting my own opinions upon the subject. They are to be viewed merely as memoranda, printed for the sake of saving trouble in the lecture-room. Neither are the arrangement and nomenclature intended to form a sketch of any system of Mineralogy. The order pursued is that which appeared best adapted, in a short and elementary course of lectures for the University, to impress some knowledge of minerals upon those who may wish to acquire it, and assist others in their desire of prosecuting the study.[2]

The *Syllabus* contains a reading list of eight standard works on mineralogy, crystallography and chemistry, three of which are in French.

We have no account from the young Professor himself as to how his very first public lecture, delivered on 15 February 1823, was received – in contrast to his predecessor's triumphant outpouring after *his* inaugural lecture sixteen years earlier (which we reported in Chapter 2). Some present must have been disappointed if they were expecting any brilliant eccentricity. Not all, however, for we have the following unsolicited testimonial from Erasmus Darwin to his younger brother Charles. Erasmus, who preceded Charles to Cambridge, as we shall see in Chapter 6, praised Henslow to his younger brother and predisposed him to sit at the feet of the Professor. Here is Erasmus to Charles Darwin in a letter dated 6 March 1823:

> Professor Henslow (on Mineralogy) has twice shown us ye experiment of ye test of Arsenic by burning it with a blow-pipe but I was so far off that I neither time smelled ye garlic odour which they describe. He has 16 guineas worth in one set of wooden crystals besides many others. The Lectures are very entertaining, & this is his first course so that he will have improved by ye time you come up.[3]

In the event, Charles Darwin never had the opportunity to attend this lecture course; Henslow delivered his final course of lectures in Mineralogy in 1827 and Darwin came up to Cambridge in the following year. We shall never therefore know whether, in the Darwins' opinion, Henslow 'improved' over the five years. Jenyns, however, loyally supports Henslow's efforts, after explaining the difficulty in following Clarke, the showman:

> Professor Henslow's style was not what would be called eloquent, but he had a good voice and a remarkably clear way of expressing himself, and of explaining in well-chosen language anything that his hearers might find difficult to understand. This is what he peculiarly excelled in as a lecturer.[4]

From Jenyns we also learn that 'the Professorship of Botany was the one to which he had been looking for some years, and for which he had been preparing himself at the time when he never anticipated that the Chair of Mineralogy would be open to his acceptance first'.[5] We hear the same from Miles Berkeley, the eminent mycologist, who recalls visits from the young Henslow before 1821:

> At that time he was making Botany a study with a view to the Regius Professorship, rather than from any peculiar predilictions for that branch of natural history above other branches. His habits, however, were so methodical that he had no difficulty in taking up half a dozen subjects at the same time . . .[6]

We can only conclude that he was prepared to apply for any vacant science Chair available to him and, since in the year of his graduation, 1818, the absentee Professor of Botany, Thomas Martyn, whom apparently Henslow had never met, was already 82 years old, whereas the very active Professor of Mineralogy, Edward Clarke, was only 48 and at the height of his powers, Henslow naturally bet upon the Chair of Botany becoming vacant first.[7] After all, his great friend and teacher, Adam Sedgwick, had not hesitated to take the Woodwardian Chair of Geology with the intention of learning the subject as he taught it, and Sedgwick strongly promoted Henslow's application to succeed Clarke, expecting that he could also learn up the necessary minimum to offer a tolerable lecture course.

The death in 1825 of Thomas Martyn, who had occupied the Chair of Botany for over 60 years, led to the long-expected vacancy for which Henslow immediately applied. There seems to have been no objection raised to the fact that he already held a Chair, and he was elected, according to Jenyns 'in the room of Professor Martyn without opposition',[8] though according to an official notice of 'University Intelligence' in the local press dated 11 June 1825, two other candidates had also come forward:

> The Regius Professorship of Botany being vacated by the death of Rev. T. Martyn B.D., three candidates have announced themselves for the office, viz. the Rev. J.S. Henslow M.A. of St. John's College, Professor of Mineralogy; the Rev. W.L.T. Garnons, B.D. Fellow of Sidney College; & the Rev. Wm. Pulling M.A. of Sidney College.[9]

It is perhaps not surprising that the other two candidates were Sidney men, as Thomas Martyn had been a Fellow and Tutor in that College, and had indeed married the Master's daughter! Of the two, William Garnons was well known to Henslow: he was a keen entomologist and a member of the Ray Club, whereas William Pulling, said to be 'a notable linguist', could hardly have been a serious candidate. Another potential applicant for the Chair was the entomologist William Kirby, whom Henslow eventually succeeded as President of the Ipswich Museum in 1850 (see Chapter 12). Hearing that Henslow was a candidate, Kirby wrote on 22 June 1825: 'I trouble you with half a dozen lines to express my wishes for your success. Had I a vote it should be at your service . . . Though I formerly felt anxious to succeed the late venerable Professor of Botany I am now too advanced in life . . .'[10] Henslow himself alludes to Kirby's interest in the post in his acceptance speech in Ipswich in 1850, explaining that his predecessor had been 'a botanist as well as an entomologist' – though his published works are entirely entomological. Clearly, Kirby's interest in the post provided yet another example where any specialist knowledge of the subject was not considered necessary.[11]

The Regius Professorship was a Crown appointment, one of three titles which Thomas Martyn had held as 'Professor of Botany'. The official archives give details of the Letters Patent of the Crown appointment dated 26 July 1825 which refer to Henslow as 'our public Professor or reader of botany', and local press references to the appointment use one or other title as if they were synonymous. Henslow exhibited the Letters Patent in the Senate House on 10 October 1825 and took oath before the Vice-Chancellor as 'King's Reader in Botany'.[12] As Jenyns explains, Henslow was not elected to the University Professorship 'owing to the circumstances of the mode of election to the University Professorship being . . . still under consideration by Sir John Richardson, to whom the question had been referred, together with that relating to the Professorship of

Mineralogy . . .'[13] Happily for Henslow, the Regius Chair carried a salary of £200 per annum, and the University Chair was devoid of any salary, so the practical effect was nil! (A further appointment, that of Walker's Reader in the Botanic Garden, came to Henslow on 25 November in the same year: this is explained in Chapter 8.)

The new Professor (or Reader) of Botany did not relinquish his Mineralogy Chair, and continued to give his old course of lectures in that subject in the Lent Terms of 1826 and 1827. During that time he was, however, much occupied with preparing the new botanical course and a syllabus to accompany that course, and he delivered his first botanical lectures in the Easter Term 1827. This must have involved him in a marathon effort of some 30 lectures first in Mineralogy, and then in Botany between the middle of February and the end of May 1827. We do not know how the first Botanical course offered in Cambridge for some 30 years was received: unfortunately, as we explain later, the set of attendance lists for Henslow's lectures that have survived do not begin till the following year, 1828, so we can only guess that they attracted as in 1828 a reasonable audience including a number of Henslow's friends and colleagues. It may, however, be significant that the official notice announcing the 1827 Botany lectures[14] states that the course 'may be attended *gratuitously* by any members of the University', which could imply that Henslow felt he had not had enough time to make a course good enough to expect people to pay! From 1828 onwards, the equivalent notice clearly states: 'Terms of attendance. One guinea'. Curiously, there is a similar uncertainty about his Mineralogy lectures: was his *first* lecture course (1823) free, and only the second one made subject to a charge? Again the archives are inadequate: we do not have a copy of the 1823 announcement, but by 1824, by which time Henslow's new syllabus was available, 'Gentlemen wishing to attend' were directed to put down their names at Messrs Deighton's [a local bookshop] where they could buy the *Syllabus*, and the 'Terms of attendance' are 'First course Three guineas' and 'Second course Two guineas.'[15] We have no notice for 1825, but for 1826 we read: 'As this Course will be limited and confined to the leading topics of the Science, it will be delivered gratis to any Members of the University who may wish to attend.' Further, we are told that there are only two (not three) lectures per week, commencing on Friday 17 February 1826.[16]

Sir John Richardson's determination cleared the way for an undisputed election to the Chair of Mineralogy, which took place in February 1828.

Henslow's colleague WilliamWhewell was elected, and held for Chair for ten years. It is not clear when, or even whether, Henslow resigned the Chair of Mineralogy he had effectively held for nearly three years. Jenyns is somewhat obscure:

> He did not formally resign the Professorship of Mineralogy till three years after [his appointment to the Chair of Botany]. This, however, was simply because Sir John Richardson's determination as to the mode of election to that Professorship had not yet been given. But he virtually resigned it, and in 1828, soon after the settlement of the question alluded to, Mr. (now Dr.) Whewell was regularly appointed to succeed him in that office.[17]

Whewell himself tells us that he is succeeding 'my very intimate friend Henslow who has since been appointed to the Professorship of Botany, and so given up that of Mineralogy'. In this passage, which contains a long eulogy of Clarke, Whewell says nothing about how Henslow performed as Clarke's successor, an omission that may be significant.[18]

By April 1827, when Henslow began his botanical lectures, any memory or influence from Thomas Martyn's days was long gone, and only the neglected Walkerian Botanic Garden and the Lecture Rooms within that Garden remained. In fact, Henslow's priorities included rescuing, sorting and ordering the considerable and diverse Museum collections, much of which had been put together by his predecessor as Professor of Mineralogy, E.D. Clarke. This material included the neglected and damaged Herbarium assembled by the Martyns, only a relatively small fraction of which he was able to rescue.

It was an exceptionally favourable time for the young, enthusiastic new Professor to make his mark. Not only had he a *tabula rasa* so far as any lecture course was concerned, but there had occurred a change in the regulations affecting the University training of medical students by which the Professor of Botany was guaranteed a minimum audience of undergraduates. Before 1831, candidates for the degree of Bachelor of Medicine were obliged to attend only the lectures of the Regius Professor of Physic and at the end of the course be examined by him. The Chair of Physic was held from 1817 to 1851 by John Haviland, an able and conscientious Professor who gave 50 lectures annually on pathology and clinical medicine. Haviland's Chair was the oldest science chair in Cambridge, said to have been founded by Henry VIII in 1540, and followed by the Lucasian

Chair of Mathematics in 1663 and then seven others with various titles and endowments throughout the eighteenth century. This unplanned growth seems not to have been accompanied by any serious attempt at devising a structured scientific curriculum until the beginning of the nineteenth century, and the majority of holders of these chairs accepted their stipends, which varied from £99 to £300 per annum, but neither gave lectures nor undertook any research. In a good many cases, they did not even reside in Cambridge.[19]

Henslow seems to have been acutely aware that he needed some guidance in planning his first course of botany. It is probable that he sought the assistance of several colleagues in other Universities, and among these we know was Robert Kaye Greville, a young and enthusiastic Edinburgh botanist who had taken his doctor's degree in law at Glasgow. One of the great advantages of Edinburgh was its very large and famous Medical School, where at this time Robert Graham was Professor of Medicine and Botany. Graham's botanical teaching to large audiences (280 in 1828!) was very successful, as can be gathered from this extract from a long, idyllic description by one of his students, Joseph Buller:

> The morning walk to the Botanic Garden, the large light conservatory looking lecture-room, surrounded by fine shrubs and beautiful flowering plants, the abundance of newly gathered flowers, with which the lectures were illustrated, and the lecturer himself, simple, unaffected, cordial, and joyous, with no dullness or tedium in him, but as fresh and healthy, and full of life, as the youths around him, remain as a permanent picture in the mind's eye, from which so many scenes have altogether faded . . . On the Saturdays he invited [the students] to accompany him on botanical excursions in the neighbourhood. These trips sometimes extended over fifteen miles; and, enjoyed alike by the teacher and the taught, they contributed in no small degree to impart a lively interest in their studies in the Lecture-Room and Garden.[20]

Though Greville did not himself attend these lectures, arriving in Edinburgh already a graduate, he obviously fell under the spell of Graham, and took part with him in explorations of the Highland flora in August and September each year. As we shall see in Chapter 8, Graham was consulted by Henslow about the Edinburgh Garden, but not apparently on his lecture course.

Greville's letter to Henslow in response to a request for help in planning his new lecture course is dated 5 January 1827. Modestly, he begins:

> I am not sure that I can give you any hints upon the subject of lectures, that you will not already have obtained. Like yourself I never attended a course of lectures on botany, and have only delivered a single popular course of 19 lectures. At the present moment however I am busied in composing a full course of 55 lectures, for next summer which I mean to give independently of my popular one, the one chiefly for medical students, the other for 'Ladies & Gentlemen'.

A 'plan' follows, in which 'Classification' [presumably of the Vascular Plants], and 'Cryptogamia – which I treat as a distinct subject' are the two last sections. Greville's comment on this order is interesting: 'I believe I differ from most lecturers in treating of classification last, but it seems to me absurd to talk of classifying bodies whose characters we are not acquainted with.' He concludes with a particularly important piece of advice on wall charts and coloured illustrations in general, which must have greatly reinforced Henslow's own ideas derived from his experience of Clarke's Cambridge lectures:

> Allow me to suggest the greatest possible assistance to a lecturer is derived from magnified colored drawings – I prepared 100 Elephant folio ones for my first course & shall have as many more. I had sap vessels as thick as my arm – the Red snow as large as Cannon balls – the flower of Campanula rotundifolia 16 inches long and the rest in the same proportions. The principal forms of leaves I also delineated, & thus illustrated in the class room, subjects which are usually studied by means of books & very dry in the lecture room. I am now going to attempt to give some idea of the geographical distribution of plants by a map of the world upon Mercators projection 10 or 12 feet square (for my lectures).[21]

Greville's later career developed particularly in studies of the lower plants, but he also wrote and illustrated *Flora Edinensis* (1824) and contributed coloured plates to *Curtis's Botanical Magazine* in the period 1824–34. In later life he specialised in water-colour scenes of the Scottish countryside.

The long friendship Henslow developed with William Jackson Hooker, the successor to Robert Graham in the Glasgow Chair of Botany, which was eventually to lead to Henslow's daughter marrying Hooker's son Joseph, seems to have begun as early as 1826, to judge from the surviving, quite copious, correspondence. By September of that year, Henslow writes to tell Hooker he has sent off a 'further packet' of dried plants, continuing an exchange that was obviously flourishing, but apologising that 'there are not many in it' and explaining that he has been 'so busy with [his] books this summer that [he has] had little time for rambling'. Preparing his first lecture course had clearly taken priority over field studies in 1826. By January 1827 he felt it necessary to ask Hooker for advice about 'the removal of our garden' (see Chapter 8) and mentions the help received from Greville: 'A friend at Edinburgh has lately introduced me to Dr. Grenville [*sic*] who has promised me his assistance & given me kind valuable hints on lecturing which I am about to commence this Spring.' This letter concludes with the following P.S. 'Are your lectures limited to the students or do Ladies and Gentlemen from the Town attend them as they do Dr. Grenville [*sic*]'[22]

From Henslow's correspondence at this period one gains the impression that he was a young man very anxious to do what his elders and peers held to be the 'correct thing' and prepared to ask advice quite widely, especially from botanists in Scottish Universities where, he had rightly concluded, some of the best teachers of natural science were to be found.

The reform of 1831, whereby medical students were required to take a much broader science course, including some Botany, might be considered to be the first step towards a greater recognition of the rôle of science in the University teaching. From 1831 all candidates for the MB degree in Cambridge were required to attend lectures and pass an examination in Botany (along with Anatomy and Chemistry), and Henslow and other science Professors benefited from this guaranteed audience. The effect, however, was rather short-lived, for in 1835 the Royal College of Physicians, despite opposition from Cambridge, abandoned their rule that only Oxford and Cambridge undergraduates could be elected Fellows of the College. The gradual effect of this removal of the Oxbridge monopoly can be seen in the figures for Cambridge MBs, which between 1842 and 1852 averaged only four, and Henslow noted in 1846 that MB students attending his lectures averaged about six.[23]

The University Archives in Cambridge fortunately contain very detailed attendance lists for Henslow's Botanical lectures, beginning in 1828 and continuing without a break until his final lecture course in 1860, delivered only months before his terminal illness. These lists give for each College the names of members attending and usually also indicate the tutors responsible for the undergraduates concerned. It is moreover possible to tell in most cases the undergraduates from graduates and Fellows because of their titles, Mr, Dr, Professor or Fellow being used for the senior members, and one can therefore obtain some idea of the proportions attending Henslow's course. These records were apparently compiled to enable each College to collect and pay the appropriate fee to the Professor giving the course.[24]

Presented as a graph (Figure 7), these statistics enable one to appreciate at a glance most of the changes in fortune during Henslow's Cambridge teaching career. The Darwin years (1829–31), when the young Charles Darwin came under the spell of Henslow's teaching, are seen to be the period of well-attended lectures, with an audience of senior and junior members. Of the total of 71 enrolled for the 1828 lecture course, no fewer than 24 were in the category of senior members attracted by the reputation of the new Professor of Botany, and although by 1829 the senior members attending had dropped dramatically to 5 (no doubt the 1828 customers were satisfied and did not return for more of the same!), the number of undergraduates had risen from 47 to 62, and by 1831, the first year of the guaranteed MB audience, had risen again to 66. The figures for 1833 are at a maximum (not in fact surpassed until the very last year, as we shall see), when 71 undergraduates registered for the course.[25]

After 1835, the last year of the Oxbridge monopoly for Fellows of the Royal College of Physicians, in which Henslow's lecture audience numbered 55, the numbers never reached 50 and indeed, for the 'lean years' between 1839 and 1851, after Henslow moved to the Rectory at Hitcham, they averaged just below 14, with a minimum of 10 in 1845. The dramatic revival after 1851, in which year the Natural Sciences Tripos came into existence, will be discussed later. The extent to which Henslow's absence from Cambridge, except to deliver his statutory Easter Term lectures, was felt by some of his former admirers, can be judged by this revealing outburst by Charles Babington in a letter dated 2 June 1846 to his colleague John Balfour, Professor of Botany in Edinburgh:

> Never was botany at so low an ebb as now in this place. A non-resident Professor, who only comes here for five weeks (as he calls it), going away on Saturday morning in each week, and returning Monday evening. I have been taking a party of our few naturalists for a short excursion on each of the last four Saturdays, but never got more than than twelve to accompany me, all of them quite beginners.[26]

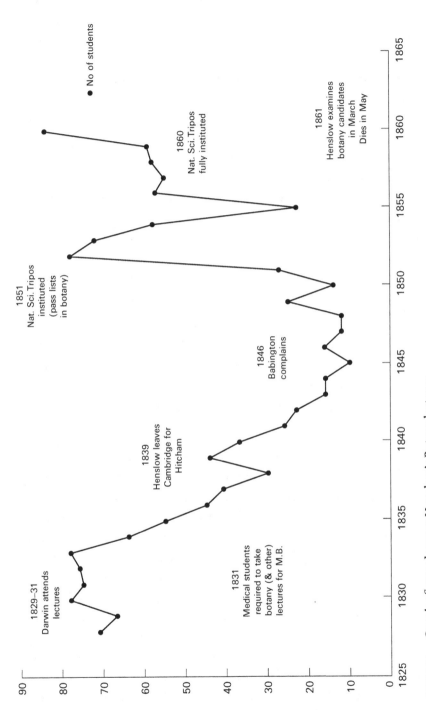

FIGURE 7 Graph of attendance at Henslow's Botany lectures.

This was the same man who was so fascinated by the young Henslow in the golden years that he attended his lectures for five successive Easter terms (1827 to 1831 inclusive), and recorded in his Journal on 2 May 1827: 'Conversed with [Henslow] after the lecture, and was asked to his house. Assisted [him] in putting his things in order before and after the lectures.' We should not judge Babington too harshly. He was, at the time, involved in the difficult task of persuading a reluctant University to find the money for the new Botanic Garden – Henslow's own project, which took shape first in 1831 and of which we tell the story in Chapter 8. Moreover, Babington was the obvious internal candidate to succeed Henslow, and was, one feels, increasingly impatient waiting for a 'dead man's shoes'.[27]

By 1828, when the detailed official attendance lists begin, Henslow is evidently a popular lecturer, drawing, as we have seen, an audience of over 70, of whom 47 were undergraduates. For those attending, Henslow had printed a small 16-page *Syllabus* of the course, which begins with a Preface of a single sentence:

> A few copies only of this Syllabus have been struck off, for the use of those gentlemen who may attend the present course; it being the author's intention to prepare another, containing more copious references, so soon as he shall have arranged further materials, and completed several more drawings for his lectures.

With such a Preface, it seems reasonable for us to treat the contents of the *Syllabus* as being only a minimal guide to the new Professor's lecture course. One thing, however, stands out immediately: the list of recommended text-books contains four titles, two of them being by the French botanist, Auguste de Candolle, and the other two by James Edward Smith. We shall see later that Henslow was greatly influenced by de Candolle's work, and indeed based his own botanical text-book, published in 1835/6, to a large extent on what he had learned from the Frenchman's publications. We might note that neither of de Candolle's recommended works was available in translation, and conclude that Henslow thought it quite reasonable to expect his students to read French. It would be nice to know how far this assumption was justified!

The English text-books recommended in this 1828 *Syllabus*, both by J.E. Smith, are the general *Introduction to Physiological and Systematic Botany*, first published in 1807 and now in its sixth edition (1827), and the four-volume

English Flora (1824–28). Smith's name nationally is permanently associated with three achievements: the purchase of the Linnaean collections in 1784, the subsequent founding of the Linnean Society of London in 1788, and the production between 1790 and 1814 of the first 36 volumes of *English Botany*, a *magnum opus* whose beautiful hand-coloured illustrations were by James Sowerby but for which Smith contributed all the text. In the history of Cambridge botany, however, Smith is remembered for the abortive attempt to appoint him as Deputy Walkerian Reader in 1819; although this incident in no way implicated the young Henslow (who was at the time still a junior member of the University), it is very relevant to the religious tensions and conflicts in the University at that time.[28]

Although there is no evidence that Smith and Henslow ever met, there can be no doubt that Smith's writings, and in particular his very successful text-book, played an important part in shaping Henslow's own ideas about what a modern botany course should contain. Smith's botanical career had begun in Edinburgh, where he had the advantage as a young medical student of attending the botany lectures given by John Hope,[29] an experience which clearly fired his enthusiasm for the subject. From the time when he bought all the Linnaean collections he increasingly saw the importance of contacts with Continental botanists, and retained the wide, optimistic international view of scientific study characteristic of Enlightenment scholars. In particular, Smith always saw the importance of plant physiology in the teaching of botany, and indeed attributed this enlightened view to his Edinburgh lectures:

> The physiological part of botany has scarcely been taught in any school, except under the late Dr. Hope at Edinburgh, where indeed this curious and interesting branch of a philosophical medical education was not always received as it deserved. The attention however which that excellent teacher merited, has been overpaid to his pupil, who is happy in acknowledging, on this occasion, how much he owes to his information and friendship.[30]

Returning to Henslow's *Syllabus*, we find no division of the material presented into separate, numbered lectures, but on the second page of text (p. 6 as paginated) we have a significant note following mention of 'flower' and 'fruit' as 'reproductive organs'. 'NB. The several parts of the flower and the method of technical description, will be explained here, to enable us to commence our "demonstrations" from living specimens.'

Jenyns gives us a glowing picture of these lecture-demonstrations:

Professor Henslow had already had three years' experience in lecturing, and he neglected nothing in his power to make his lectures attractive and popular, without departing from a plan of study that could alone give the students a proper grounding in Botany as a science. One great assistance he derived from his admirable skill in drawing. His illustrations and diagrams ... representing all the essential parts of plants characteristic of their structure and affinities, many of them highly coloured, were on such a scale that when stuck up they could be plainly seen from every part of the lecture-room. (Plate 3) He used also to have 'demonstrations', as he called them, from living specimens. For this purpose he would provide the day before a large number of the specimens of some of the more common plants, such as the primrose, and other species easily obtained, and in flower at that season of the year, which the pupils, following their teacher during his explanations of their several parts, pulled to pieces for themselves. These living plants were placed in baskets on a side-table in the lecture-room, with a number of wooden plates and other requisites for dissecting them after a rough fashion, each student providing himself with what he wanted before taking his seat. In addition to these were rows of stone bottles containing specimens of all the British plants that could be procured in flower, the whole representing, as far as practicable, the different natural families properly named and arranged.[31]

It seems that Jenyns himself did not actually attend any of these lectures, though his acquaintance with Henslow began as early as 1820, when Jenyns tells us they both attended lectures in Chemistry given by Professor Cumming. By the time that Henslow took the Chair of Botany they were firm friends, sharing a passion for entomology, conchology and other branches of field zoology, and Jenyns would undoubtedly have learned from friends in Henslow's botanical audience just how good the lecture-demonstrations were. As we shall see, these acquaintances included Charles Darwin himself and, in compiling the biography of Henslow, Jenyns eventually consulted Darwin and other celebrated former pupils about the influence the young Professor had, and incorporated some of this material in his book.

Before we look in any more detail at Henslow's teaching of botany, some comparison could usefully be made with the botanical courses given by his predecessor, Thomas Martyn, more or less continuously from 1764 to 1796. It is, of course, true that the gap of 30 years without any botanical lectures meant that very few if any in Henslow's first audience would have been in a position to make direct comparisons, but we (and they) have access to the pamphlet which Martyn published in 1764 entitled *Heads of a Course of Lectures in Botany*. Dedicating this work to Richard Walker, Vice-Master of Trinity College, the benefactor whose generosity had enabled the University to set up its first Botanic Garden in 1762, Martyn writes:

> The following Work may claim your Patronage with some Propriety, since it is to you that it owes it's [*sic*] Birth. Had it not been for your Munificent Design in founding a Botanic Garden, the Author would scarce have thought of attempting to restore the Study of Botany, which was almost lost among us. To teach this extensive Science without a good Garden is next to impossible . . .'

This dedication is dated Jan. 1, 1764. Unfortunately for the new Professor, Richard Walker died on 15th December of the same year, and Martyn lost an important ally within the counsels of the University. We return to the varied fortunes of the Walkerian Garden when we discuss Henslow's New Botanic Garden in Chapter 8. What is important for the present purpose of comparison is that both Thomas Martyn and Henslow saw the provision of a good Botanic Garden as a *sine qua non* for effective teaching of their science in the University. In this respect they both agree with John Hope and his successors in Edinburgh, and indeed Martyn explicitly commends Hope's energy in his Preface, with these words:

> The Botanic Garden at Edinburgh already begins to flourish under the Conduct of Professor Hope, who seems intent at the same time upon searching for the almost neglected Plants of Scotland . . .

Hope in Edinburgh and Martyn in Cambridge were both young, newly-appointed Professors at around the same time beginning to teach their science in the University. Hope was successful, establishing a wide European reputation and launching on to important scientific careers students such as James Edward Smith. Martyn had far less influence. The main reason for the difference seems to lie in the size of the medical faculty, which provided most of the students who took the botanical courses. In Edinburgh this Faculty of the University, as we have seen, had already achieved national – and even international – fame by the time of Hope's appointment, and throughout the 1760s and 1770s there were 50 to 60 students attending the botany lectures, almost all in the Faculty of Medicine, with occasionally two or three from the Faculty of Divinity. Not only did this guarantee Hope a gratifyingly large audience, but it also increased his earnings significantly from the course fees. The very different picture presented by medicine in Cambridge is of a series of weak attempts throughout the eighteenth century to provide medical training in a University dominated by mathematics and the Newtonian tradition. The number of students taking the MB degree in Cambridge were: 1720s 67, 1730s 46, 1740s 35 and 1750s 15; and by 1759 the Cambridge-trained physician Richard Davies was complaining that 'the arts subservient to Medicine have no appointments to encourage Teachers in them. Anatomy, Botany, Chemistry and Pharmacy have been but occasionally taught.'[32]

For his lecture course in the Easter Term of 1833, Henslow had prepared an eight-page printed pamphlet. Unlike the very provisional *Syllabus* for 1828, this definitive *Sketch of a Course of Lectures*, which was obviously intended to be used by all the participants, indicated clearly the division between 'Demonstrative' lectures held 'on Tuesdays and Thursdays' and the 'Physiological' ones on 'Mondays, Wednesdays and Fridays'. Assuming that the 1833 course (and those of subsequent years) occupied five consecutive weeks, then 15 lectures, more than half the course, were devoted to 'plant physiology' as it was then broadly defined, and by interspersing these more theoretical lectures with his 'Demonstrations', Henslow seems to have succeeded in relating the study of form (what we would call morphology, anatomy and taxonomy) to that of function. Let us look more closely at the outlined content of the 'Demonstrative' lectures. Towards the right-hand margin of the pages we find a general title for the grouped subjects. The Introduction, dealing with 'external characters of the several organs', is called 'Glossology' (our Morphology), and is followed by 'Organography' (our Anatomy), which first discusses 'cellular tissue', 'woody fibre' and 'vascular tissue' and then proceeds to consider 'compound' organs (e.g. stomata, hairs, glands) and 'complex' organs. At this point Henslow puts in a N.B: 'As the first few minutes of every Lecture will be occupied with the demonstration of some British plant, the various forms of the complex organs will be best explained at these times.'

The 'complex organs' to be discussed are divided into 'conservative and fundamental' (root, stem and leaf), and 'reproductive', subdivided into inflorescence, flower and fruit. A list is given of 28 Orders from which 'most probably' the material used for the demonstrations will be chosen:

26 of these are recognisably modern Families of flowering plants, and the remaining two are Ferns and Mosses. This section of the syllabus is marked 'Phytography' – a term we rarely use today, meaning Plant Description – and what follows is marked 'Taxonomy'[33] and deals with the 'Artificial System of Linnaeus' and the 'Natural System' (not explicitly attributed to de Candolle, but obviously his). Pages 4 and 5 of the pamphlet set out, on facing pages, the 'Artificial System of Linnaeus' and an 'Artificial Analysis of the Natural Orders' attributed to John Lindley, whose *Introduction to the Natural System of Botany* (1830), and *Introduction to Botany* (1832), both published since Henslow's first lectures, appear in the reading list. This list differs significantly from the 1828 one: Lindley's text-books have replaced those of Smith, and de Candolle's three-volume *Physiologie Végétale* (1832) is added to the two earlier works by the Frenchman.

Henslow's use of Linnaeus' 'Sexual System' is apparently seen by him as a convenient device for identification:

> 'After the demonstration of each specimen, its Genus and Species will be ascertained by referring to its Class and Order in the Artificial System of Linnaeus. When the name of the plant is discovered, its right position must be sought for in the Natural System. But after some little progress has been made, this may be ascertained without at all referring to the Linnaeus System, by merely consulting some artificial analysis of the Natural Orders, like the one given by Professor Lindley in his *Introduction to the Natural System*.'

Anyone who has tried to teach Angiosperm classification may be forgiven for thinking that Henslow was rather over-optimistic. It would indeed be very revealing if any of his pupils had recorded details of what and how they learned from Henslow's lecture-demonstrations, but nothing seems to have survived. One might hazard a guess that the keen and intelligent student would have put together some knowledge of common British plant families, genera and species from the combination of the lectures and the field excursions – just as they do at the present day! The significant factor was undoubtedly Henslow's enthusiasm, manifested, as Darwin testified, in both his 'Demonstrations' and the field excursions that he ran.

Henslow's move from the text-books of Smith to those of Lindley obviously reflects his conviction that the 'Natural System' of the French School is inherently 'better', and no doubt he elaborated this view in the course of his lectures. His 1829 *Catalogue of British Plants arranged according to the Natural System*, which was a check list of the British Flora 'primarily intended for students who attend the Botanical Lectures at Cambridge' – to use Henslow's own words from the Preface – makes his conversion abundantly clear, for the Linnaean sexual system is completely omitted from the book. This slim book seems to have filled a gap in the literature, achieving a second edition in 1835.[34] To make the book really useful to the assiduous student attending his lectures, all those species recorded for Cambridgeshire are provided with a letter 'c'. A table also gives data for the numbers of genera etc., from which one can learn that Henslow counts the British vascular plants (Phanerogams) as numbering 1495 species, of which Cambridgeshire has 836.

By 1833 Henslow must have been busy writing his own text-book, which appeared in 1835 under the title *Principles of Descriptive and Physiological Botany*. The eight-page Introduction to this book is admirably clear; the material is 'conveniently arranged . . . under two heads . . . Descriptive and Physiological'. The 'Descriptive' section, the author explains, is

> devoted to the examination, description and classification of all the circumstances connected with the external configuration and internal structure of plants, which we consider in much the same light as so many pieces of machinery, more or less complicated in their structure; but of whose several parts we must first obtain some general knowledge, before we can expect to understand their mode of operation, or to appreciate the ends which each was intended to effect. In the 'Physiological', which is the other department, we consider these machines as it were in action; and we are here to investigate the phenomena which result from the presence of the living principle, operating in conjunction with the two forces of attraction and affinity, to which all natural bodies are subject.

The two Parts of the book are roughly equal in length: 145 pages for the first 'Descriptive' part, and 159 pages for the second 'Physiological' part. Those critics and writers who have assessed Henslow's botany as being 'mainly concerned with systematics' seem to have grossly misunderstood the man.[35] It is unfortunate that the 50-page review of de Candolle's three-volume *Physiologie Végétale*, which Henslow wrote in 1832, was published anonymously in the *Foreign Quarterly Review* 22. This essay, almost a book in itself, should certainly be read by any scholar interested in the rise of biology in the nineteenth century: it is obvious how much the 'Physiological' part of Henslow's text-book is based on de Candolle. Of course, it is abundantly clear that the development of modern plant physiology owed little or nothing to Henslow, to whom, for example, the fixation of carbon and the relationship of this phenomenon to the green colour of plants was at the time he wrote his text-book a great mystery. As we shall see in Chapter 11, however, in later life Henslow put much effort into keeping abreast with the rapidly-developing subject, especially as it affected agriculture, and at no stage does he show any sign of thinking that Botany is, or should be, only a descriptive science.

The extent to which accurate botanical description of detailed plant structures much preceded any understanding of what these structures 'were designed to do' is, of course, obvious from the whole history of Angiosperm taxonomy, based as it is on comparison of the flower and fruit. It comes as a surprise to see how much detailed comparative floral morphology is presented at a time when the function of flowers was still unclear. Excellent illustrations of flower structure are, for example, provided by John Martyn, second Professor of Botany in Cambridge, in his booklet entitled *First Lecture of a Course of Botany* published in 1729, almost a century earlier than Henslow's first lecture. Henslow himself seems to have made little effort to see the detail of floral structure as revealing, to use his own words in the passage quoted above, 'their mode of operation, or ... the ends which each was intended to effect', though there are impressive examples where he *does* relate form and function, as in his study of a hybrid *Digitalis* (see Chapter 6). This is in spite of an accumulating body of published material on the relationship between insects and flowers, by far the most important of which was C.K. Sprengel's book, *Das entdeckte Geheimnis der Natur im Bau und in der Befruchtung der Blumen*, published in 1793. As Proctor et al. explain:

> Sprengel's work made little impact for over half a century, although his ideas seem to have been quite widely known and discussed, perhaps more by entomologists than by botanists.[36]

Eventually, it fell to Charles Darwin to make the next important contributions to the function of flowers, in a series of publications from 1857 until the culmination in the book entitled *The effects of Cross- and Self-fertilisation in the Vegetable Kingdom* in 1876. We can see in Darwin's eventual contributions to the study of floral biology some of the fruits of Henslow's teaching.

Even if we may feel that Henslow's stated desire to try to relate form to function is oddly undeveloped in terms of Angiosperm flowers, we should also remember how relatively late was the growth of recognisably modern

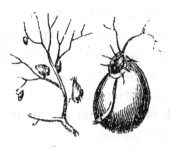

FIGURE 8 Bladderwort, *Utricularia*, from Henslow's text-book.

plant physiology. This important branch of our science was held back by the absence of anything we would now call 'organic chemistry' and also of any general understanding of what came to be known by the middle of the century as the 'cell theory' of the structure of organisms. In Cambridge the eventual break-through of plant physiology was a more painful process than it might have been because Henslow's pupil and eventual successor, Babington, as we have already seen, was waiting too long for his former Professor to die . . . a sad commentary on the state of unreformed Cambridge.

Sometimes, however, we find that Henslow *does* relate form to function, attempting to answer the questions 'how does this structure work?' and even more daringly 'what is it for?' In one particular case we can appreciate the special link between teacher and pupil: this concerns the remarkable submerged carnivorous plant *Utricularia* or Bladderwort. On page 41 of his text-book Henslow describes, illustrates and briefly discusses the structure of this aquatic plant, which he knew well in the Cambridge area, and which he demonstrated on his excursions in the wetlands with his students (Figure 8). No doubt he did what generations of Cambridge teachers of botany have done: in front of an admiring group of students at the edge of the fenland ditch he pulled out from the water a length of Bladderwort with its attractive orange-yellow flower-spike, and asked the assembled party to hold the dripping mass up to their ears to listen to the popping of the bladders. Here are Henslow's words that accompany the illustration:

> The roots of certain aquatics belonging to the genus *Utricularia* are furnished with appendages in the form of little membranous bladders which are partially filled with air, and serve to float the plant, in order that it may be enabled to flower above the surface of the water.

There is no mention of its trapping small water-fleas and other aquatic animals, though Henslow describes both *Dionaea*, the Venus Fly-trap, and

Drosera, the Sundew, later in the same book, accepting somewhat cautiously as 'the only plausible conjecture' that the plants are actually feeding on their prey.

Henslow's drawing of the *Utricularia* bladders certainly shows accurate observation of form. His statement that the bladders are 'partially filled with air' is, however, only true *when the plant is lifted out of the water* and air is drawn in to the many tiny bladders with an audible popping noise; so we award him half-marks for 'how does it work?' But the conclusion he draws is then quite wrong: the specific gravity of these long intricately-branched submerged shoots is *not* reduced by air-filled bladders, and the ability to bear a flowering stem erect is in no way dependent on bladders. In fact, many other submerged aquatics such as *Myriophyllum* and *Hottonia* manage to float near the surface and flower above it without any obvious adaptation, as Henslow must have observed.

Look now at Charles Darwin's book on *Insectivorous Plants* published in 1876, well after Henslow's death. A whole chapter is devoted to *Utricularia*, which he must have met, possibly for the first time, on Henslow's excursions. We learn here that Darwin 'was led to investigate the habits and structure of the species of this genus . . . more especially by Mr Holland's statement that 'water-insects' are destined for the plant to feed on'. A footnote tells us that Holland published his paper in 1868. Darwin's own careful observations confirmed that indeed small crustaceans were often found inside the bladders, and that the special glands on the inner lining could have a digestive function, but, interestingly, even he could not detect the remarkably precise trigger mechanism that drew in the microscopic prey. It was not in fact until 1911 that the functioning of the Bladderwort traps was finally described.[37]

Ironically, the publication of Henslow's text-book coincided with the ending of the 'Oxbridge' monopoly on Fellows of the Royal College of Physicians which led to a great reduction in undergraduates attending his lectures, and we can only conclude that the sale of the text-book, which cost six shillings (a relatively large sum in 1835), would be very small in Cambridge. No further Syllabuses of the actual lecture courses were, apparently, printed until 1848, and, as we shall see, this is a suggested (rather than an actual) course for the proposed new Pass Degree. So Henslow does not include his own excellent text-book in any reading-list for his students. The book actually ran to a second edition and, according to Jenyns, 'a third was in preparation at the time of his death, with extensive modifications'.[38] Jenyns' opinion of the book as 'long considered the best manual of structural and physiological botany in the English language' may be a little exaggerated, but it certainly achieved some general acclaim and a reasonably wide audience. A wide audience was, indeed, the explicit aim of the book. That section of the Introduction (pp. 4–5) entitled 'Advantages of our pursuit' is clearly aimed at the rapidly-expanding educated middle class and makes delightful, if somewhat verbose, reading. Here is an extract:

> All who feel an unaccountable delight in contemplating the works of nature; who admire the exquisite symmetry of crystals, plants and animals; and who love to meditate upon the wonderful order and regularity with which they are distributed; possess a source of continued enjoyment within themselves, which is capable of producing a most beneficial effect upon their temper and disposition, provided they do not abuse these advantages by making such studies too exclusively the objects of their thoughts and care.[39]

By 1838, Henslow was, however, beginning to share his enthusiasms with a much wider audience, and struggling with the problem that most of that audience could barely sign their own name, let alone read and understand such perorations! We shall see in later chapters how he tackled these problems.

Although Henslow's main schoolboy enthusiasm was for entomology, there seems little doubt that, like many keen naturalists with a love of the English countryside, he had developed an early interest in 'herborizing'. After all, some knowledge of the flora and vegetation is essential to any worth-while entomology as soon as it develops into something more than mere specimen-collecting, because a particular host plant is essential for many insect species. But it was obviously more than that. It seems very probable, therefore, that the young Henslow was in the habit of collecting and pressing botanical specimens long before the Chair of Botany became his, but none of this activity seems to have been more than a small part of his general collecting. Certainly the influence of Leach and Stephens on the schoolboy ensured that his organised, systematic study was zoological, and principally expressed in mounted collections of insects and shells; Jenyns' account of the early stages of his acquaintance with Henslow in Cambridge confirms this picture.

From about 1818 onwards we have evidence that Henslow began making specimens of unusual plants in the vicinity of Cambridge, presumably keeping these in his own collection until after 1825. An example is provided by the surviving specimen of *Ornithogalum nutans*, a garden escape recorded for the first time in Cambridgeshire by Henslow as being 'abundant but accidental between Trumpington and Cambridge, 2 May 1821.' Commenting on this first record, Babington in his *Flora of Cambridgeshire* (1860) writes: 'Abundant for one season (1821) between Cambridge and Trumpington, but did not occur again!; H[enslow]. Not a native of this county'. [The exclamation mark is the convention used in this Flora to indicate that the author had seen the specimen.]

From 1825, when Henslow began the considerable labour of rescuing the damaged Herbarium of John and Thomas Martyn, he obviously made a more organised effort to build up a properly-documented local collection, and had made enough progress by 1827 to begin to publish a series of Annual Reports on the Botanical Museum. Dated 25 March 1828, the second report briefly explains the rescue of the old Museum and Library materials and continues:

> The Senate having passed a Grace during the last year, for allowing an annual stipend for supporting and increasing this Museum, there need be no apprehensions of any further neglect in this department and Professor Henslow therefore ventures to solicit the assistance of his botanical friends in procuring materials for supplying the following collections, which he has commenced:
>
> 1. Herbarium of the Plants of Cambridgeshire
> 2. Herbarium of the Plants of Great Britain
> 3. Herbarium for a general collection.
>
> N.B. All local and rare specimens will be acceptable for the first two collections, and any native exotics or cultivated specimens for the third. It is requested, where possible, that the exact habitat, <u>distinctly written</u>, and time at which each specimen was gathered, may accompany it [There follow further categories to cover lower plants, bulky fungi, fruits and seeds, etc.][40]

An excellent example of Henslow's care in creating the basis of the University Herbarium we have today is provided by a sheet of the Safflower, *Carthamus tinctorius*, which contains two specimens of this Composite, nowadays not uncommon in southern England as a bird-seed alien species. The larger specimen carries a handwritten label with a quotation from the Linnaean description and 'Mus. Martyn': it is obviously one of the few sheets rescued by Henslow from the Martyn Herbarium. The smaller specimen adjacent carries the label (in Henslow's writing): 'H. Bot. Cantab 2 Sept. 1827 J.S. Henslow'. Thus in a single, well-preserved herbarium sheet we have evidence of Henslow's thorough, organised approach: he preserves the historically interesting specimen, the provenance of which we are quite ignorant about, with a new, accurately labelled specimen taken by him from his Botanic Garden on a day in September 1827.

From 1826, when Henslow began regular 'herborizing excursions', his systematic documentation of the local flora is evident in a number of publications, among them what we might now call printed 'check lists' of plants collected on a particular excursion. Two lists cover local excursions in and around Cambridge on 8th May 1828. The first of these lists the 'plants gathered near Cambridge in May 1828', with an indication 'C' for those found in 'deserted chalk-pits near Cherryhinton' and 'S' for those found at 'Shelford Common'. This list obviously selects the more 'interesting' finds, omitting, for example, all grasses, but listing six *Carex* species, a fact that reminds us that Henslow had obviously spent time learning his way about the sedge family Cyperaceae. In fact, we have surviving herbarium specimens of the sedge he called *Carex fulvus* (nowadays *C. hostiana*) collected at Shelford Common in 1825 and 1826, together with one of his water-colour paintings showing the same species, which was probably part of the same collection (Figure 9).

This check list is also interesting in emphasising the land use changes since Henslow's time. Shelford Common no longer exists: like other ill-drained boggy pastures in the vicinity of Cambridge, it was the home of several species which became rare or extinct as the nineteenth century proceeded, and which are now represented by very few relatively undrained habitats. On the other hand the Cherry Hinton chalk-pits have survived – the site is now a Local Nature Reserve – and all Henslow's 'special plants' are still to be seen there, with the sole exception of the Burnt Orchid, *Orchis ustulata*.

The check list for the following week's 'herborizing', entitled 'Plants gathered near the Banks of the Cam . . . on 15th May 1828' is much more extensive, and seems to include all vascular plants seen, adding for good measure four mosses and four fungi – 138 species in all. Similar printed lists survive for 1829, and in 1830 Henslow printed a consolidated list, showing all plants seen on five local excursions held that year. No further

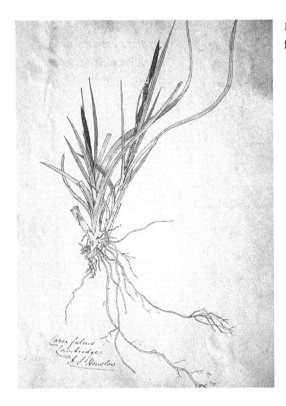

FIGURE 9 Sketch of 'Carex fulvus' by Henslow.

lists survive, and it seems likely that he abandoned this particular piece of documentation.[41]

A favourite excursion, sometimes undertaken as one of the 'official' list, was a walk to 'Whitwell Farm and the Petrifying Spring at Coton'. Henslow became aware of the botanical interest of this site as early as 1818, the year of his graduation, for we have a specimen of *Paris quadrifolia* collected by him at 'Whitwell near Coton' on 1 April 1818. A specimen of the aquatic moss *Fontinalis antipyretica*, collected by Henslow at Coton in 1821, is obviously from the Petrifying Spring (Figure 10). Visiting this Petrifying Spring had developed in the eighteenth century as a pleasant local tourist attraction, and no doubt that is how Henslow first found there the remnant of woodland in which there were relatively rare plants, including Herb Paris. The species list for 30 April 1830 contains 71 vascular plants, three mosses and one fungus, and includes *Orchis morio*, *Helleborus viridis* and *Stellaria holostea*, all now very rare in the Cambridge area. Henslow's interest in Herb Paris developed into a pioneer study of what we would now call 'meristic variation'. Noting that the normal 4-symmetry of the leaf and flower is not completely regular, he began to collect samples with the help of 'two

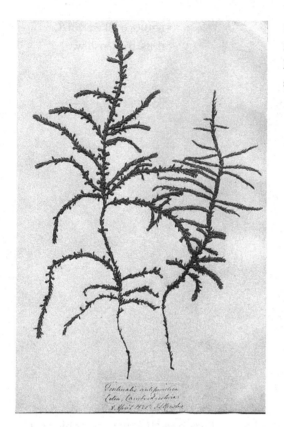

FIGURE 10 Henslow's specimen of the aquatic moss *Fontinalis* from the Coton springs.

friends, Messrs. Babington and Downes, in a habitat near Cambridge'. On each of 1500 flowering specimens [!] he, with the help of his student 'friends', carefully scored the number of foliage leaves in the whorl and for the same plant the number of floral parts. The results Henslow published in 1832 in a remarkable pioneer study which must rank as one of the earliest statistical studies of meristic variation in any plant (Figure 11). It was in one sense too early, for the development of statistics, and in particular ideas and treatments for what we would now call the 'normal curve', were in their infancy.[42]

Another example of Henslow's application of quantitative methods to botany is the paper he gave to the Philosophical Society in 1834 on the spiral arrangement of the cone-scales of Spruce (*Picea excelsa*). Again this seems to be something of a pioneer investigation into a subject that developed later and became known as phyllotaxis, a term first used in 1857 (Figure 12).

Many of Henslow's records, accompanied by herbarium specimens, are of course of interest to students of the Cambridgeshire flora; details of

a. flowering stems   *b* foliaceous whorl that crowns the barren stalks
c, d, e filament bearing two anthers   *f, g*   stigma becoming branched
h   Trillium: an allied genus

FIGURE 11 Illustrations accompanying Henslow's *Paris quadrifolia* paper.

some of these are given in Appendix 5. Before we leave the topic of Henslow's herbarium activities, however, we should mention his *Fasciculus.* In 1834, Henslow prepared for sale a set of 50 British plants, 'price 4s/-', most of which were collected locally. Such sets of plants were frequently made for sale at the time, and indeed many of the larger Herbaria in later Victorian times were accumulated through purchase or exchange of sets of plants (often called 'Exsiccatae'). In the accompanying text, Henslow tells us that 'the only object in preparing these *fasciculi* is the desire of assisting beginners in botany in acquiring a knowledge of our native plants, many of the species have been selected from the more obscure or difficult orders, but no attempt has been made to procure exclusively such as are rare'. This statement is borne out by the fact that no fewer than 16 of the plants are grasses, mostly very common ones – though there are classical rarities,

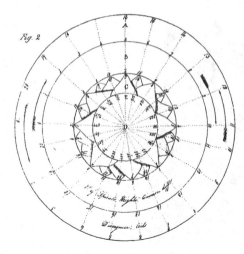

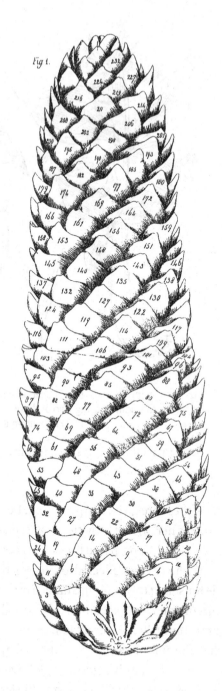

FIGURE 12 Illustration prepared by Henslow for his paper to the
Cambridge Philosophical Society on the cone-scales of Spruce (1834).

some (such as the sedge *Rhynchospora alba* collected at Gamlingay on 29 July 1833) now extinct in Cambridgeshire. Although the *Fasciculus* is labelled 'no. 1', no later sets seem to have been prepared, and we can only conclude that Henslow was disappointed by the reception.[43]

6   Educating Charles Darwin – and others

In her excellent biography of the young Charles Darwin, Browne writes:

> Meeting [Henslow] was rightly considered by Darwin as the one circumstance 'which influenced my career more than any other'. No other man had so immediate an impact on his developing personality, nor did any other figure have so important a role in directing the course of his early scientific experiences.[1]

No-one who has attempted a biography of Darwin has seriously questioned Henslow's primary role; indeed, it might be said that Henslow's formative influence remains one of the few assessments that survives unchallenged all kinds of critical reworking of the story of Darwin's life and work.[2]

The first Darwin–Henslow meeting was in no way an accidental, serendipitous event. Darwin's elder brother Erasmus, with whom Charles had spent much of his first year (1826) in Edinburgh University, had painted for him a glowing picture of the qualities of the Cambridge Professors and the University in general, based on Erasmus' own undergraduate days in Cambridge from 1822 to 1825. In particular, Erasmus had greatly appreciated the lectures, accompanied by demonstrations of chemical experiments, given by the Professor of Chemistry, James Cumming (to whom Henslow had earlier acted as an assistant), and had also attended the course given by Henslow, the young, enthusiastic Professor of Mineralogy, in 1825. As Browne explains, Henslow liked 'the gentle cultivated Erasmus, one of the earliest Cambridge "Apostles", full of intellectual promise', and must have been 'just as inclined to be as agreeable to Darwin as Darwin was to him'.[3]

In spite of this considerable preparation of the ground, however, Darwin did not attend Henslow's botany lectures in his first year (1828), though his name does appear on the botanical lecture lists for the three successive years 1829 to 1831. In fact Henslow's lectures were the only ones given in any course that would now be recognised as 'natural science' attended by Darwin during his whole Cambridge undergraduate career. Darwin gives his own explanation of his behaviour in the *Autobiography*:

Public lectures on several branches were given in the university, attendance being quite voluntary; but I was so sickened with lectures at Edinburgh that I did not even attend Sedgwick's eloquent and interesting lectures. Had I done so I should probably have become a geologist earlier than I did. I attended, however, Henslow's lectures on Botany, and liked them much for their extreme clearness, and the admirable illustrations; but I did not study botany. Henslow used to take his pupils, including several of the older members of the University, on field excursions, on foot, or in coaches to distant places or in a barge down the river, and lectured on the rarer plants or animals which were observed. These excursions were delightful.[4]

We should not think that there was anything particularly unusual in the young Darwin's student experience in late Georgian Cambridge. Newtonian mathematics still dominated the teaching, and there was no Natural Sciences Tripos until much later.[5] Unlike his teacher, who was a more than averagely competent mathematician, Darwin showed no particular aptitude for the subject – though he seems to have enjoyed Euclidean geometry. His Tutor in Christ's College, John Graham, later Master of the College, saw in the young Darwin nothing but 'a typically idle "poll man" destined for the Church, dawdling away his Cambridge days with his horse and his gun until thrown into a panic by the approaching examinations'.[6]

Darwin was officially registered for an ordinary BA degree, which was the usual preliminary to entering the Church, and his syllabus included some elementary mathematics, Paley's *Evidences of Christianity*, and a few classical Latin and Greek texts. In later life, Darwin's own comments on this syllabus are illuminating:

> During the three years which I spent at Cambridge my time was wasted, as far as the academical studies were concerned, as completely as at Edinburgh and at school . . . In my second year I had to work for a month or two to pass the Little Go [the Previous Examination], which I did easily. Again in my last year I worked with some earnestness for my final degree of BA, and brushed up my Classics together with a little Algebra and Euclid, which latter gave me much pleasure as it did whilst at school . . . In order to pass the BA examination it was, also, necessary to get up Paley's Evidences of Christianity and his Moral Philosophy . . . The logic of this book [*Evidences*], and as I may add of his Natural Theology gave me as much delight as did Euclid. The careful study of these works, without attempting to learn any part by rote, was the only part of the Academical Course which,

as I then felt and as I still believe, was of the least use to me in the education of my mind.[7]

Without Henslow, what would the young Darwin have made of his career? Would he, under parental pressure, have settled for ordination and a comfortable country living? The climate of the times would certainly have encouraged such a solution, and his doubts and reservations about Christianity, as we shall see in Chapter 9, do not seem to have been much more serious than Henslow's own until much later in his life. Of course, we cannot answer the question. We can, however, say in what ways Henslow would have left his mark on this pupil, even if the voyage of the 'Beagle' had never happened.

It was, apparently, the Henslow soirées that gave Charles Darwin his opportunity to meet and appreciate the man who 'knew every branch of science' and whom he was 'prepared to reverence'. He was invited through the good offices of his cousin William Darwin Fox, with whom he cemented a great friendship in his first year at Christ's College. Although it appears that Darwin was not a regular attender at the soirées until his second year,[8] it seems likely that he had fallen under the spell of Henslow before the end of 1828, for in January of that year Darwin had begun his Cambridge undergraduate career, and Henslow began his regular soirées, held on Fridays in Full Term, in February. Browne is sure that

> the last two terms at Cambridge – the ones after passing his degree examination in January 1831 – were the months [Darwin] really came to know Henslow closely. During those final weeks he took long walks with him and joined the family dinner table, becoming such an obvious favourite that he was called by some of the other dons 'the man who walks with Henslow'.[9]

We have no detailed eye-witness account of a soirée at the Henslows', but we have several witnesses, including Darwin himself, to the success of these informal gatherings of 'no more than ten or fifteen people eased by the homely surroundings of a family man and by Mrs Henslow serving tea'.[10] The most familiar and detailed account of the soirées and their social and educational success is to be found in the Memoir of Henslow by his brother-in-law Leonard Jenyns who was, of course, in a uniquely favourable position to appreciate the hospitality of his sister and her husband. It is fortunate for us that Jenyns went about his task of writing the Memoir very methodically, consulting a wide range of friends and colleagues to supplement his own considerable knowledge of his brother-in-law's life, works and character. All the replies he received are preserved[11] and so one can go back to the originals, to check and see what Jenyns used and omitted. Apart from Darwin's own testimony, the most interesting tribute, worth quoting *in extenso*, came from Miles Berkeley, whose contribution to the study of mycology and plant diseases in the mid-Victorian period was quite outstanding. Berkeley writes:

> My more intimate acquaintance [with Henslow] was from 1821 to 1829. I had for some years before I went up to Cambridge been much attached to Botany and Henslow called on me at the request of some relations in Blackheath who had in former days been acquainted with certain members of his family. At that time he was making Botany a study with a view to the Regius Professorship, rather than from any peculiar predilictions for that rather than for any other branch of Natural History. His habits were so methodical that he had no difficulty in taking up half a dozen subjects at the same time, and he had so many little plans and schemes for economisng time and communicating efficiently with others, that he could get through more work in a given time than perhaps any other of his contemporaries. During my residence in Cambridge [1821–25] I could scarcely be said to be on very intimate terms with him, but after I took my degree, I went two or three times to visit him, once I recollect well with my friend Lowe, and profited very much by our intercourse. I had the pleasure of being present at one or two of his soirées which he contrived to make extremely agreeable and interesting, and he not only gave young men of a scientific turn opportunities of beneficial intercourse, but he had especial meetings of more intimate friends, for improvement in French, at which the Marquesa Spinosa [*sic*] were I think usually present.[12] Somewhere about this time he was engaged in a history of Hybrid Digitalis, which is a model of patient investigation. He had a good microscope, though of rather early date, and was skilled in its use.'[13]

The 'history of Hybrid Digitalis' referred to was a study undertaken by Henslow on a plant that had occurred unexpectedly in his garden in Cambridge: 'chance having favoured me with a hybrid Digitalis during the past summer (1831), in my own garden, I employed myself, whilst it continued to flower, which was from June 19 to July 22, in daily examining its character and anatomizing its part of fructification'[14] (Plate 4).

We can imagine how the young Darwin would have been fascinated by this study, involving as it did very careful and accurate use of the microscope. His friend John Maurice Herbert had presented to him in May 1831 a fine 'Coddington microscope'. As Browne tells us:

> The present was the most useful scientific gift Darwin could have hoped to handle. Every free moment was now filled with exploring the miniature delights of a natural history world formerly denied to him. His last two months at university were spent studiously gazing through the new instrument.[15]

Darwin recalls in later years a 'trifling incident which showed [Henslow's] kind consideration'.

> Whilst examining some pollen-grains on a damp surface I saw the tubes exserted, and instantly rushed off to communicate my surprising discovery to him. Now I do not suppose any other Professor of Botany could have helped laughing at my coming in a hurry to make such a communication. But he agreed how interesting the phenomenon was and explained its meaning; so I left him not in the least mortified, but well pleased at having discovered for myself so remarkable a fact, but determined not to be in such a hurry again to communicate my discoveries.[16]

We are not told in the Autobiography when this 'trifling incident' occurred, but it seems reasonable to suppose it was around the time that Henslow himself was doing his careful microscopic investigation of the *Digitalis* hybrid. This, and similar events, must surely have been the basis for Darwin's assessment:

> Another charm [of Henslow], which must have struck everyone, was that his manner to old and distinguished persons and to the youngest student was exactly the same: to all he showed the same winning courtesy. He would receive with interest the most trifling observation in any branch of natural history; and however absurd a blunder one might make, he pointed it out so clearly and kindly, that one left him no way disheartened, but only determined to be more accurate the next time. In short, no man could be better informed to win the entire confidence of the young, and to encourage them in their pursuits.[17]

The admiration, even hero-worship, shown by the pupil Charles Darwin for Henslow seems to have been at times excessive, and must have elicited snide or even adverse comment from his student acquaintances. John Medows Rodwell, for example, a student at Gonville and Caius College and contemporary with Charles Darwin (they attended Henslow's lectures together in the years 1830 and 1831), recalled, in reminiscences sent to Francis Darwin, a memorable occasion of a visit to Gamlingay, a favourite acid heath and bog area which generations of Cambridge naturalists have visited:

> We once had a very amusing excursion to Gamlingay heath in search of Natter-jacks [*Bufo calamita*, a rare native toad, long extinct at Gamlingay]. Darwin was very successful in detecting the haunts of these pretty reptiles and catching them. He brought several to Prof. Henslow who said laughingly – 'Well, Darwin, are you going to make Natter-jack pie?'[18] (Plate 5)

This was probably Henslow's annual student excursion to Gamlingay, one of a series that he held in the Easter Term in connection with his lecture course, and seems almost certainly the occasion laconically recorded by another pupil, Charles Cardale Babington, in his journal for 19 May 1831: 'Henslow's party to Gamlingay'.[19]

It was Rodwell who also recalled in after life the story of Darwin and the Bladderwort *Utricularia* which we have already met in Chapter 5:

> In one of Professor Henslow's botanizing excursions to Bottisham Fen I well recall an amusing incident which befel Darwin. In order to clear the ditches we were provided with several jumping poles with which we had to find the Utricularia a specimen of which caught his keen eye, and in order to secure it he attempted to jump the ditch on the opposite side of which it grew. Not however having secured sufficient impetus for the leap, the pole stuck fast in the middle in a vertical position of course with Darwin at the top. Nothing daunted however, he coolly slid down, secured the prize, and brought it, all besmirched as he was to the amused Professor.[20]

Another contemporary recalling for Francis Darwin what he could remember of Darwin's student days was William Allport Leighton, who had been at school with Darwin:

> '. . . and we never met again until we went to Cambridge. Here we were pupils of Professor Henslow and attended his botanical lectures and his weekly conversaziones and occasional herborizations. Darwin's manner

was still reserved and proud but it was patent to all that his mental and intellectual powers were greatly developed and he seemed to be pondering in the mind some good results. Professor Henslow was a good botanist and geologist and a most patient instructor answering every simple question with as much gravity and kindness as if it had been of greatest importance. Thus his pupils were never discouraged but thoroughly confided in him and made him their friend thro' life. Darwin hung upon the Professor's lips and words and no doubt he influenced him very much as to his voyages and explorations in the Beagle. I remember that the Professor in his concluding remarks at the close of lectures said he hoped his teaching had influenced many to perseverance – certainly he knew it had influenced <u>one</u>.[21]

Some tension and jealousy was inevitable, and was almost certainly exacerbated by Darwin's own temperament, which has inevitably been discussed by many writers. It was nowhere more obvious than in the tensions between Babington and Darwin. Both claim to have been Henslow's chosen assistant to prepare and clear away the materials used in his lectures, and no doubt the Professor's tact was often used to steer a tranquil course. Ludicrously, the Darwin–Babington tension surfaced almost violently in competition for rare beetles. The young Babington, destined eventually to succeed Henslow as Professor of Botany, was a passionate collector of beetles – so much so that he earned the simple nickname of 'Beetles' Babington. Darwin's own collecting passions included beetles; as Browne explains:

> Neither man bothered to hide his dislike of the other: theirs was a long-standing difference originally rooted in mutual competition for water-beetles, and matters came to a head when Darwin discovered that one particular natural history supplier – a Mr. Harbour – was giving Babington first choice of the insects coming in. Harbour was 'a damned rascal', Darwin exploded, and he vowed never to buy from the man again.[22]

One other quality in Henslow's teaching must have been of enormous importance, not only to Darwin but to other keen and able students. This was the breadth of the knowledge Henslow possessed. It is tempting to suppose that there was nothing unusual in Georgian and early Victorian times for a man to be an all-round naturalist, and it is of course true that many a country vicar, with the comparative leisure and comfort of a Gilbert White if not his talent, could collect birds, beetles, shells and flowers and keep a naturalist's diary recording the first burst of song in January from the song-thrush or the time of flowering of the primrose. Jenyns, indeed, did exactly this when he became Curate and later (1828) Vicar of the Cambridgeshire parish of Swaffham Bulbeck. But Henslow's achievement was rather different. He was, it is true, a typical collector: as we shall see, throughout his life he recorded, drew and preserved an unwieldy and apparently undisciplined range of natural objects; but he had an ordered mind which could, at least in the early days, take in and give out large quantities of information in a pleasantly accessible form. He could also, as we have seen in Miles Berkeley's tribute, work very methodically, 'so that he had no difficulty in taking up half a dozen subjects at the same time'. These qualities, when combined with a lively social conscience and a good sense of humour, must have made him an irresistible influence. So it is in field and laboratory demonstrations rather than in the lecture course *per se* that Henslow's success was most obvious. Indeed, though Darwin himself speaks warmly in later years about the Henslow lectures, commending them for their clarity, one suspects that he was, for once, exaggerating the value to him of this part of 'the Henslow experience'. We have already quoted Jenyns' own assessment of Henslow as a lecturer: essentially, no eloquence, but a good voice, and an ability to present facts clearly, and to explain difficult points. Anyone who has experienced University lecture courses as a student, and later been in the position of delivering essentially the same kind of lecture course, is likely to decide that there is a place for eloquence and panache and also a place for ordered presentation. Each contrasting method can play its part in the complex business of imparting a full, rounded education. What does *not* succeed is obscure inaudibility. Perhaps Henslow's good voice and well-planned material commended him to the majority of his student audience, a large proportion of whom were, during the years 1831–35, medical students for whom attendance was practically compulsory.

Henslow's acquaintances in the world of natural history were not by any means confined to the strictly academic strata of society. He had the open, enthusiastic temperament and the social position needed to make any guest feel at home, especially at the Friday evening soirées. One of the more colourful of these guests, who has left us a fascinating account of a week in Cambridge in 1828, was the famous American naturalist and painter of birds, John James Audubon (1785–1851). A letter from Henslow to Jenyns, dated 6 March 1828, is worth repeating in full for its charming enthusiasm:

My dear Leonard,

You must contrive to come over here either on Friday, or Monday next – Mr Audubon has brought some engravings of his work on American birds, more splendid that you ever beheld – Elephant folio, highly coloured, 2 guineas per No – each No containing 5 plates – Birds natural size – Groups of birds & flowers where the subjects are small – 26 years in the woods of America – losing 70 per cent by publication, from a want of subscriptions – determined to continue it at all costs – subscriptions increasing rapidly – The Public library has subscribed and I hope the Fitzwilliam [Museum] will. I have also planned a subscription of 5/- per head per annum for a copy for the Phil. Soc., there being 5 Nos = 10 Gs. published annually, Of course I shall put your name down.

The man is very amiable, very zealous, and deserves to be extensively patronised. The paper for each plate costs him 2/-, so that you have engraving and colouring for 6/-. He is to bring his drawings here on Friday evening – & I intend to summon a meeting of the Council [of the Philosophical Society] on Monday for the purpose of mentioning the subject & one or 2 other things I have to say. Perhaps you wd prefer coming on that day – I have secured a copy of the Flora Danica 10 vol. fol. for 10 [shillings]

Yours ever sincerely

J S Henslow[23]

The young Henslow is so enthusiastic that all punctuation is thrown overboard in this outpouring! Whether it was Henslow himself, having perhaps met Audubon during his stay in London the previous winter, who invited this remarkable American 'backwoodsman' to visit Cambridge, is not clear, but we know from Audubon's own diary[24] that he duly came to Cambridge, arriving on 3 March

at this famous University town at half-past four this afternoon, after a tedious ride of eight & a half hours from London, in a heavy coach which I entered at the White Horse, Fetter Lane, & I am now at the Blue Boar, & blue enough am I . . . The driver held confidences with every grog shop between London & Cambridge, & his purple face gave powerful evidences that malt liquor is more enticing to him than water.

Audubon is a delightful diarist, and it is tempting to repeat the whole entry for each of his six days in Cambridge, culminating on Sunday 9 March, when he is overcome with awe at the solemn beauty of the place and its religious observance. His contact with Henslow, it would seem, though obviously enjoyable, does not stand out from that with the other 'learned men' to whom he had brought letters of introduction and, disappointingly, he gives no account of the dinner on Friday 7 March – the Henslow soirée to which Jenyns was also invited. The entry for 5 March incudes: 'Professor Henslow invited me to dine on Friday, & just as I finished my note of acceptance came in with three gentlemen.' There follows an account of his dining in Hall that evening at Corpus Christi College, his host apparently being 'the Rev. H. Greenwood'.[25] The description is so good that it deserves a verbatim extract:

> . . . & then we went into the 'Hall', where dinner was set. This hall resembled the interior of a Gothic church: a short prayer was said, & we sat down to a sumptuous dinner. Eating was not precisely my object, it seldom is: I looked first at the *convives*. A hundred students sat apart from our table, & the 'Fellows', twelve in number, with twenty guests constituted our 'mess'. The dinner, as I said, was excellent . . . We then went to the room where we had assembled, & conversation at once began; perhaps the wines went the rounds for an hour, then tea & coffee, after which the table was cleared, & I was requested to open my portfolio. I am proud <u>now</u> to show them, & I saw with pleasure these gentlemen admired them. I turned over twenty-five, but before I had finished received the subscription of the Librarian for the University, & the assurance of the Secretary of the Philosophical Society that they would take it. It was late before I was allowed to come away.

Though it is obvious from his journal that Audubon was greatly impressed by the kindness and generosity of his University hosts, he never loses sight of the main purpose of the visit, which was to sell as many copies of his *Birds of America* as possible to the rich patrons he found throughout Britain.[26] Today it is possible to see in Cambridge libraries the fruits of Audubon's 'patience and industry'. The success of the convivial evening at Corpus is to be seen in the University Library: a fine first edition of the *Birds of America*. Similarly the Fitzwilliam Museum houses its *first* edition, the origin of which can be traced to the Henslow soirée day – Friday 7 March.

> After breakfast I went to the library, having received a permit [presumably the University Library], & looked at three volumes of Le Vaillant's 'Birds of Africa', which contain very bad figures. I was called from here to show my

work to the son of Lord Fitzwilliam who came with his tutor, Mr Upton. The latter informed me the young nobleman wished to own the book. I showed my drawings, & he, being full of the ardor of youth, asked where he should write his name. I gave him my list; his youth, his good looks, his courtesy, his refinement attracted me much, & made me wish his name should stand by that of some good friend. There was no room by Mrs Rathbone's [a Liverpool Quaker who with her husband enthusiastically supported Audubon], so I asked that he write immediately above the Countess of Morton, & he wrote in a beautiful hand, which I wish I could equal, 'Hon. W.C. Wentworth Fitzwilliam'. He is a charming young man, & I wish him bon voyage through life.

We assume that Henslow himself did the necessary persuasion of the Council of the Philosophical Society on the Monday following the Audubon visit. Their first edition, described as 'the five volumes of Plates and the five volumes of text of Audubon's *Birds of America*' in the Society's Council Minutes of 16 October 1922, was eventually sold by the Society after careful soundings with libraries and booksellers about its value.[27] What would the 1827 set of the five parts containing the first 25 Plates fetch in the open market today? A complete first edition (1827–38) is variously estimated as worth half a million to four million pounds. Even allowing for inflation, perhaps the Philosophical Society should have hung on to their set, even though it was duplicated in other Cambridge libraries. Henslow himself, we assume, made nothing on the deal. He did, however, earn a native North American bird named after him: Henslow's Sparrow, *Ammodromus (Emberiza) henslowii* (Plate 6). Strictly speaking, this bird is a Bunting rather than a Sparrow, but its common name remains in the modern American ornithological literature, and ensures that the fourth Professor of Botany in Cambridge is eponymously recognised by American naturalists.[28] Audubon writes 'In naming it after the Rev. Professor Henslow of Cambridge, a gentleman so well known to the scientific world, my object has been to manifest my gratitude for the many kind attentions which he has shown towards me.'[29]

It is, perhaps, not inappropriate that Henslow's limited fame in the New World should be transmitted largely through an inconspicuous little bird. Although it is true that Henslow was not an ornithologist, he was pre-pared to 'collect' birds by shooting or trapping them and having them stuffed to present to the Philosophical Society's Museum,[30] and was emphatically a naturalist whose early enthusiams were undoubtedly for animals rather than plants. His field excursions betrayed that continuing enthusiasm and he would never have captured the young Darwin's affection if he had been narrowly and pedantically limited to collecting and naming plants.

The extent to which the young Henslow pursued in Cambridge invertebrate zoology in the field is made very clear by Leonard Jenyns who writes:

> Zoology . . . seemed then [1820] to be his favourite pursuit. Botany had been scarcely taken up . . . though, within a twelvemonth from that time, we commenced together the formation of a collection of British plants. He would often make excursions in the fens or down the river, where he was the first to notice many of the local species of insects & shells that are to be found in that district.

Jenyns goes on to describe how:

> there is one little creek running into the Cam, at a place called Backsbite [sic: it is 'Baitsbite' now] about two miles below the town of Cambridge, where Henslow discovered 'upon floating weeds, two specimens of an insect which, from its extreme rarity, has been met with only in one or two instances in the country. This species, the *Macroplea equiseti* of entomologists, is figured in Curtis *British Entomology*, from the identical specimens taken by him on the above occasion; and the volume in which it occurs is dedicated to the captor, by its talented author . . .[31]

Henslow and Jenyns pursued together with equal enthusiasm their studies of conchology, the fruits of which can be seen in Jenyns' monograph on a group of small bivalve molluscs, published in 1832.[32] Some of the flavour of these joint collecting expeditions in the 1820s can be distilled from their correspondence: thus, on 4 October 1826, Henslow writes as follows:

> Dear Leonard,
>   You will find the minute ciliated helix (★) at the bottom of the pill box in your tin case. It seems to me very curious – perhaps the one you mentioned as found near Dartford. The Acari are in the bottle of spirits with a few minute beetles etc which were in the Madingley moss. Turn them all into a saucer, & be careful in searching for the Acari as some are very minute & the

FIGURE 13 Sketch of *Helix Madingleyensis* by Henslow.

spirits rather muddy. Be careful also in opening the pill box, where you will find a little hairy insect with long horny jaws, which I found crawling on my writing paper one day (*). I am naming the *Peziza* – [?] if it may not turn out a young specimen of the large one I found in the plantations last year.

Yours ever,

    J. S. Henslow[33]

This letter is embellished by tiny sketches at the points indicated by *, and an enlargement of the 'minute ciliated helix' on the opposite page, inscribed 'Helix Madingleyensis' (Figure 13). No such *Helix* name exists in a published form, and we must assume that Henslow was being light-hearted in so 'christening' his find.[34]

Cooperation between Henslow and Jenyns on the study of the local molluscan fauna takes a somewhat unexpected turn in the story of the Roman Snail, *Helix pomatia*. In an unpublished Ms written by Jenyns and dated 1869 there is a mass of information originally assembled for publication as a joint project with Henslow:

> It was originally intended to have been a much more detailed work; the late Professor Henslow and myself having jointly, very many years ago, conceived the idea of getting together materials for a complete Fauna Cantabrigiensis. – This design however, was frustrated by the Professor being early called away from the University to reside elsewhere, and myself also from the county a few years later. – It was only quite recently that it appeared to me that such notes as I professed on the subject, if fairly written out and deposited in the New University Museum, – might be of service to

any naturalists who took an interest in the Natural History of Cambridgeshire, a county which, through drainage and enclosures, has lost of late years so many of its rarer species of birds and other animals.[35]

Under the genus *Helix* the following entries appear:

16. H. pomatia; Linn. –

This species is not a native of Cambsh., so far as I know. Many specimens, however, were turned out formerly in Cherry-Hinton Chalk pits by the late Professor Henslow, and some of these were found there alive in 1826. – Others were turned out by myself on the Devil's Ditch and in Burwell Chalk-pits, in Sept. 1822. –

17. H. aspersa; Müll. –

Abundant every-where, especially in old gardens. – Occurs of a large size in Burwell Chalk-pits, some of the specimens measuring in diameter full an inch and a half. –[36]

So Henslow and Jenyns 'turned out' Roman Snails in Cherry Hinton Chalk Pits, on Devil's Dyke, and in Burwell Chalk Pit in the 1820s, and Jenyns treats the snail as an introduced, not a native, species. This is surprising, because the later writers on Cambridge natural history seem to have been unaware of Jenyns' 'confession', and have insisted that one local site in a chalk-pit represented the most northerly *native* locality for the species.[37] The other large species of *Helix*, H. aspersa, which was clearly as abundant in gardens in Jenyns' time as it is today, was always accepted as native. Modern opinion, however, shaped by the absence of authenticated pre-Roman records in any British archaeological site of either species, is inclined to treat both the Roman and the common garden snail as Roman introductions as food.[38]

Jenyns' statement that Henslow 'turned out many specimens' of *Helix pomatia* at Cherry Hinton reminds us that Henslow's interest in zoology from his childhood persisted throughout his life, and involved him in keeping all sorts of animals as 'pets' in cages. The Roman Snail, being a virtually domesticated animal for food in France, would have been an obvious choice which he could easily obtain from commercial sources, or from his French naturalist contacts. Although Henslow himself does not tell us, we can reasonably surmise that the verification of Roman Snails persisting in Cherry Hinton until at least 1826 was part of the Henslow–Jenyns study of what we would now call the ecology of *Helix pomatia*. Henslow in particular was showing an interest in the factors determining the distribution of plants and animals, an interest that was of course taken up and greatly developed by his star pupil, Darwin.

Perhaps the most fruitful year of the Henslow–Jenyns field studies followed Henslow's marriage to Harriet in December 1823. To quote Jenyns: 'We naturally, therefore, associated much together, and often made expeditions for collecting about Cambridge, and in the neighbourhood of my father's house at Bottisham Hall, where he was a frequent visitor.'[39] The year 1824 saw Jenyns as Curate at Swaffham Bulbeck, and launched on his career in the Church, and Henslow happily married to Jenyns' sister, so it is not surprising that it was an ideal season for their field studies. Indeed, with joint visits to Cherry Hinton, the Gog-Magog hills, the Devils Dyke, Bottisham Fen, and especially their pioneering two-day exploration of Gamlingay on 24 and 25 August, they laid the foundation that was to become Henslow's course of excursions. In the Memoir, Jenyns plays down the importance of these excursions for their botanical rewards, perhaps naturally (since that was his own enthusiasm) stressing the zoological content, and it is necessary to go to Jenyns' unpublished 'Natural History Journals' to get a better picture.[40] The following extract makes clear that the pioneering visit to Gamlingay had a primarily botanical motive, and it is reasonable to suppose that Henslow already had in mind the value of an annual excursion there when, as he undoubtedly hoped, he would become Professor of Botany.

> 24.8.1824 Early this morning, I started in company with Prof. Henslow, on an expedition to Gamlingay, for the purpose of Botanizing, &c. Our route lay thro' Barton, Comberton, Toft & Bourn, which last village soon after we had passed, we crossed the London road, & so on to Gamlingay . . .
> Gamlingay is situated quite at the extreme South West point of the County of Cambridge, & is particularly interesting to a Botanist, from its variety of plants. This is oweing to the soil, which consists of a fine silty, heathy sand; differing in this respect from all the other parts of the county: for some few miles before you arrive at the village going from Cambridge, the Geologist takes leave of the Chalk, and finds himself on the iron sand formation which shews itself in a very conspicuous way, in several places in the neighbourhood of Gamlingay, particularly in a large sandpit to the left of the road leading from the village to Eltisley bar; – where regular seams of ironstone may be seen running horizontally . . . .

The whole of today was spent in examining Gamlingay Heath & Bogs which are on it, & Whitewood. Our Botanical treasures were as follows.

[There follows a list of 27 species, containing many of the famous acid-loving Gamlingay rarities, mostly now extinct.]

Their most remarkable zoological find was the Natterjack Toad which, as we have already seen, was to excite the young Darwin on his first visit seven years later.

Responsibilities and cares of family life, together with professional and clerical duties, crowded in on Henslow in his thirties and, by 1831, when Darwin was in his final year at Cambridge, Henslow must have abandoned any real hope of exploring the natural history of exotic lands himself which was his childhood ambition. Vicariously, however, young Darwin provided at least a partial satisfaction. He had begun in the Lent Term of 1831 to recruit among his teachers and fellow-students a party to make an expedition to Tenerife in the Canary Isles during the coming Summer vacation. On 11 July, Henslow received a letter from Darwin, written from his Shrewsbury home, which conveys the enthusiasm of the young pupil finally converted to the field study of natural history on a wider scale than the Cambridgeshire countryside. Opening with some talk of his preparation for accompanying Sedgwick on his geological field trip (see below), Darwin returns to his overweening current enthusiasm:

> . . . and now for the Canaries – I wrote to Mr. Ramsey, the little information which I got in town – But as perhaps he had left Cam. I will rehearse it. – Passage 20£: ships ? & return during the months of June to February . . . I hope you continue to fan your Canary ardor [*sic*]. I read and reread Humboldt, do you do the same, & I sure nothing will prevent us seeing the great Dragon Tree [*Dracaena draco*, the famous endemic tree of the Atlantic Isles, of which a huge specimen existed in Tenerife until it was blown down in 1868] . . . I am very anxious to hear how Mrs. Henslow is – I am afraid she will wish me at the bottom of the Bay of Biscay, for having been the first to think of the Canaries. I am going now to trouble you with several questions . . .

The letter finishes:

> And now for a troublesome commission. Would you be kind enough to exert your well-known judgment & discretion in choosing for me a Stilton

Cheese, fit for eating pretty soon – Would you have it directed to
Shrewsbury & I will pay the man when I come up in October . . . Excuse all
the trouble I am giving you, .&. Believe me my dear Sir
Yours ever most sincerely,
    Chas. Darwin[41]

It seems that Darwin had hopes that both Henslow and his friend
Marmaduke Ramsay, Fellow and Tutor of Jesus College, who had been
undergraduate friends and had taken their BAs together in 1818, would
accompany his party to Tenerife. The whole tone of this letter, from a 'star
pupil' to his teacher, is remarkably free from inhibition, and surely the
request for the cheese reinforces this feeling!

    In the event, nothing came of the planned Tenerife expedition. Ramsay
died tragically in Perth on 31 July, a great shock to Henslow. The letter to
Darwin from Henslow on 24 August that begins with a reference to
Ramsay's death is the famous one in which Henslow, acting on a request
from his friend George Peacock, Professor of Astronomy and Fellow of
Trinity, to recommend a naturalist to accompany Captain Fitzroy on the
voyage of the 'Beagle', tells Darwin that he will be receiving this offer, and
urges him to accept. In his letter to Henslow, Peacock had in fact suggested
Leonard Jenyns, but if not 'is there any person whom you could strongly
recommend?'[42] Henslow himself, it seems, was briefly tempted, but saw it
was not possible for him. Jenyns, however, seems to have given the invitation much more serious thought before turning it down.

    We have Darwin's own account of this in a letter to his sister Susan,
written in Cambridge on 4 September 1831:

My dear Susan
    As a letter would not have gone yesterday I put off writing till to day. – I
    had rather a wearisome journey, but got into Cambridge very fresh. – The
    whole of yesterday I spent with Henslow, thinking of what is to be done. –
    & that I find is great deal . . .
    Henslow will give me letters to all travellers in town whom he thinks may
    assist me.
    Peacock has sole appointment of Naturalist the first person offered was
    Leonard Jenyns, who was so near accepting it, that he packed up his
    clothes. – But having two livings he did not think it right to leave them. – &
    to the great regret of all his family. – Henslow himself was not very far from

accepting it; for Mrs Henslow, most generously & without being asked gave her consent, but she looked so miserable, that Henslow at once settled the point.[43]

Jenyns, according to Browne, 'always regretted his unimaginative decision & in future years found it difficult to listen to Darwin's stories about the voyage, even more so when fame as a scientific traveller & writer was heaped on him'.[44] Reading Jenyns' *Memoir*, one is indeed struck by the fact that, although he *begins* his list of Henslow's pupils 'now in the first rank of living naturalists' with Darwin's name, he makes little of Darwin's outstanding reputation, already established by the time Jenyns is writing (1862). It is true that Jenyns does explain later in the same book Henslow's reaction to the publication of the *Origin* and, to some extent, reveals his own view on the controversy, but his comments seem to be muted.[45]

The most important outcome of Darwin's planning for the Tenerife expedition was his enthusiastic conversion to field geology. Again, it seems clear that Henslow provided the initial stimulus: he 'taught Darwin how to use technical instruments like a clinometer, & explained the trigonometry that lay behind calculations of the inclination of rock beds'.[46] Even more importantly it was Henslow who, recalling how much he personally had benefited from a summer geological expedition with Adam Sedgwick in his own undergraduate days, persuaded Sedgwick to take Darwin for part of his summer expedition in 1831. Neither Sedgwick nor Henslow, it should be remembered, approved of Lyell's *Principles of Geology* (1830–33) – though, characteristically, Henslow recommended to Darwin that he should obtain a copy of the first volume of this work, published in July 1830, and take it with him on the 'Beagle' – but 'on no account to accept the views therein advocated'!

The impact of Lyell's *Principles* on Darwin was enormous. As Darwin himself put it in a letter to a colleague in 1844: 'The great merit of the *Principles* was that it altered the whole tone of one's mind, & therefore that, when seeing a thing never seen by Lyell, one yet saw it partially through his eyes'.[47] Lyell and Henslow were contemporaries, born within a year of each other, and it was inevitable that Henslow would be required to assess this remarkable new work. After all, Henslow had held the chair of Mineralogy for five years from 1822, and had made for himself a reputation as a competent geologist with his paper on Anglesey in 1821. We shall return in Chapter 9 to Henslow's reservations about the conclusions Lyell seemed to be reaching about the history of the earth and its living beings, fossil and recent. For the moment, it would be sufficient to emphasise how significant it was that Henslow recommended this new and potentially revolutionary scientific work to Darwin at the right time.

Although nothing came of the Tenerife plans, Darwin did accompany Sedgwick for two weeks of his geological field trip to Wales.[48] Afterwards Darwin was to take the opportunity of stressing how important the experience had been: 'Tell Professor Sedgwick he does not know how much I am indebted to him for the Welch [sic] expedition – it has given me an interest in geology, which I would not give up for any consideration.'[49]

The story of Darwin's final acceptance of the 'Beagle' offer, which involved overcoming his father's initial objections, is too familiar to be repeated here.[50] In the period between August and the final sailing of the 'Beagle' on 27 December 1831, Henslow's role had become the relatively minor, essentially practical one of recommending to Darwin several reliable naturalist colleagues of his in London who could advice him on equipment and how to collect and document his specimens.

But Darwin expected much more than this from Henslow as he awaited the departure of the 'Beagle', as evidenced by this extract from a letter he wrote to Henslow on 3 December 1831, when he thought departure was imminent. (They actually sailed on the 27th.)

> I am very much obliged for your last kind & affectionate letter. – I always like advice from you; & no one whom I have the luck to know, is more capable of giving it than yourself. – Recollect, when you write, that I am a sort of protégé of yours, & that it is your bounden duty to lecture me. – I will now give you my directions: it is, at first, Rio; but, if you will send me letter on first Tuesday (when packet sails) in February, directed to Monte Video, it will give me very great pleasure.

Henslow rose to the occasion and did not disappoint his 'sort of protégé'. He even got his letter off in February so as to catch the 'packet' boat sailing for Rio de Janeiro. This chatty letter dated 6 February 1832 begins a correspondence with Darwin which covers the whole five years of the 'Beagle' voyage.[51] In the first letter, Henslow tells his former pupil, *inter alia*, about the results of the January Senate House examinations, including the 'plucking' [failure] of 'Lord Sandwich and a nephew of Lord Grey's at Trinity', and comments 'You see we are becoming quite radical.' He also voices his concern that the young Darwin should enjoy life, rather than aim at worldly fame out of any sense of duty: 'Much therefore as [I] should like to see you return laden with the spoils of the world yet if you do not find yourself *content* I should much rather see you sooner than I hope to do.'

Darwin received this when the 'Beagle' reached Rio de Janeiro on 4 April 1832. He obviously appreciated the news of the fate of his colleagues in examinations and other gossip: 'we arrived here on April 4th; when amongst others I received your most kind letter: you may rely on it, during the evening, I thought of the most happy hours I have spent with you in Cambridge'.

This letter, which he began on 18 May but did not finish and post until 16 June, is full of his experiences since sailing from Plymouth on 27 December 1831. It is clear from the letter that Darwin was collecting with considerable enthusiasm as instructed by his teacher. He confesses, however, that 'geology and the invertebrate animals will be my chief object of pursuit though the whole voyage' – plants, as we shall see, he relegates to the bottom of the pile. He also reveals his homesickness:

> I suppose you are at this moment in some sea-port, with your pupils – I
> hope for their & your sake, that there will be but few rainy mathematical
> days – How I should enjoy one week with you: quite as much as you would
> one in the glorious Tropics.[52]

Darwin was quite right: Henslow received the letter, sent on from Cambridge, in Weymouth in July, where he had taken his wife and family, together with a couple of pupils, for a summer 'holiday'. It is obvious from Henslow's second letter, not posted until 21 January 1833 (but dated in error 15 Jan. 1832), that he is finding life very demanding, but his excuse for not writing is that he had heard (in December) that Darwin's box was on its way, and had decided to wait: 'It is now here & everything has travelled well.' The letter was started on 15 January, but was interrupted by the fact that Henslow is 'examiner in Paley[53] & in one hour [has] to attend at the Senate House'. He finds the necessary 'few more minutes' to finish and post the letter six days later!

The 'meat' of this letter is typical of Henslow: it contains very practical criticism of Darwin's collections and packing, with detailed comments on how well or badly particular specimens have travelled. Darwin's few botanical specimens come in for special comment:

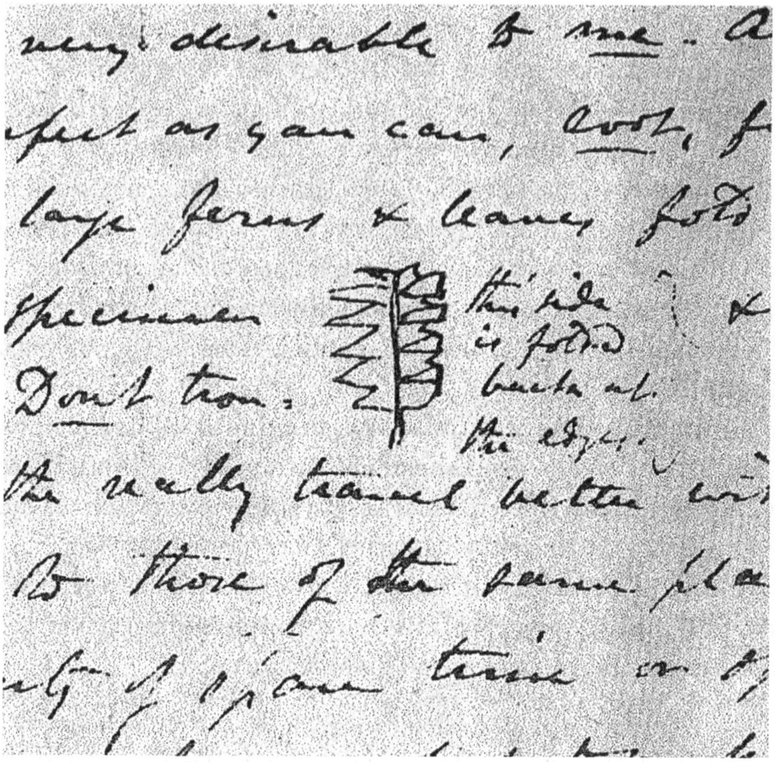

FIGURE 14 Henslow's sketch of the correct method of pressing a leaf or fern.

Most of the plants are very desirable to me. Avoid sending *scraps*. Make the specimens as perfect as you can, *root flowers* & *leaves* & you can't do wrong. In large ferns & leaves fold them back upon themselves on *one* side of the specimen & they will get into a proper sized paper. *Don't* trouble yourself to stitch them – for they really travel better without it – and a single label *per month* to those of the same place is enough except you have plenty of spare time or spare hands to write more.

The explanatory sketch, showing exactly how to press a fern or other large leaf so that, when mounted on a sheet for the Herbarium, it is possible to see both the upper and the lower side, is very typical (Figure 14). The results of his instruction can be seen in the later Darwin 'Beagle' specimens (Figure 15). Henslow follows this with a very practical comment on how the specimens have travelled:

– And now for the Box – Lowe[54] *underpacks* Darwin *overpacks* – The latter is in fault on the right side. You need not make quite so great a parade of tow &

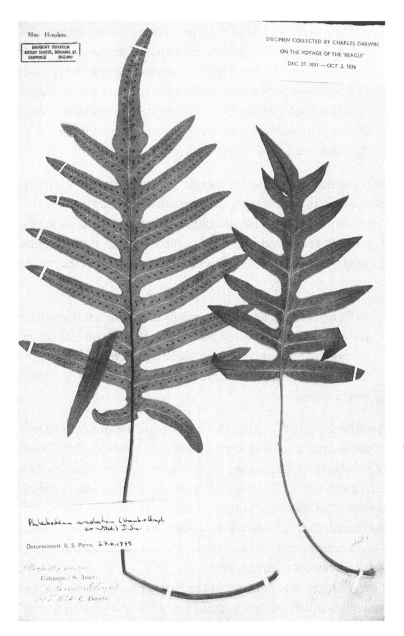

FIGURE 15 A Galapagos fern collected by Darwin. Note the carefully turned-back pinnule as instructed by Henslow.

paper for the geolog. specimens, as they travel very well provided they be each wrapped up *German fashion* & closely stowed – but *above all things* don't put tow round *any thing* before you have first wrapped it up in a piece of thin paper – It is impossible to clear away the fibres of the tow from some of your specimens without injuring them – An excellent crab has lost all its legs, & an Echinus ½ its spines by this error. Another caution I wd give is to place the number on the specimen always inside & never outside the cover. The moisture & friction have rubbed off one or two – & I can't replace them. I shall thoroughly dry the different perishable commodities & then put them in pasteboard boxes with camphor & paste over the edges, & place them in my study or some very dry place.[55]

Darwin's second letter from the 'Beagle' to reach Henslow was posted in Montevideo on 15 August 1832, and contains a revealing passage about how uncomfortable he feels collecting *botanical* specimens (as opposed to zoological and geological ones). After explaining that 'the box contains a good many geological specimens' and some discussion of that material, he says:

> As for my plants, 'pudet pigetque mihi'. All I can say is that when objects are present which I can observe & particularize about, I cannot summon reso- lution to collect where I know nothing. It is positively distressing, to walk in the glorious forest, amidst such treasures, & feel that they are all thrown away upon one.

On the other hand, he does feel that he has made a useful collection in the Abrolhos [a small coastal archipelago between Bahia (modern Salvador) and Rio de Janeiro]: 'as I suspect [my collection] nearly contains the whole flowering vegetation'. It is easy to sympathise with Darwin. Clearly the richness and luxuriance of tropical forest vegetation overwhelmed him, and he literally did not know where to start. Where, however, he had a few days on a relatively arid island he could aim to collect everything in flower, and generally seems to have done so.

Henslow's letter of 21 January 1833 reporting on the contents of the first of Darwin's boxes of specimens took over two years to reach Darwin. This was a long time, even allowing for the slow and uncertain communications by sea and the fact that the 'Beagle' itself was moving around. Throughout the years 1832 and 1833 Darwin continued to write to his teacher, and received several letters from friends and family in England, especially his sister Caroline, so that he was kept in touch, but it is obvious from his replies to these correspondents how he misses letters from Henslow. Writing to his cousin W.D. Fox from Buenos Ayres on 25 October 1833 he complains:

I hope you will write to me, & as I do, give (but a longer) a history of your-self. Excepting my own family I have very few correspondents; & hear little about my friends. – Henslow even has never written to me. I have sent several cargoes of Specimens & I know not whether one has arrived safely: it is indeed a mortification to me: if you should have happened to have heard, whether any have arrived at Cambridge, do mention it to me. – It is disheartening work to labour with zeal & not even know whether I am going the right road. – How is Henslow's family & what is he himself doing? I should so enjoy receiving ever so short a letter from him. – But patience is a fine virtue & there does not lack opportunity of exercising it.[56]

For more than two years the 'Beagle' was on the east coast of South America, arriving at Salvador ('Bahia') on 29 February 1832, and leaving the Straits of Magellan in June 1834. From June 1834 to September 1835 they were on the west coast, reaching the Galapagos Islands on 15 September and spending just over one month there. After that the 'Beagle' was on its way home by circumnavigating the globe (Figure 16).

So Darwin had to wait until near the end of his voyaging down the east coast of South America to receive any report from Henslow on the condition and value of his boxes of specimens. To make things worse, he received Henslow's report on the 'second cargo' first, in a relatively short letter dated 31 August 1833 when he arrived in the Falkland Islands in March 1834, but had to wait until the 'Beagle' docked in Valparaiso before he read Henslow's two earlier letters, one of which is obviously the letter Darwin was so anxiously awaiting – that of 21 January 1833 containing the detailed comments on the first 'cargo'.

In this, one of the longest letters to Henslow, Darwin reveals his anxieties, his concerns lest his annotated collections should prove grossly inade-quate, and most importantly his illness, which began on 20 September and lasted until late October. This illness caused Fitzroy to delay leaving Valparaiso until 10 November 1834, by which time Darwin had sufficiently recovered to finish and post his long letter, begun on 24 July with its ecsta-tic opening sentence:

A box has just arrived, in which were two of your most kind and affectionate letters; you do not know how happy they have made me – one is dated Dec. 12th 1833, the other Jan: 15th *of the same year!* By what fatality it did not arrive sooner, I cannot conjecture; for it contains the information I most wanted about manner of packing etc . . .[57]

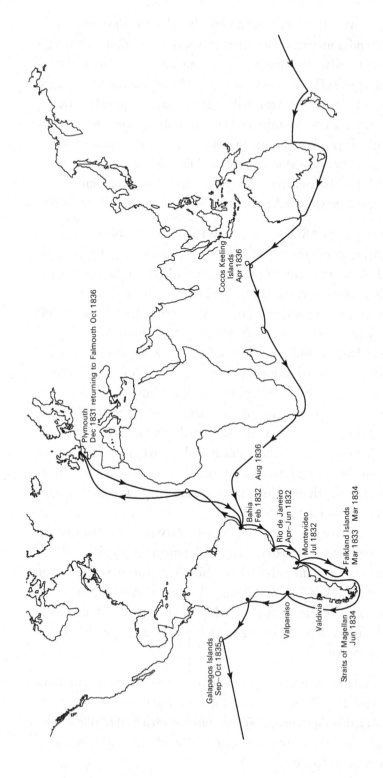

Plymouth
Dec 1831 returning to Falmouth Oct 1836

Cocos Keeling
Islands
Apr 1836

Bahia   Aug 1836
Feb 1832

Rio de Janeiro
Apr–Jun 1832

Montevideo
Jul 1832

Falkland Islands
Mar 1833 Mar 1834

Valparaiso

Valdivia

Straits of Magellan
Jun 1834

Galapagos Islands
Sep–Oct 1835

FIGURE 16 Map of the 'Beagle' voyage.

This illness has been much discussed by biographers of Darwin and others, who have speculated that he may have contracted Chagas disease, and that the effects of this may have persisted throughout his life. Most recent commentators, however, discount this theory, and look to more mundane causes for his indisposition. Browne gives a useful summary of current opinion:

> Sour new-made wine seems as good a reason as any for his disorders in Chile, and his story of pushing onwards towards Corfield for eight days easily accounts for his feverish prostration. Acute food poisoning followed by overexertion and then . . . doses of toxic emetics would have incapacitated the hardiest.[58]

The winter of 1834–35 was a difficult time for Darwin and also for Fitzroy, captain of the 'Beagle'. What seems to have redeemed it, certainly for Darwin and possibly also for the captain, was the fact that they were on the spot for a large volcanic eruption and even more impressively for an earthquake which they experienced in the small Chilean port of Valdivia and later saw the destruction caused in the nearby town of Concepcion.[59]

Poor Darwin's distress at not having heard from him has got through to Henslow by the summer of 1834, and he duly discharges his duty to his pupil in a letter from his living in Cholsey, dated 22 July 1834. Fox seems to have conveyed how Darwin was feeling:

> – Fox and his wife spent a day with us at commencement – He tells me that you are very irate at not having heard from me – which I don't exactly understand, as I should have thought that you ought to have received two letters at least from me by the time I heard from you – That I have not written so often as I ought I will readily admit for I never do any thing as I ought – but really & truly I have written & I trust that you have had positive proof of it before now. I don't know that I have much local news to tell you which is likely to be of any interest.[60]

This letter seems to be the last from Henslow to Darwin during the voyage. So the correspondence after Valparaiso is totally one-sided. Darwin continues to keep Henslow informed of his thoughts and deeds, but all his collections after leaving the South American mainland were stored on board the 'Beagle'. These include the famous Galapagos plants, eventually delivered to Henslow early in November 1836, and announced by Darwin himself who personally supervised their transport from the 'Beagle' to Cambridge.

In a letter from Lima, Peru, to Henslow dated incorrectly 'July 12th 1835'[61] Darwin anticipates his joy at their next destination, the Galapagos archipelago. From this and other letters to his friends and family it seems clear that it is predominantly the *geological* interest of the Galapagos that excites him. As Browne explains:

> Much of his anticipation was admittedly the effect of having left South America for good. He was longing for a change of scenery and different activities to occupy his time. But his scientific interest was fully engaged as well because the islands promised a new kind of geological situation ... The group was of volcanic origin and, he thought, relatively recent.[62]

The second volume of Lyell's great work had already reached Darwin, and its discussion of island floras and faunas greatly stimulated him, and challenged him to explore this unique volcanic archipelago.

The rest of the 'Beagle' voyage, circumnavigating the world, took about one year. So far as Henslow is concerned, the only botanical collection made by Darwin that his teacher dealt with completely was made on the Keeling Isles (modern Cocos Isles) on 1 to 12 April 1836.[63] Henslow received these collections together with the Galapagos plants and much other material, in November and December of the same year, after Darwin had supervised the unloading of his precious cargo of boxes. He begins a long letter to Henslow, dated '30–1 Oct 1836' from his rooms in London:

> My dear Henslow
>
> I have delayed writing, as I daily expected the Beagle would arrive, and I should be better able to tell you how my prospects go on. – I spent yesterday on board at Greenwich, & brought back with me the Galapagos plants; they do not appear numerous, but are I hope in tolerable preservation. – Tomorrow I will procure a box & will send them to Cambridge.

Darwin had already paid a short visit to Cambridge, seeing both Sedgwick and Henslow; he was, characteristically, in a quandary as to whether he should make his base in Cambridge or London: further on in the rather rambling letter we read:

> I have often thought of your really most kind offer of talking with Mrs. Henslow about my taking up my quarters with you. Few things could give me more happiness, and at the same time do me more real good. But I fear I

should me[*sic*] much in the way, & I have been thinking of another plan which would be better for the work; that is to take lodgings with two sitting rooms & a bedroom, (which I daresay could be procured), in one of which my servant could work & it would at the same time serve for a warehouse for the skins &c &c &c. – Perhaps my servant might live in the house. – In College I should only have one room, which, although a large one, would be inconvenient; and it may turn out more advisable not to remain a whole year, it is a great expence to buy furniture crockery &c &c &c for any shorter period. If I subsequently live in London, I shall follow my brothers plan take the whole of an unfurnished house, excepting the shop or office, then furnish two rooms & keep the others for lumber.[64]

In the event he stayed a few days with the Henslows then moved into a house in Fitzwilliam Street[65] until March 1837 when he moved to London. The brief stay in the Henslows' house was from 13 to 16 December, and is the visit recorded by Romilly when he had tea at the Henslows and met Darwin (Chapter 4).

Henslow does not tell us why he selected only the two Galapagos *Opuntia* species (Figure 17), and the Keeling Island flora,[66] on which to publish. Undoubtedly, Darwin's presence in Cambridge would be a factor, and he may have looked for relatively small tasks to placate his former pupil. Darwin would not give up the pressure, as this extract from a letter to Henslow in November 1839 reveals:

I am delighted to hear you have taken some of my plants with you to Hitcham. – I believe you have received a message I sent you saying that Humboldt in a letter to Me expresses at great length his **vivid regret** that M. Henslow has not been able to describe the species, or even characterize the genera of the very curious collection of plants from Galapagos. – Do think once again of making one paper on the Flora of these islands – like Roxburgh on S. Helena, or Endlicher on Norfolk Isld. – I do not think there will often occur opportunities of drawing up a monograph of more interest. – if your descriptions are frittered in different journals, the general character of the Flora never will be known, & foreigners, at least, will not be able to refer to this & that journal for the different species. – But you are the best judge. –

Not even an exhortation from the great Humboldt succeeded, and it must have been a great relief to Henslow when the younger Hooker finally

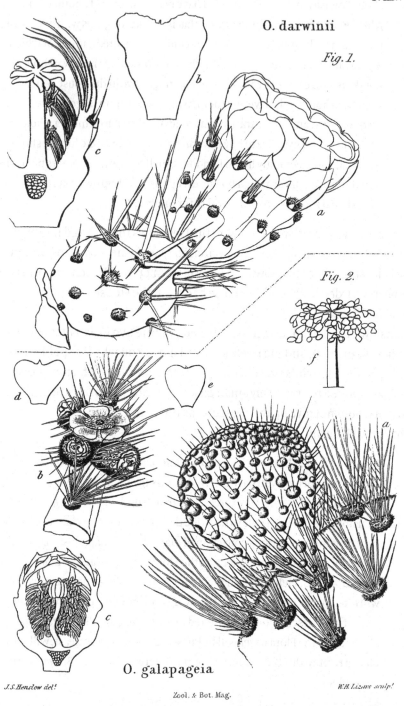

P. XIV.

O. darwinii

Fig. 1.

Fig. 2.

O. galapageia

J.S.Henslow del.

Zool. & Bot. Mag.

W.H.Lizars sculp.

FIGURE 17 *Opuntia* plate drawn by Henslow for his paper of 1837 describing Galapagos plants.

undertook the study of all the Galapagos material, and lifted the burden off the conscience of the older man, who just could not cope.

It is no part of our book to repeat the story of the shaping of Darwin's great work *On the Origin of Species*, though we shall consider in a later chapter Henslow's own reaction to his pupil's ideas as they developed. We leave the story of Henslow, Darwin and the 'Beagle' with a picture of a rather harassed and over-worked Professor not coping too well with his promise to receive, care for and even study his pupil's 'cargos' of boxes of specimens.

7 The middle years:
politics, policing and publication

We have already seen how conscientiously Henslow tried to fill what he saw as the proper role of a Professor of Botany, and how his remarkable influence on the man we now call his star pupil, Charles Darwin, gradually developed. We now turn to other aspects of Henslow's Cambridge life in the 'middle years', and in particular try to assess the pressures and difficulties he faced. Undoubtedly one of the severest pressures was financial. In the ten years between 1825 and 1835 six children were born (one of whom died in infancy) and the Henslows moved from their first house, known as 'Gothic Cottage' and situated in Regent Street, 'to a larger terrace house overlooking Parker's Piece'[1] (Figure 18). Jenyns tells us that

> during the chief part of his period of residence in Cambridge, Professor Henslow was obliged to take pupils, from the straitness of his income as a married man. This employment took up much of his time, and it is a marvel that under such disadvantages he could find leisure to do so much in other ways.[2]

As Professor of Mineralogy, Henslow had received £100 per annum, a meagre sum and of course totally inadequate to keep a wife and family in a house in Cambridge with the usual domestic staff that a University man would be expected to employ. The children came quite quickly in series: Frances Harriet on 30 April 1825, Louisa Mary on 6 October 1826, John Jenyns (who died before his first birthday) on 17 August 1828, Leonard Ramsay on 19 June 1831, Anne on 23 June 1833 and finally George on 23 March 1835. In addition to the Professor's salary, which rose to approximately £200 for the Regius Chair of Botany, Henslow's 'perpetual curacy' at Little St Mary's Church brought him some money,[3] and another source was the fees paid individually by the men attending his lectures. In total this was still inadequate, and Henslow found it necessary to take in a wider range of pupils, especially 'poll men' whose only concern was to pass the Senate examination so as to qualify for a BA degree. That the 'poll men' were an irksome chore becomes clear in a revealing autobiographical passage in Henslow's 1846 *Address to the Members of the University of Cambridge* . . . an impressive statement of the value of his new Botanic Garden, to which we

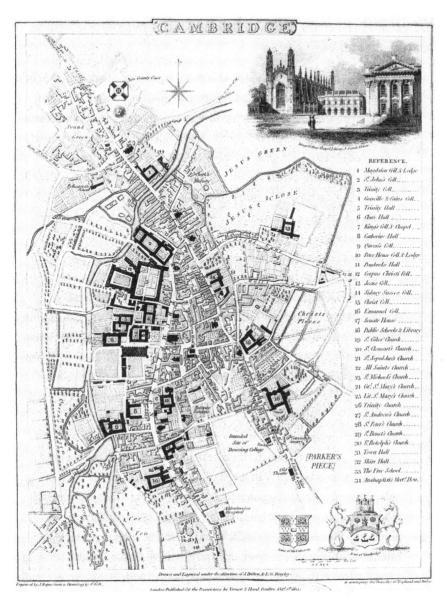

FIGURE 18 Map of Cambridge in 1810.

refer in the following chapter. In this passage Henslow seems to be reacting to criticisms made, not least by his former pupil and eventual successor as Professor of Botany, Charles Cardale Babington, that the 'absentee Professor' appearing only to give his Easter Term lectures was not pulling his weight. It is a unique outburst, and reads as 'coming from the heart'. We shall see in Chapter 10 that the years 1843–46 were unusually demanding for Henslow in Hitcham, and the following passage must be read against that background.

> I shall here take a slight review of what was the position of the Professor of Botany whilst I resided in the University. I do this, not with any desire of justifying my own insufficiency for advancing the science, or of exculpating myself from any merited rebuke for having neglected whatever facilities I possessed for so doing: whether they were offered me by the present Garden, or by many valuable works on Botany in our Public Library. But I do this in order that the public may be able to understand more thoroughly than they do at present, what are the peculiar disadvantages under which Botany and other Natural Sciences are to be upheld in our University, as well as the peculiar advantages we there enjoy for promoting them. I possess that sort of love for Natural History which makes the pursuit of any branch of it a delightful occupation to me. When appointed to the Botanical Chair I knew very little indeed about Botany, my attention having before been devoted chiefly to other departments; though I probably knew as much of the subject as any other resident in Cambridge. There had been no lectures delivered in Botany during the last thirty years of the protracted life of my predecessor, who had not resided for several years before I came to Cambridge, and whom I never saw. After entering on my office I was deprived of the opportunity of paying an undivided attention to its duties, by the necessity under which I then lay of improving my income beyond the very scant stipend which is attached to our Professorships. Five or six hours a-day devoted to cramming men for their degrees, is so far apt to weary the mind as to indispose it for laborious study: especially if we happen not to be gifted with talents or energy sufficient to overcome such an obstacle. Attention to the mere arrangement of an extensive herbarium also takes up a vast deal more time than persons not acquainted with the subject are at all aware of. Botany is now so very extensive a subject, that it needs the devoted study of many years to enable any one to master what has been done in it by others; and it requires the continued attention of a whole life to qualify a man for taking the lead in any one of its departments.[4]

Some alleviation of the family's financial straits came in the summer of 1832, when the Lord Chancellor, Lord Brougham, offered to Henslow the living of Cholsey-cum-Moulsford in Berkshire, an offer which he accepted without hesitation. Jenyns tells us the living was worth '£340 per annum ... a considerable addition to his income'. Very fortunately Henslow reveals his own feelings about accepting the living in the letter he wrote to Darwin on 21 January 1833 from which we have already extensively quoted in the previous chapter. It is part of the preliminaries in which he gives some local news and gossip:

> When at Oxford [for Henslow's first British Association meeting] I received a letter from the Ld. Chancellor giving me a small living in Berksh: about 14 miles from Oxford. Of course I do not reside as I never mean to quit Cambridge without something very extraordinary should happen. I never mean to leave for lucres sake.[5]

In the light of his eventual acceptance of the rich living of Hitcham this statement is quite revealing. It suggests that when the offer of Hitcham came his way in 1837, Henslow accepted it, feeling that he could do for Hitcham what he had been doing for the four years from 1833 to 1836 in Cholsey – namely continue to live in Cambridge and reside merely for the period of the Long Vacation – the months of July and August – 'amongst his parishioners' as Jenyns puts it (Figure 19). We shall see in Chapter 10 how and why this did not work in Hitcham, with the consequential decision to reside in the parish and give up his Cambridge home.

Another Henslow–Darwin letter during the 'Beagle' voyage gives us a glimpse of how the Vicar of Cholsey views his summer residence amongst his parishioners. In the 'gossip and news' part of the letter, written from Cholsey and dated 22 July 1834, Henslow says:

> I am at present rusticating for the Vacation at my living – & enjoy the change from a town to a country life most exceedingly – There are no immediate neighbours & I am not bothered by morning visits – My parish abounds in poor, & small farmers who leave every thing to the parson without attempting to assist him – However I am quite satisfied with my visit, the only drawback being the long distance which I have to bring my family – about 100 miles.[6]

Henslow was doing what many other Oxbridge College men did when they obtained a country living: employed a curate to perform the necessary duties in his parish for most of the year, visiting or residing himself only for limited periods. His first curate was Thomas Cottle, son of the previous vicar of Cholsey, Wyatt Cottle, and later on he employed Richard Broome Pinniger whom he took with him to Hitcham. If we are to accept what Henslow tells Darwin at its face value, he seems to have felt little concern to

FIGURE 19 Cholsey Church.

alter anything he found in Cholsey, and complains of nothing except the distance from Cambridge. Financially the net gain from the Cholsey living after allowing for paying a curate could have brought a doubling of his salary to over £400. If we accept that '£300 marked the boundary between keeping up a respectable middle-class appearance and having to struggle', as Knight suggests,[7] then Henslow's main financial worries may have been largely over by 1834.

He would, however, have remained very interested in the possibility of obtaining a comparable benefice nearer to Cambridge. Such an opportunity presented itself in 1836, when he was offered an exchange of parishes by which he could obtain the living of Histon, just three miles from the centre of Cambridge, and ideally suited to his academic life. This offer came to him through the good offices of William Brougham, the younger brother of the Lord Chancellor, Henry Brougham, who had offered Henslow the Cholsey living. William Brougham was an undergraduate contemporary of Henslow, a Jesus man who took his BA degree in 1819; it is from a surviving letter of his to Henslow dated 16 June 1836 that we are able to piece together the story.[8] The vicar of Histon was Thomas Pennuddock Michell,[9] appointed to the living in 1821; the tone of Brougham's letter to Henslow suggests strongly that either Brougham himself or some third party had persuaded Michell to offer to exchange Histon for Cholsey. That Henslow received such an offer, and accepted it with pleasure, is clear from his correspondence with Hooker.[10] His letter of 11 July 1836 included the following:

> ...I am truly happy to say that I have every prospect of exchanging my living for one within 3 miles of Cambridge which will keep me here in the summer for the future and add considerably to my stock of botanical time for though I should take on fourteen of the Sunday duty and be constantly in the parish I should certainly keep a curate resident on the spot – This would afford me leisure for the duties of my Professorship whilst at the same time I should have the satisfaction of being in close communion with my parish.

Alas, this was not to be: we find Henslow writing again to Hooker on 17 May 1837, and having to report: 'I am sorry to say that the gentleman who agreed to exchange with me broke off at the last somewhat in the style of a lady who changes her mind after the wedding clothes are purchased.' This flippant tone must have disguised a severe disappointment. It is surely not too much to claim that had Michell not backed out, and Henslow become Vicar of Histon, where he could have combined with a clear conscience his parish duties with his academic ones, the whole history of Cambridge botany would have been radically different. In the event, Michell remained at Histon for 35 years, from 1821 to 1856.

One aspect of Henslow's academic life that deserves attention is his lack of any strong connection to a particular College. This, of course, contrasts very obviously with both Sedgwick and Whewell, who were, quite simply, Trinity men. As we have seen, Henslow seemed to cement no strong friendship with his contemporaries in his own College, St John's, and indeed, as Jenyns tells us, 'among his most intimate acquaintance, and with whom he seemed most to associate, were Calvert, then tutor of Jesus, Dawes, of Downing, [Henry] Kirby, of Clare, – and Ramsay of Jesus . . .'[11] Of course, Jenyns himself was a Johnian: but he was four years younger than his brother-in-law, and there is little evidence that their common College played any part in their relationship, which arose primarily through a shared passion for field zoology. Henslow seems to have been a University, rather than a College, figure, foreshadowing the considerable changes that he himself was welcoming in his last years: the removal of the celibacy restriction on College Fellowships, the rise of joint University rather than College responsibility for Science courses in particular, and the growth of political power in the hands of University officers.[12] The fact that the Henslow soirées were social occasions where men and women mixed freely at a private house in the *town* and not in College rooms must have done much to advance the image of reformer amongst many academics and their wives who were not necessarily interested in any aspect of natural science. Often, indeed, they were accomplished in music, languages and the visual arts: in this respect we should see Henslow as encouraging the artistic accomplishments of his wife and daughters, whilst not being particularly so inclined himself.

Richard Dawes, three years older than Henslow, who came up to Trinity in 1813 and graduated (as Fourth Wrangler) in 1817, was very important in Henslow's Cambridge life. He was made Fellow and Bursar of Downing College in 1818 and Tutor there in 1822. To quote Pettit-Stevens:

> In Dawes' time the Downing Combination Room acquired a social and convivial celebrity second to that of no other college in the University. Here it was the delight of such men as Whewell, Romilly, Peacock, and Sedgwick (the admirable *raconteur* and the man of marvellous memory), to gather round the hospitable board of their old Trinity associate.[13]

This was the social group to which Henslow was attracted, a group of Trinity men who found the refreshing, aspiring College of Downing so congenial. The College, with its new, colonial-style campus and handsome buildings, had received its royal charter in 1800, and admitted its first undergraduates in 1819, and many had seen in Downing a new pattern that could reinvigorate the old Collegiate system. Perhaps Downing could have led the way, but unfortunately Dawes, who was immensely popular in the College and the obvious choice as Master when this post became vacant in 1836, was not elected because, together with Henslow and 62 other members of the Senate, they had declared themselves in favour of the abolition of the religious tests preventing Dissenters from taking degrees in Oxford and Cambridge. Dawes married, gave up his Fellowship and began an influential career in the Church.[14]

We get the first evidence of Henslow's increasing attachment to the Dawes circle in Downing in another of his gossipy paragraphs in a letter to Darwin dated 6 February 1832: 'Worsley has resigned the Tutorship of Downing – He was absent all this term & I acted for him as Chaplain & dinner-eater on Sundays & other days when Dawes was absent – He thoroughly disliked the duties of a College life & has taken a wise step.'[15]

This association with Downing was to prove particularly valuable to Henslow after he moved to Hitcham. It gave him a *pied-à-terre* in Cambridge ideally situated near to the new Botanic Garden, and seems to have been used by him during the weekdays of the Easter Terms when he delivered his lectures and conducted the field excursions. It was also useful to Henslow in that he could not only dine himself in Downing, but invite his colleagues to join him for dinner there. Babington records two such occasions in his Journal. In 1840, he tells us, Henslow commenced his lectures on 6 May, and on 15 May the entry reads: 'Dined with Henslow at Downing, and met Prof. Starkie, the candidate for the Town. We went with Starkie to a meeting in King Street, and I made a speech in his favour.' In the following year, Babington records: 'May 27. Dined at Downing; it being Henslow's last day in Cambridge.'[16]

Dawes had a distinguished career in the Church, and was ultimately Dean of Hereford. Like Henslow, he devoted himself in his parish work to the cause of popular education, developing local schemes to improve the lot of the poor, and he and Henslow kept in touch by correspondence, as we see in Chapter 10.

As an established academic with a Cambridge house and an apparently very promising future, Henslow soon involved himself in politics. We can piece together some of his activity because, by chance, two boxes of letters addressed to him have survived in the archives, letters which are entirely concerned with his role as Chairman of a local Committee for the re-election of Lord Palmerston as a University MP.[17] A short letter, dated 19 December 1825, is from Wood, the Master of St John's, calling together a meeting to form this Committee: it must have been at this meeting that Henslow was elected Chairman, and began to accumulate a considerable file of correspondence. The 'Palmerston Committee' ran its canvass from

the Sun Inn, a hostelry, now demolished, which was opposite the Main
Gate of Trinity. Henslow's Committee was successful, and Palmerston and
the then Attorney-General, Sir John Copley, were elected to the two
University seats.

From this activity arose the cooperation between Henslow and John Lamb,
Master of Corpus Christi, who was six years senior to Henslow, and a brilliant scholar who became Master of his College at the early age of 32.
Henslow and Lamb published a pamphlet containing a 'recommendation'
to the Vice-Chancellor, supported by over a hundred members of the Senate,
to make illegal the payment of expenses claimed by 'out-voters' from the
candidate they voted for in University Parliamentary elections. This political
question caused a great stir in the University, and undoubtedly brought the
young Henslow much local publicity. Most resident members of the
University supported the 'recommendation' but, perhaps understandably,
'with non-residents [ie. the 'out-voters'] the case was far different, to judge
by the literature which the movement called forth. Nor was their wrath
appeased by the extraordinary conduct of two members of the Senate, who
on the day of election insisted that the oath against bribery should be admin-
istered to each voter as he came to the Vice-Chancellor's table'.[18]

This early taste of local political controversy, which first reveals that side
of Henslow's character of which we shall see more in later chapters in the
confrontations with the local farmers in Hitcham, seems to have developed
quietly in the years of national political turmoil that culminated in the
defeat of the Tory administration under Wellington and the passage of the
Whig Reform Act of 1832. Henslow as a young man was a Conservative
voter, but when Palmerston changed sides, and joined the Whig admin-
stration in 1830 as Foreign Secretary, Henslow followed his example and
became a Whig. The Darwin–Henslow correspondence affords us a rare
glimpse of Henslow's political interests at this time, and incidentally
reminds us of the extent of quite violent 'politicking' that seems to have
prevailed not just in Cambridge, but throughout the country in the 1830s.[19]
In a letter to Darwin dated 25 October 1831,[20] Henslow writes:

> The day after tomorrow our County election begins. Mr. Jenyns is
> Chairman of Capt. Yorke's Committee, and proposed at one time to sleep at
> my house during the week, by which manoeuvre I had calculated upon
> getting my windows smashed! [Yorke was the Tory anti-Reform Bill
> candidate.]

We are not told what actually happened on the fateful day: presumably George Leonard Jenyns (Henslow's father-in-law) slept elsewhere, and the Henslow windows remained intact! A few months later, Henslow again refers to politics: writing to Darwin on 6 February 1832, after Darwin had set sail on the 'Beagle', he allows himself a little levity and offers some advice and support to his favourite pupil:

> Pray are you yet a Whig? – for I heard from Wood that your Brother told you it was impossible to touch pitch & not be defiled [an editorial footnote says 'a reference to Robert Fitzroy's staunch Toryism']; Whatever you become I know it will be from honest conviction, and therefore tho' I shan't change my principles myself I shall be quite content to allow you to change yours without thinking the worse of you for so doing – not that I suppose you will care if I should – only as we used to agree in these matters we will now agree not to let our disagreement (if it should turn out so) trouble us.[21]

In January 1835 there was a general election. The two Cambridge borough seats were contested by three candidates: Thomas Spring Rice (later Lord Monteagle) and George Pryme, both Liberals, and J.L. Knight, a Conservative. Henslow, who supported the Whig cause, was convinced that bribery had been used by Tory supporters, and published his *Address to the Reformers of the Town of Cambridge* in an attempt to persuade, as he put it, 'weak and wavering minds of those whom I know to be with [the reformers] at heart, but were hesitating, from some deficiency of moral courage, between obedience to the dictates of conscience and submission to the powerful influence in operation against them'. Initially, as Jenyns explains, 'his interference was restricted to private reasoning with parties whom he considered as having been unfairly influenced in the decision of their votes and to publicly putting forth his views on the subject in the pamphlet'.[22]

We can take up the story from the local press, when the bribery charges were aired in the Cambridgeshire Lent Assizes before Lord Abinger on Thursday 19 March 1835. The *Cambridge Independent Press*,[23] a supporter of the Whig reformers, reports Mr Gunning opening the case, as he outlines his account of Henslow's part in the affair:

> Several of the most respectable citizens of Cambridge came to a determination to prevent the practice of bribery. They made no secret of this resolution. Attention was paid to certain individuals and amongst other facts were discovered those which became the subject of the present action.

FIGURE 20 'Henslow Common Informer' inscription on Corpus Christi College.

They then consulted who should prosecute. It was necessary that someone should become the informer. It was found that many actions had failed, from the suspicions attached to the persons who prosecuted; and Professor Henslow, acting under those principles which he had always professed, felt it was his duty not to shrink and to become the informer against Mr. Fawcett.

So Henslow became a 'common informer', an action which brought him much local notoriety, with his political opponents organising gangs of unruly men to shout slogans against him, and to paint insulting graffiti on Cambridge walls. Extraordinarily that adornment of the wall of Corpus Christi College with the clear legend in large capital letters 'HENSLOW COMMON INFORMER' survived well over a century into the 1960s (Figure 20).[24]

Like other prominent University men, Henslow acted on occasion as a Proctor who patrolled the main streets of Cambridge to apprehend 'wrongdoers'. In Georgian and Victorian times the Proctors had the right to arrest women suspected of prostitution, a right that was finally extinguished as late as 1894,[25] and details of these activities to protect the 'young gentlemen' of the University are given in the Spinning House Committal Book.[26] The first of Henslow's terms of duty covered February to June 1830, in which period he was responsible for 22 arrests. The infamous 'Spinning House' to which the arrested women were taken was a building in Regent Street quite near to the Henslows' own house, and next door to the Fountain Inn ('founded 1749') which is still there; the Spinning House itself was pulled down in 1901 to make way for a Police Station. Detail of the story is given by Holbrook (1999). Holbrook mentions in particular Adam Sedgwick as an assiduous Proctor: 'one imagines him a tall black stalking figure scouring the high streets industriously seeking out sin', and Henslow is a minor figure in her story.[27] We can analyse Henslow's cases to see how lenient he was. Of the 22 arrests of women made by Henslow in the 1830 period of duty, nearly half (10) were discharged the following morning, sometimes with an 'admonition' not to 'sin' again. These ten included all four 'first offence' cases. In his second period of Proctorial Service in 1834, he arrested ten women between 12 February and 31 March and discharged four, including one Mary Ann Webb who had been arrested no fewer than 24 times, on grounds of her 'ill health & promise to go home'. There is no evidence from the records that Henslow was significantly more diligent (or more lenient) in discharging his duty than the other Proctorial Officers.

A glance of the list of Henslow's publications in the Appendices will reveal that most of his natural history output occurred between 1828 and 1837. Between these dates Henslow had 30 papers and notes published, 22 of which were botanical, four zoological and four others. In an excellent, well-researched paper on the rise and fall of specialist journals in Britain catering for aspects of natural history, Allen (1996) has set out the story covering Henslow's period:

> The first true journals in the field of natural history were almost all non-commercial enterprises, sponsored by one or other of the loftier bodies as those successively came into existence. With the Royal Society and its *Philosophical Transactions* as their model, they each early saw it as desirable, as well as a matter of prestige, to have an outlet for at least the more important material read before their members at their meetings. The Linnean Society's *Transactions* were the first to start, in 1791, to be followed by the Geological Society's in 1811, respectively three and four years after the founding of the parent body. The interval was shorter still in the case of the Cambridge Philosophical Society (which had the advancement of natural history as one of its declared aims on its founding in 1819), while the

Wernerian Society launched its *Memoirs* in the very same year, 1808, that it broke away from the Natural History Society of Edinburgh to pursue an independent existence.[28]

Henslow was enthusiastically in at the beginning of this rapid development through his role in the formation of the Cambridge Philosophical Society, and was equally a pioneer in the next phase associated with the name of John Claudius Loudon. Again, to quote Allen:

> As so often with major new departures, it required an outsider to demonstrate the possibilities that this series of developments had begun to open up. John Claudius Loudon, the person who now arrived on the scene to fill this role, was a man of quite exceptional industriousness even by nineteenth-century standards. A landscape gardener by training, his never-idle pen and a bubbling fascination with novelty soon led him into architecture on the one hand and into the compiling of encylopaedias on gardening and farming on the other. Keenly attuned to the interests and tastes of the newly-comfortable middle classes, in 1826 he exploited the gap in the market for a periodical devoted to the practical side of horticulture. The resulting *Gardener's Magazine and Register of Rural and Domestic Improvement* (to give it its full title) quickly proved a resounding success ... Brimming over with confidence now and with a gardener's interest in botany already, Loudon saw a *Magazine of Natural History* as the appropriate next addition to his stable. Brought out in the spring of 1828, this was boldly made a bi-monthly from the first, the aim being to go monthly once it was established. The publishers, Longman & Co., ingeniously retained stereotyped plates, thereby allowing extra copies to be produced on demand.
>
> Intially, this magazine too proved a conspicuous success, the first few issues selling over 2,000 copies each. Once again, it seemed, Loudon had hit upon exactly the right formula ... The style, deliberately, was informal and chatty.[29]

To this successful new magazine Henslow was a founder-contributor. In a letter to Henslow dated 7 March 1828, Loudon thanks the 'Reverend and Dear Sir' for an 'offer of assistance' in launching his new journal:

> I beg leave to offer you my best thanks for your letter of the 4th instant. The offer of assistance on the part of yourself and the Reverend L Jenyns, I should have highly valued at any time but I particularly prize it at the

commencement of a work much of the value of which I feel will depend on the general cooperation of naturalists. – I shall state at once what I think you and Mr Jenyns could do for me.

An account of the origin, rise, progress and present state of the Museum for Natural History attached to the Philosophical Society of Cambridge, including notice of the latest purchases. Minute details need not be gone into, but leading dates, facts, and features, so as to give a general interest to the thing.

A paper on the Natural History of Cambridge and its neighbourhood, in which might be introduced an historical account of the taste for natural history at Cambridge (meaning among the learned men there) from the earliest times to the present, including notices of the different institutions, museums, lectures, Botanic Garden etc to which this taste has given rise, short notices of the eminent naturalists which have studied at Cambridge or lived there and a disquisition on the influence of the study of natural history on the other studies pursued at Cambridge – shewing how much it would add to the enjoyments of clergymen and others destined to live in the country to have a taste for Natural History &c &c So extensive a subject would occupy a series of papers: but being so serious in its objects, it admits of being both interesting and instructive.

Occasional short notices on any point or of meetings or Transactions at the Museum or in any way connected with Natural History, Account of public lectures &c &c, and in short a monthly letter containing scraps of information suited for various heads enumerated in the prospectus, might I should think be readily furnished . . . between you and Mr Jenyns, and for which I should be singularly obliged to you, and as a mark thereof send you the work regularly as it comes out.

I have thus Reverend and Dear Sir expressed to you very fully what you can do for me and have only to repeat my best thanks to yourself and the Reverend L Jenyns and to add that I hope you will let me have something in time for the first number

I remain Reverend and Dear Sir

Much obliged and Very faithfully

J C Loudon[30]

Henslow lost no time in preparing a suitable contribution to meet the first of Loudon's requests, namely a short account of the Cambridge Botanical Museum which was duly published in the first volume, and accompanied by a note by Henslow, dated 'April', on his discovery of *Althaea hirsuta* 'plentifully in several cultivated fields about the point of the junction of the three parishes of Cobham, Cuxton and Strood, flowering throughout July'.[31] Evidence of much botanising by Henslow in the county of his birth and childhood is rather scarce, but there is the particularly interesting case of *Fumaria vaillantii*, one of a group of small-flowered annual fumitories, detected by Henslow as being 'a variety of *F. parviflora*' and confirmed by Lindley as being a species new in Britain. Henslow communicates this find as a note to the *Magazine of Natural History* on 16 September 1831. It seems probable that his sharp eye had noted this 'variant' *Fumaria* 'on Chatham Hill, Kent' on the same occasion when he recorded *Althaea hirsuta*.[32]

It is not appropriate to comment on all Henslow's publications in Loudon's journal, but we should note that he used it both for short notes and for substantial papers, such as the important study in 1830 of the native *Primula* species, which we discuss in Chapter 9. To give the flavour of his eclectic output in the journal we have chosen a few examples from the more or less complete list given in the Appendix. Henslow the precise observer and recorder is exemplified by the following three-line correction of a previous author's effort:

> *Calendar of Nature in England for* 1830. – In the table, p. 168., the columns representing the rainy and snowy days are evidently misplaced in reference to the headings they bear. – J. S. Henslow. *Cambridge, April* 9.[33]

Under the section 'Retrospective Criticism' the journal published comments on previous papers, and several of Henslow's contributions are to be found here. 'Humming in the Air' is a particularly attractive one, involving King's College Provost's Lodge and Leonard Jenyns:

> *Humming in the Air* (p. 110.). – As corroborating O.'s conjecture, I would refer you to an account transmitted to you last year by Mr. Jenyns, of an extraordinary swarm of minute flies, which settled in the Lodge of King's College. Although they were exceedingly minute, yet, upon our listening with very moderate attention, their aggregate humming was very distinctly audible. The explanation of this phenomenon, mentioned by White, seems to me to be satisfactorily given by O. (p. 110.) – J. S. Henslow. *Cambridge, Feb.* 14. 1832.

The Editor follows this with the very detailed account of the phenomenon by Jenyns, of which only the introductory part is given here:

We here insert the account supplied by the Rev. L. Jenyns, adverted to above by Professor Henslow.

An extraordinary Swarm of Flies. – During the month of September last, a small dipterous insect, belonging to Meigen's genus Chlorops, and nearly allied to, if not identical with, his C. Læta, appeared suddenly, in such immense quantities, in one of the upper rooms of the Provost's Lodge, in King's College, Cambridge, as to render the fact worthy of being recorded. The same species of fly, or one closely approaching to it, is not uncommon in most houses, at least in Cambridgeshire, towards the decline of the summer; but in this instance their numbers were so great, and their appearance so sudden, as to surpass any thing of the kind I had ever before witnessed.[34]

Mammals find their place also in Henslow's contributions. Following several contributions on the Stoat, and its white coat in winter, we find this brief Henslovian note:

The Stoat (p. 77.). – In this county, Cambridgeshire, the stoat does sometimes change its coat, and assume the perfect dress of the ermine, as two fine specimens in the museum of our Philosophical Society will testify.
– J. S. Henslow. Cambridge, February 14, 1832.[35]

A natural history observation including an advertisement for his beloved Philosophical Society Museum!

Finally, in this selection, we give Henslow's account of how the 'great and the good' in the University are pursuing their hobby and building up their collections. There is a rather sour note in this contribution concerning the state of the University's own collections, which were badly housed at the time.

Collectors and Collections in Natural History in the University, Town, and County of Cambridge. – Sir, I send you a list of such collections of natural history as I at present recollect in this neighbourhood, and will endeavour to increase it for you by further enquiry. The four collections belonging to the university are unfortunately little accessible even to the members of the university, and still less so the public at large. These are, in fact, of no real utility, except to the professors themselves; and, were it not for the prospect of better days, in which the force of public opinion will at length compel the university to provide some better establishment for her museums, it would really be quite disheartening to go on labouring day by day in the accumulation and arrangement of specimens which are destined to remain concealed, and perhaps doomed to rot, without one particle of benefit being derived from their existence. However, there is at length some hope of our obtaining museums which will be placed on as liberal a footing as our public library.

Four university collections
{
Zoology and comparative anatomy, Rev. Professor Clark.
Geology, Rev. Professor Sedgwick.
Mineralogy, Rev. Professor Whewell.
Botany, Rev. Professor Henslow.
}

General natural history (containing a good collection of British birds, &c.),
  Philosophical Society.
British birds, Rev. Dr. Thackeray, Provost of King's.
  [followed by a list of thirteen 'gentleman' indicating their specialist
  knowledge or enthusiasm] [36]

The resounding success of Loudon's *Magazine of Natural History* was relatively short-lived, and by 1832, after Loudon had delegated editorship to his young assistant, John Denson, the terminal decline had begun. For some later papers, Henslow transferred his loyalty to another journal, *Magazine of Zoology and Botany*, which had recently been launched with Edinburgh as its base. The most significant of these papers we have already mentioned: Henslow's description of the two Galapagos *Opuntia* species, which appeared in the first volume. From 1837, when the first volume appeared, the rise and fall of natural history journals is a complicated story which we can ignore.

So far as Henslow is concerned, publication of notes and longer articles was obviously a 'club' activity which he enjoyed, and from which he derived a wide circle of naturalist gentlemen acquaintances. In this respect he was, again, a pioneer of the pattern of learned journals containing both amateur and (increasingly after Henslow's death) professional members which we still enjoy today. Nor would the presence, even preponderance, of ladies in some of the societies which still flourish today have seemed in any way undesirable to him. The modern *Botanical Society of the British Isles* would, one feels, have pleased him greatly!

A less happy association of Henslow with a publisher is commemorated in a journal called *The Botanist*, one of several journals emanating from a remarkable local botanist, Benjamin Maund, of Bromsgrove,

Worcestershire, who also ran a chemist's and a bookshop. How the original association with Maund began is not clear, but Maund persuaded Henslow to help him with material for his journal. How Henslow viewed this association can be gleaned from this short comment from a letter to Hooker dated 17 May 1837: 'You have perhaps seen my name attached to a humble work called the Botanist which Mr Maund puffs off much more than I like. I have had very little time indeed to spare for him & have described 3 or 4 plants myself besides the dictionary part...' Henslow's contributions are mostly in the first volume, and include '*Cactus speciosus*', of which his original illustration survives.[37] (Plate 7)

The 'dictionary part' was a botanical glossary, illustrated with wood-cuts, and originally published incompletely in parts in Maund's journals, but then eventually published separately as a book of 218 pages.[38] Entitled *A Dictionary of Botanical Terms*, it is a tribute to Henslow's meticulous scholarship.

So much for Henslow's publications during the Cambridge years. In the later chapters we see how active he continued to be in Hitcham, but using the local press, and concerned to reach a very different audience via, for example, the *Gardener's Chronicle* and the *Journal of the Royal Agricultural Association*.

During the 1830s Henslow developed an interest in what Jenyns called 'antiquarian researches'.[39] He was present at the opening of the first great barrows at Bartlow on the Cambridge–Essex border on 21 April 1835, together with Sedgwick, Whewell and others, and again in April 1838 and April 1840. More importantly, when farm-workers accidentally discovered a Romano-British burial chamber at Eastlow Hill, Rougham in July 1843, Henslow undertook in September of that year a careful description of the contents of the next of these barrows to be excavated at Rougham, and published in two papers with his own careful drawings an account of this site. Edward Martin, an expert on East Anglian archaeology, assesses these papers as the first 'modern' scientific illustrated accounts of any archaeological dig in East Anglia.[40]

One feature of special interest to Henslow was the presence of leaves and twigs of box, *Buxus sempervirens*, in and around the objects found in the burial sites. This material was discussed by Gage in his paper reporting the results of the Bartlow dig of 1838. His remarks reveal that the help of Robert Brown was enlisted to confirm the identification:

Leaves were found adhering to the bottom of the cinerary urn, from which it would appear that some had been thrown in before the urn was deposited: while round the handle of the lamp a wreath would seem to have been entwined.

'These vegetable remains,' remarks Mr. Brown, F.R.S., who has had the kindness to examine them, 'appear to consist of the epidermis of leaves and ultimate branches of box, the vascular part and the parenchyma being in most cases entirely removed! I judge the leaves to belong to Box (*Buxus sempervirens*), from their insertion as indicated to the ramuli, from their outline, size, thickened margin, and arrangement, and form of stomata, which in most cases, however, are removed, leaving round apertures of the form and size of the whole stoma.'

Professor Henslow informs the writer of this memoir, that a skeleton was lately found in or near Chesterford churchyard, together with a Roman vase, and that box leaves lay loose in the soil near the skull and vase. Some of the leaves are in his possession, and they are similar to those found at Bartlow.[41]

Henslow's annotated specimen of box leaves and twigs from 'a Roman tomb at Chesterford' fortunately survives as a specimen sheet in the University Herbarium in Cambridge (Figure 21). His discoveries and later ones have been discussed in recent literature in connection with evidence for the native status of box in Britain; generally the view is held that the presence of box foliage in Romano-British burial sites cannot be taken as evidence of local wild origin, because trade within the Roman Empire could easily have included imports of Mediterranean box, or box could have been planted and even cultivated on Romano-British estates.[42]

Towards the end of his life, Henslow renewed his interest in a particular archaeological controversy – the nature and significance of flint implements or 'celts' found in East Anglian deposits. We deal with this controversy in our final chapter.

Before we leave the subject of Henslow's publications in the 'middle years' we should note *Le Bouquet des Souvenirs: a Wreath of Friendship*, an odd and rather mysterious work that carries on its title-page as author only Henslow's name, as responsible for the 'botanical portion'. This book of about 200 pages, with 25 coloured plates, appeared in 1840: it has a short anonymous Preface, from which we learn that '[the volume] was a year's occasional amusement to three friends' who express the hope that 'the addition of the botanical descriptions may render it more generally interesting'.

Box leaves

from a

Roman tomb

at

Chesterford.

Twigs & Leaves of Box (Buxus sempervirens) from a Roman Tomb at Chesterford (either within or below a cinerary Urn). N.B. A similar deposit occurred in one of the Bartlow Tumuli, of which I possess specimens.

J. S. Henslow

FIGURE 21 Henslow's specimen of box leaves from a Romano-British
burial site.

Following the Preface is a 'letter' from Henslow to his 'Dear Sister' which
begins:

> In complying with your request 'that I would furnish you with a popular
> description of the several plants figured in "A Wreath of Friendship,"' I
> must risk my reputation as a Botanist with scientific readers, and hazard
> my character, by being thought pedantic, with those who are unaccus-
> tomed to the use of technical terms.

Which of Henslow's four sisters who survived him, namely Ann,
Charlotte, Eleanor and Louise, is the one referred to remains a mystery.
Jenyns, apart from including the book in his 'List of Publications' by
Henslow, makes no comment on it. Nor is it clear who the 'three friends'
were – though we are perhaps at liberty to assume that it was a small group
of talented literary and artistic ladies, one of whom was 'Miss Henslow'.

8 The Botanic Garden:
Old and New

Following his election to the Regius Professorship in 1825, Henslow was also appointed Walkerian Reader in the small Botanic Garden that had been given to the University by Richard Walker, Vice-Master of Trinity College, in 1762. It was in his capacity as Walkerian Reader that the new Professor was able to set out energetically to rescue what he could of the collections in the old Botanical Museum, and to clear space for his course of lectures, with their accompanying 'demonstrations' and practical classes, that he began to deliver annually in 1827. So far as the Botanic Garden itself was concerned, Henslow must already have been aware that this part, at least, of his new empire was quietly continuing to function under the able Curatorship of Arthur Biggs, who had been appointed in 1813 on the death of his more famous predecessor, James Donn.[1]

The Walkerian Garden was laid out from 1760, when Walker bought the Old Mansion House on the site of the ancient monastery of the Augustinian Friars, situated on nearly five acres of land bordered by Bene't Street, Free School Lane and Pembroke Street. Walker himself had appointed Thomas Martyn, 'now Titular Professor of Botany', to be the first Reader, and Charles Miller the first Curator. Miller was the second son of the famous Philip Miller, Curator of the Chelsea Physic Garden and author of the best-selling Gardeners Dictionary; he and Thomas Martyn were childhood friends from Chelsea, and it is therefore entirely understandable that the new Walkerian Garden in Cambridge was closely modelled on the Chelsea Physic Garden with its largely formal design and rectangular flower-beds.[2] Martyn and Miller seem to have made an enthusiastic start: in April 1761, in a letter to his friend Richard Pulteney, Martyn writes:

> Our Garden begins to flourish; shrubs and trees are already planted; plenty of seeds, both tender and hardy, are sown; a stove is building; & Stone is preparing to raise the superstructure of a greenhouse on the foundation which was raised last year.[3]

To Thomas Martyn, whose election to the Chair of Botany in 1762 on his father's retirement looks – and almost certainly was – a nepotistic affair,[4] botany was a gentleman's hobby. He himself admitted this; writing in later life he tell us that he was first 'engaged in academical studies & afterwards in those of the professions I had determined to adopt. Botany was rather the amusement of my leisure hours, than my serious pursuit.'[5] To such an academic, a small, formal Physic Garden would seem to offer all he needed to build a modest interest among the young gentlemen together with any other University men who might wish to learn a little of the world of plants. How modest was his aim is obvious from this further extract from the letter to Pulteney quoted above:

> All this, I hope, will increase the number of botanists among us. Indeed, we already begin to grow considerable, for I never had more than one companion before this Spring, but now I have three, & expect soon to have two or three more converts.

The state of this Walkerian Garden just before Henslow's appointment can be judged by the report of a visit made by Josef August Schultes, the Austrian botanist, who was at the time Professor of Botany at Landshut in Bavaria. Schultes' long letter to his friend Count Steinberg, sent from London at the end of a short visit to England in late August 1824, gives an entertaining account of a journey that began in Harwich, took in a visit to James Edward Smith in Norwich, and included days in London, Cambridge and Oxford. Schultes' impressions are conveniently available in translation in an edited version by W.J. Hooker entitled *Schultes's Botanical Visit to England*, and our quotations are extracts from this paper.[6] James Edward Smith at home made an exceedingly favourable impression:

> Sir J. E. Smith displayed to us the treasures of his collection (in reality the only one of its kind) with a courtesy & kindness which are peculiar to great & well-educated men; and which in this truly noble person are heightened by such charms of gentleness and affability, as cannot fail to attract to him most forcibly even such individuals as have but once enjoyed the privilege of his society.

After such an experience, strengthened by Schultes' disgust at the rejection of this paragon of virtue by the University of Cambridge some years before on the grounds of his being a 'Dissenter', it is not surprising that the Cambridge visit sounds to have been rather disappointing.

> We hired a postchaise from Newmarket to Cambridge, which is situated in a rather bleak neighbourhood . . . I had hoped here to meet my late friend

Dr. E. D. Clarke, Professor of Mineralogy, who once spent an evening with me at Landshut, on his return from Egypt, & had invited me in return to see him & his Garden at Cambridge [sic]. He knew not that he was asking me to come and see his effigy, when he gave me the invitation; – the marble bust which the University has placed in his honour in the library is all that was left of my friend. I was told that Dr. Clarke's death was occasioned by the irritation that an insect gave rise to, and which was drawn into his nostril by smelling of a flower.'

No mention of Clarke's successor, though Henslow had already been in the Chair of Mineralogy for a couple of years. Perhaps this is not surprising, since Schultes' visit was so short, but it does reinforce our picture of the young Henslow as having, before becoming Professor of Botany, no reputation that would have alerted an eminent foreign botanist to seek him out. When Clarke invites his botanical host to visit 'him and his Garden at Cambridge', we realise that Clarke was probably the only academic in the period between 1810 and 1820 who would have had any interest in the Garden: as we have seen in Chapter 2, Thomas Martyn had made over his Lecture Room and his own botanical collections to Clarke for his use in the building within the Garden, so that the Professor of Mineralogy would at least be a frequent visitor there.

The Schultes report continues rather bluntly: 'The Garden at Cambridge contains about five acres of very bad ground', but follows with

> & there are from five to six thousand species of plants, the greater part of them cultivated in beds. It does not present so pleasing an appearance as the Dutch botanic gardens, but is, however, kept very neat & is well arranged . . . The care of the Garden is committed to Mr. Biggs whom we did not find at home. The stoves are well built, & they may have been hitherto large enough; but the progress of the Science will soon cause their size to be insufficient, as they extend only to 216 feet. A building was erected some years ago, for the lecture-rooms of the Professors of Botany, Chemistry, Mineralogy & Mechanics. The Alpine plants, among which are some rare species from the Scotch Highlands, are very properly cultivated in small pots, & placed during winter under glass. The assistant-gardener, who conducted me through the grounds, was not able to tell me the annual expenditure of the institution. The work-people receive two shillings a day.

FIGURE 22 The Old Botanic Garden (Ackermann, 1815).

In an age of telephone, fax and e-mail it is perhaps easy to forget that Henslow and his contemporaries had no way of discovering at a distance whether a visit would be successful or not. Schultes was lucky to find J.E. Smith at home in Norwich, but very unlucky in the Cambridge visit. Nevertheless we can obtain a picture of a small, well-tended botanic garden with something of a speciality in its 'alpine' collection. We do not have to believe the figure of 'five to six thousand species', which perhaps came from the 'assistant gardener' anxious to impress a foreign visitor, but the praise for handling 'alpines' seems genuine enough. It is a small tradition established by James Donn, the Curator whom Biggs succeeded. Donn had his early training under William Aiton, first Curator of the new Botanic Garden at Kew established in 1759, and Aiton had trained under Philip Miller at Chelsea; all this practical expert knowledge came with Donn to Cambridge and provided him with personal links to both Chelsea and Kew.

The well-known print of the Walkerian Garden[7] (Figure 22) shows the Garden looking north, with a skyline of St Benet's Church tower, King's College Chapel and Great St Mary's Church tower; it is a quiet, attractive

FIGURE 23 Entrance to the Old Botanic Garden (Le Keux for Cooper's *Memorials ...*).

scene, but the individual, exotic-looking trees must owe something to the artist's imagination. The view of the 'entrance to the old Botanic Garden' showing the fine wrought-iron gates with a group of academic gentlemen and their ladies entering the Garden is not dated, but must have been during Thomas Martyn's reign[8] (Figure 23). These gates, which survived in Downing Street long after the Walkerian Garden was abandoned, were dismantled and re-erected as the formal entrance to the present Botanic Garden in 1909. A third print, dated 1800, shows the lecture-rooms as they would have been when Henslow was appointed, with a broad north–south walk through the Garden[9] (Figure 24).

   Detail of the layout of the Garden can be seen on the plan[10] (Figure 25). Space for trees and shrubs is very restricted around the margins. It is emphatically a Physic Garden, catering for the growing of a range of herbaceous plants, in a direct line of descent from the medieval herb gardens where medical students could be taught about the medicinal use of herbs. Having said this, we ought to give some credit to Thomas Martyn, whose original intention, in his first lecture course given in 1763, is clearly to teach

FIGURE 24 Lecture-room in the Old Botanic Garden (Harraden, 1800).

botany on a world scale as a science in its own right and no longer as the handmaid of medicine, but this vision and enthusiasm did not long survive.[11]

What was wrong with the site – the 'five acres of very bad ground' inspected by Schultes in 1824? Perhaps we tend to forget what gross pollution accompanied the burning of coal (and peat) fires in the cities and towns of Georgian and Victorian England. The Walkerian Garden was surrounded by smoky chimneys on three sides, and was only 'open' on the narrow south side in Pembroke Street. An anonymous author ('J.D.') describes in 1833 in alarming detail a peculiar hazard affecting Arthur Biggs' attempts to keep a scientifically respectable, accurately-labelled collection in the old Walkerian Garden:

> Jackdaws are comparatively numerous at Cambridge. The Botanic Garden
> there has three of its four sides enclosed by thickly built parts of the town
> and has five parish churches and five colleges within a short flight of it.
> The jackdaws inhabiting (at least for a certain time in each year) these
> and other churches and colleges had, in the years 1815 to 1818 . . . discov-
> ered that the wooden labels placed before the plants whose names they
> bore in the Botanic Garden would well enough serve the same purpose
> as twiggy sticks off trees, and that they had the greater convenience of

GROUND PLAN OF THE BOTANICAL GARDEN, 1838.

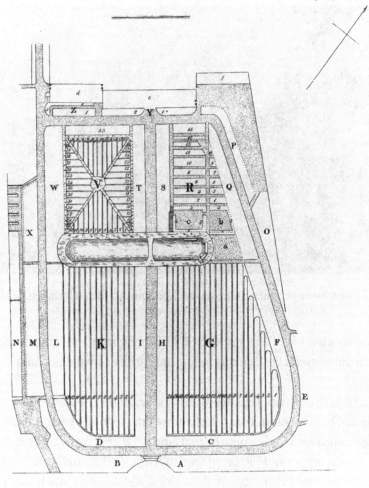

Metcalfe & Palmer, Lithog: Cambridge

DESCRIPTION OF THE PLAN.

The dotted space is gravel walk 9 ft. 6 in. broad.

d Stove in length 58 ft.	*e* Greenhouse in length 104 ft.
f Greenhouse ..... 50 ft. 4 in.	From *d* to the S. wall... 61 ft.
From 53 to W. wall 75 ft. 4	53.15 pits in breadth ..... 9 ft.

Y a space in length 75 ft.   The central space and paths of turf.

The pond is 22 ft. broad and 158 ft. long.

K, G contains beds each 169 ft. long and 4½ ft. broad, and 1½ ft. apart.

DCF a bed 10 ft. in breadth.   *AB* another 20 ft. broad.

The spaces designated by the middle sized capitals are flower-borders.

R contains a number of sheltered partitions.   *E* the Botanical Museum and Lecture Rooms.

Between *A* and *B* is the principal entrance which is not kept open.

FIGURE 25 Plan of the Old Botanic Garden (Smith, J.J. ed. *Cambridge Portfolio*, 1840).

being ready prepared for their use and placed very near home. A large proportion of the labels used in this Garden were made out of deal laths and were about nine inches long, and about an inch or more broad . . . and although of this size, as they were very thin when dry, pretty light. To these the jackdaws would help themselves freely whenever they could do so without molestation, and the time at which they could do this was early in the morning before the gardeners commenced work for the day, and while they were absent from the Garden at their meals, and the jackdaws would sometimes fetch away labels during the gardeners' working hours from one part of the garden, when they observed the gardeners occupied in another, as was often the case in their attending to the plants in the greenhouse, etc. . . . Those who are aware how closely some species of the grasses, garlic, umbelliferous plants, etc., resemble each other and also how needful it is to prefix labels to them, as remembrances of their names, will readily perceive that much inconvenience arose from the jackdaws appropriating some of the labels; and this especially when they removed, as they sometimes did, the labels from sown seeds, as the plants arising from those seeds must in some species grow for a year or more before their names could be ascertained. I cannot give a probable idea of the number of labels which the jackdaws annually removed, but have been more than once been told by persons who had ascended the tower of Great St Mary's Church and the towers or steeples of other churches that wooden labels bearing botanical inscriptions were abounding in these places. The house of the late Dr Kerrick, in Freeschool Lane, was close beside the Botanic Garden; and the shaft of one of the chimneys of his house was stopped up below, or otherwise rendered a fit place of resort for jackdaws. From this chimney shaft Dr Kerrick's man-servant got out on one occasion eighteen dozen of the said deal labels; and these he brought to Mr Arthur Biggs, the Curator of the Botanic Garden. I saw them delivered and received . . . This number of labels and the fact of the occurrence of plant labels on other buildings about the town prove that in general terms the aggregate of labels lost from time to time could not be inconsiderable.[12]

Such a quiet backwater, gradually becoming more and more neglected by the University, starved of funds for any fundamental improvement, and suffering from atmospheric – and avian! – pollution, was emphatically not what Henslow wanted. He made repeated representations to the Governors of the Garden that the University should take its responsibility for the future of botanical science more seriously, and particularly that it would be necessary to remove the Garden to a larger, more open site out of town. After five years of political pressure within the University, Henslow succeeded, and the University acquired, by a special Act of Parliament dated 30 March 1831, the '38 acres and 23 perches' on which the present Botanic Garden stands.

An Act of Parliament was necessary because the acquisition involved sale of the land by Trinity Hall, and all sales of land by Cambridge Colleges until the middle of the nineteenth century required a special Private Act of Parliament. In the preamble to the Act, the case for re-siting the Garden – which was obviously Henslow's, though the wording is cumbersome and legal – is clearly stated:

> although the site of the said Botanic Garden was at the time of the . . . indenture of release and assignment near the outskirts of the town of Cambridge, yet by the great increase and extension of the same town the said Botanic Garden is now clearly surrounded by buildings, whereby the free circulation of air is impeded and restricted . . .

The case made by Henslow, in his own words, is fortunately set out for us *in extenso* in a twenty-page pamphlet that he had privately printed in 1846. We shall see later why there was a fifteen-year delay involved in the production of this admirably clear *Address to the Members of the University of Cambridge on the expediency of improving, and the funds required for remodelling and supporting, the Botanic Garden.*

An extract from the *Address* sets out the case for a much larger Botanic Garden:

> It is indeed true enough that one man with half-a-dozen flower-pots may do more towards advancing Botany than another will feel inclined to attempt with twenty or thirty acres of garden at his command: but it may very safely be asserted, that the larger the number of living species that are cultivated in a Botanic Garden, the greater will be the facilities afforded to us all; not merely for systematic improvement, but for anatomical and other experimental researches essential to the progress of general physiology. It is impossible to predict what particular species may safely be dispensed with in such establishments, without risking some loss of opportunity which that very species might have offered to a competent investigator, at the exact moment he most needed it. The reason why a

modern Botanic Garden requires so much larger space than formerly, is chiefly owing to the vastly increased number of trees and shrubs that have been introduced within the last half century. The demands of modern science require as much attention to be paid to these, as to those herbaceous species which alone can form the staple of the collections in small establishments. The considerable portion of the ground which would be devoted to an Arboretum may be kept up at very much less expense than the rest, but would add very greatly to the ornamental as well as to the efficient character of the Garden.[13]

There are three new ideas here. Most importantly, the provision of as wide a selection of plants as possible is given as a primary aim, with the intention that the teaching of Botany, and the 'anatomical and other experimental researches' which should be an essential part of University Science, can be carried out most efficiently. Note that there is no longer any mention of the particular requirements of medical teaching as a justification: the image of the Physic Garden is at last laid aside. The second, quite revolutionary, idea is that a modern Botanic Garden needs far more space because it should accommodate 'the vastly increased number of trees and shrubs that have been introduced within the last half century', and that 'a considerable proportion' of the Garden should become 'an Arboretum'. Thirdly, the cost of upkeep of the Arboretum would be much less that the more traditional, very labour-intensive formal systematic beds and glasshouses, and would moreover 'add very greatly to the ornamental . . . character of the Garden'.

This vision of the purposes of the New Garden was clearly an optimistic one. Henslow consistently supported the movements to reform the University's teaching, and in particular anticipated those fundamental changes which were just being introduced before he died.

We now return to the story of Henslow's new garden in chronological order, using as a basis the Minutes of the Botanic Garden Trustees and Syndicate in the archives of the University Library.[14] His lobbying was clearly gaining support through 1827, so much so that a Grace was submitted to the Senate on 21 March 1828 to appoint a Syndicate 'to report on the proper measures to be taken for the removal of the Botanic Garden'. By the end of the following year, a second Grace, passed on 9 December 1829, appointed a Syndicate 'to consider the best means of removing the Botanic Garden . . .': significantly, this Syndicate was a larger body, including the

FIGURE 26 Site of the New Botanic Garden, 1809.

Trustees of the Botanic Garden and William Hildyard, Tutor of Trinity Hall. The outline of a land deal with Trinity Hall was beginning to emerge. Biggs, as Curator of the Walkerian Garden, was officially asked his opinion on the suitability of the projected new site, and his report (not, alas, surviving) was approved at a meeting of the Syndicate on 8 February 1830, at which it was 'unanimously agreed that the most eligible site for a new Botanic Garden is the ground adjoining Mr Pemberton's land, at the entrance into Cambridge from the London Road'. At this meeting, therefore, the Syndics requested that 'the Vice Chancellor would ascertain the Terms on which the Master & Fellows of Trinity Hall, & Mr Bullen, the lessee of the land would be willing to part with twelve or fourteen acres.'

Here, for the first time, we meet a member of the Bullen family. The land on which our Botanic Garden now stands (Figure 26)[15] was leased by Trinity Hall to the Revd John Bullen of Barnwell, Vicar of St Andrew the Less, and the lease passed on his death in 1822 to Catherine, his wife. A new lease was drawn up to run for 21 years from Michaelmas 1823. Her eldest son George, a farmer living on Trumpington Road, was presumably farming this land, but George also had a separate lease of about 7 acres of land from the University. It is this portion of land, on the north-east edge of the present Garden, that was eventually transferred to Trinity Hall as part payment for the main site. It appears that, faced with the complication of the sitting tenant, the Botanic Garden Syndicate, with Henslow's full agreement, was envisaging a situation where they would be content with 'twelve or fourteen acres' for immediate development. However, at a further meeting on 20 February, 'It was resolved that it was highly expedient that the University should purchase the whole of the allotment of land belonging to Trinity Hall consisting of 38a, 0r and 23p and that a communication to this effect should be made by the ViceCh to the Master of Trinity Hall.' Things were moving fast, and Henslow must have been looking forward to developing his new Garden and holding his classes there, when a totally unexpected complication arose: the death on 27 February of George Bullen, who, we must assume, had been playing a central part in the negotiations as a tenant of the University and sub-tenant and adviser to his mother, who held the main lease. A draft agreement, not sealed and therefore not legally binding, dated 2 November 1830, stated that the Vice-Chancellor agreed to pay Catherine Bullen £560.1s. for full and immediate possession of the land.[16] It seems that Catherine was prepared to settle for the sum of £628.21s.11d, a figure put forward by Messrs Watford & Nockolds, whom both parties had agreed should make the assessment, but she was advised not to settle by a lawyer son-in-law, John Watson. In the meantime, however, the Act of Parliament authorising the sale and transfer of land between Trinity Hall and the University received the Royal assent on 30 March 1831.

Assuming that all was going well with the planned acquisition of the site, the Vice-Chancellor of the day, William Chafy, Master of Sidney Sussex College, in a letter dated 9 July 1830, had invited a Mr Edward Lapidge, a London architect, to tender for the contract to plan the new Botanic Garden. The postscript to the letter: 'You may perhaps recollect meeting me at Mr. Henson's at Shelford' suggests that it was a chance meeting and conversation between Lapidge and Chafy that got Lapidge the job. Lapidge responded quickly, accepting the commission, and stayed in Cambridge from 26th to 29th July to see and measure the new site; he also reported to the Vice-Chancellor, and had a conference with Henslow, at which, presumably, Henslow was able to agree with him the main features of the new

Garden. During August and September Lapidge worked on his plans, and attended a meeting of the Botanic Garden Syndicate during a second visit to Cambridge between 4th and 6th October. At this meeting the Syndicate approved Lapidge's plans, and he again saw both Henslow and the Vice-Chancellor.

Lapidge's plans, accompanied by a detailed manuscript report, have survived in full.[17] His report begins with a strong recommendation to the University that they should purchase 'the whole of the field & also the small intersecting property of Ashby & others'. The reason he gives sounds very familiar – indeed, quite prophetic!:

> ... future inconvenience of the Establishment would be prevented; it being certain that the ground left unpurchased by the University will become valuable for building, as soon as the botanic garden shall be established. In making this purchase as a precautionary measure, to prevent future annoyances to the Garden, the University would only occupy the space required for the garden, & lay down the remaining land as a Meadow, over which it would possess the control.[18]

Lapidge's report proposes that the Garden proper should occupy the centre of the roughly rectangular site, with a formal entrance on the South side, and the handsome 'Conservatory' building at the end of the main gravelled walk (Figure 27). He takes the view that the 'several compartments or quarters' which he makes a central feature in front of and behind the Conservatory 'would probably be appropriated by the Professor for the following purposes as he may judge to be the most suitable in his classification'. The list that follows, which it is reasonable to assume was drawn up after the conversation he had with Henslow in late July, is worth giving in extenso, because it shows the extent to which Henslow had already in his mind the kind of botany he wanted to teach using living material. The list runs as follows:

> 'Four acres in the main Area of the Garden for a Hortus Linnaeensis & a
> Hortus Jussieuensis
> Four paved areas adjoining for the reception of the Greenhouse plants in
> Summer
> A quarter for native English, Irish and Scotch plants
> –   Twining, creeping & climbing plants, shrubby & herbaceous
> –   Florist's Flowers, as pinks, carnations, etc

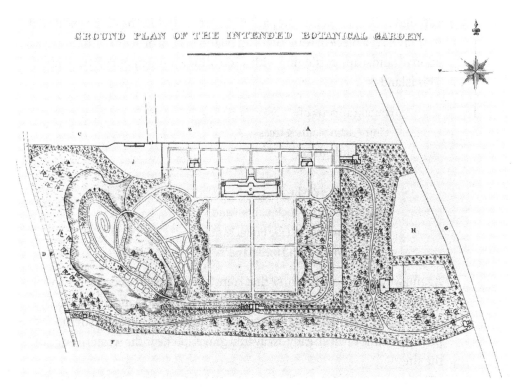

GROUND PLAN OF THE INTENDED BOTANICAL GARDEN.

FIGURE 27 Lapidge plan of the New Botanic Garden, 1830.

– Annual border Flowers
– Biennials
– Bulbs
– Roses of species and varieties
– Choice flowering Plants
– Variegated plants
– Shrubs – hardy foreign
– Alpine plants
– Plants used in Medicine
– Hardy poisonous plants
– Culinary Vegetables
– Different sorts of fruit trees & vines which grow in the open air in England
– Specimens of grafting
– Agricultural plants, including the different kinds of Hedges
– Grasses & Clovers

Lapidge envisages 'in the low part of the ground which has been excavated for gravel . . . a piece of water with an Island in it, in order to combine a variation of landscape gardening with a scientific arrangement of the plants'. This island is

> to contain all sorts of:
>> American plants & trees
>> Bog earth plants
>> Ferns
>> Marsh trees & shrubs
>> a Rock work for rockplants – and
>> Vaults underneath for Fungi & Mosses
>> an aquarium round the Island[19]

Again, we can reasonably infer that Henslow had expressed a desire to develop these special collections associated with the island in the lake, building on the vision that Lapidge as a landscape architect had of the potential value of an already excavated gravel-pit near the western edge of the site.

The importance of 'American trees' appears for the first time in this reference. Henslow would have been well aware of the spread of interest in Britain for new hardy plants, especially trees and shrubs, of North American origin, a movement that grew rapidly in the succeeding decade. By the time, fifteen years later, that it was possible to begin the detailed development, the case for an Arboretum of North Temperate trees and shrubs was very much more obvious to Henslow. How far already in 1830 Henslow had seen the full potential of Lapidge's design is not easy to determine: in particular we should perhaps not assume that the peripheral systematic arrangement of the hardy trees that is one of the most impressive features of 'Henslow's Garden' today was in his mind when the Lapidge plan was accepted. All Lapidge says about an Arboretum in his report is the following:

> On the exterior of the wall [separating off the central portion which he planned to be the Garden proper] a general Arboretum of
>> deciduous Forest trees
>> Evergreen trees
>> Firs and Pines
>> Willows

Duplicate flowering Shrubs & flowers in the borders of the Plantations which skreen [sic] the Garden from the Wind.'

We do not know how far Henslow made any detailed contribution to Lapidge's design for the main Conservatory. The impression given by Lapidge in his report is that he went to some trouble to acquaint himself with the Botanic Gardens of Glasgow, Edinburgh, Dublin and Brussels, and that he was in no way dependent on others for detailed information. He says: 'I have considered the extent of Glass in the Botanic Gardens of Glasgow & of Edinburgh' . . . and follows this with figures of sizes of the glasshouses there. We should remember, however, that Henslow himself had already sought and received quite detailed information from Hooker in Glasgow and Graham in Edinburgh on their respective Botanic Gardens. In a letter to Hooker as early as 11 January 1827, we read:

> I reproach myself in trespassing upon your time but as the cause of Botany is at stake perhaps you will excuse me – we are (& have been for some time) planning the removal of our garden, & I wish to gain some information respecting other gardens in the kingdom, by which we may in some degree guide ourselves – I should therefore feel much obliged if you could without much trouble to yourself procure me an answer to the following questions, respecting the Glasgow garden.
>
> 1st No of acres enclosed
> 2d   No & extent of the glasshouses
> 3.   Salary of the curator or head gardener
> 4.   No of men under Curator
> 5.   Whole annual expenditure
> 6.   Whence the funds are devised[20]

Hooker obviously acted quickly, for Henslow writes again on 23 January: 'I am extremely obliged to you for the information you have given me.' We do not have this 'information' from Hooker, but that from Graham on 6 March in the same year has survived, and obviously follows the order set out by Henslow in the letter to Hooker. It is quite detailed, but cannot have been entirely cheering to the recipient, especially on the finances of the Edinburgh Garden: 'The whole allowances made to me [to run the Garden] are £444 per annum and we are starving, tho. I subscribe to prevent our going to ruin, abt. £200 per annum.'[21]

The intransigence of John Watson finally becomes clear from a meeting of the Botanic Garden Syndicate in 1833 when a letter from him was considered and in effect rejected: the Syndics simply reiterated their offer of a sum of money based on their previous offer three years earlier. It is a bleak exchange, and the whole matter seems from this point to have been shelved.[22] We have traced no explicit reference by Henslow as to how he viewed another ten years or so before the lease was due to expire and the deal to acquire the land could go through; as we have seen in Chapter 5, he was very busy in the early 1830s with teaching, writing papers and especially preparing his text-book for publication, and it is surely characteristic of the man that he should spend no time 'crying over spilt milk'. He was certainly interested in what Biggs was growing in the Walkerian Garden during this period. For example, he published around 1840 a description of the tropical American '*Gesnera douglasii var. verticillata*', together with a coloured illustration saying 'the present plant was grown in the Botanic Garden at Cambridge'.[23]

In 1840 Henslow published a paper describing the history of the Walkerian Garden and its present state, and explaining the 1831 purchase of land and the new site. Interestingly he makes no comment on the lack of action, restricting himself to facts and eventual intention:

> The site proposed for a new Botanic Garden lies a little out of the town on the London Road. It is a field of 30 acres, in the middle of which it is intended to lay out 4 or 5 acres as an herbaceous ground and to arrange the rest as ornamental walks, which will afford an opportunity of growing all trees capable of standing our climate. We subjoin the plan proposed by Mr. Lappidge [sic] who was employed by the University for this purpose.[24]

It is from this paper that we take both Lapidge's original plan and the plan of the Walkerian Garden.

Lapidge himself eventually breaks the silence, writing a courteous letter to the Vice-Chancellor to remind the University that his bill, which he had submitted over ten years earlier, had never been paid. He obviously knew what the problem was, for he reminds the Vice-Chancellor that 'the tenant in possession of the land [had] demanded an extravagant sum for his interest in it'. Judging the time to be ripe, he continues:

> You will doubtless recollect my drawings for the proposed Botanic Garden in Cambridge, which have lain dormant these ten years ... That I should feel very great interest in the work you will rapidly suppose, my plan having

been approved by the Syndicate on the 5th October 1830, and my bill of £131.5.0 having remained in abeyance ever since that period. I conceive the tenants term to be nearly expired, that the subject could now be advisedly and successfully revived in anticipation of that event.[25]

His plea was successful; the then Vice-Chancellor paid his bill, an act requiring a special Grace on 19 February 1841.

Perhaps it was the Lapidge letter which caused the first stirrings in the University to reactivate the plan, for by December 1842 we find Henslow replying to a letter from Whewell on the subject, in which he says: 'though a few acres will be enough for shrubs and herb plants it would be a thousand pities to deprive us of any portion of the 30 [acres] if we should ever contemplate possessing anything like an arboretum'. Obviously the Henslow–Whewell discussion of the New Garden continued, for in a further letter, dated 1 March 1844, Henslow is expanding his ideas on using 'the whole 30 acres'.[26] The appointment of a special Syndicate on 7 February 1844 'to consider whether & what steps should be taken towards changing the site of the Botanic Garden' was the first formal action to plan for the release of the land acquired for the New Garden in 1831.

Henslow's correspondence with William Hooker at this period reveals how much he felt the need of external advice about the design of the new Garden. His letter to Hooker of 1 December 1844 encloses a sketch plan which marks a square 10 acres of ground in the middle of the north side of the new site as the 'proposed site of the Garden', and the Arboretum with the word 'seminary' [?], in the area south of this. On the west side is 'low ground, gravel dug – ? aquarium' and there are two entrances, one on the north and the other at the north-east corner, with a third, marked ?, at the south-west corner. The existence of this sketch, which antedates the appointment of Andrew Murray[27] from a list of four candidates discussed in some detail with Hooker during March and April 1845, clearly shows that Henslow had tried to accommodate the Lapidge original design to the political requirements that only a relatively small part of the available site of nearly 40 acres should be developed initially as the Botanic Garden proper (Figure 28).

Henslow writes again to Hooker on 9 August, after Murray's appointment, saying how pleased he is with Murray, and subsequent letters reveal that Henslow and Murray are discussing differences in their respective plans. An exchange between Henslow and Hooker in early September

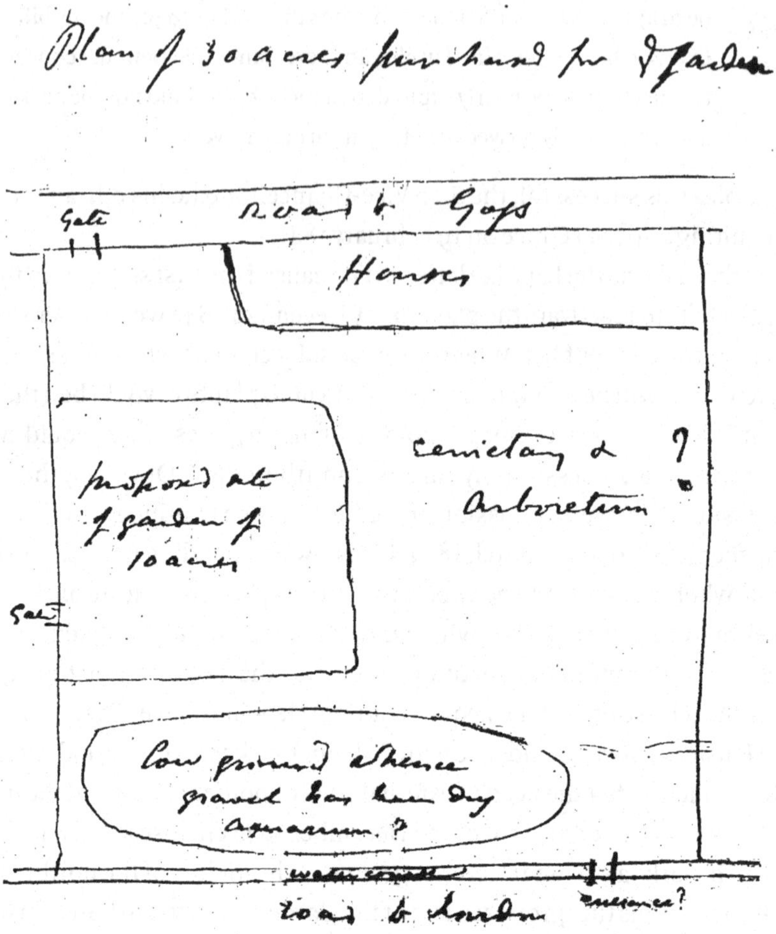

FIGURE 28 Henslow's sketch plan of the New Botanic Garden.

concerns the pros. and cons. of appointing a 'professional man', Henslow suggesting, and Hooker denying, that the latter had recommended dispensing with a 'professional'. Here is part of Hooker's letter of 9 September 1845:

> ... I do not recollect however at any time advising you to lay out the Garden without applying to a professional man. My advice I think was the very opposite to that & I felt the more confident on that point – when I found now large the Garden was to be – 20 acres. Make it ever so good you will find some that will find fault with it & it will be something to be able to mention in reply the name of an eminent landscape gardener whose design it was. Certainly I considered Murray's first attempt a failure and I

did not scruple to say so. Your little plan is in many respects a great improvement but I do not at all like the arrangement of the herbaceous portion wh. would look frightfully in the winter nor of the [ ] mass of what I assume to be gravel immediately around the great range of houses. The form of a terrain ought to correspond in a measure with the form of the range & there ought to be a good deal of well kept lawn near it. I fear however you would find me more inclined to find fault than able to mend. Still I shall be very happy to see any plan of yours or Murray's & to express my views respecting it.[28]

Henslow, writing on 11 September, presumably in reply to the above, says that he hopes to visit Hooker 'once more about our scheme' and adds: 'I will write to the V.C. and get him to send me the disapproved plan which was furnished by Lapidge about 12 years ago and bring it with Murray's New Ideas . . .'[29]

We have only a little evidence about the cooperation between Henslow and Babington in the establishment of the New Garden. Babington's own Journal is quite surprisingly devoid of any references to the Garden for the years in question, though he does record, for 7 February 1844, being 'placed on a Syndicate to consult concerning the removal of the Botanic Garden', and also tells us that the new Syndicate met on 10 February, and again on the 26th of that month when he dined in Trinity with Whewell.[30] It is obvious that he is acting informally as a deputy for Henslow who was, of course, present only for the Easter Term five weeks unless there was urgent University business. It can hardly have been an accident that Henslow's name was not put forward immediately for the new Syndicate – though he *was* added in May 1844.

A single letter preserved in the University Archives provides us with the best evidence of Babington's initial estimate of the new Curator, Andrew Murray. The main purpose of the letter dated 27 May 1845 and addressed to Hitcham is to acknowledge a draft paper and deal with some matters concerning the finances of the British Association which was due to meet in Cambridge in June. The second paragraph continues:

Mr Murray is come and I have had some conversation with him today. I like him well. We are to have a meeting with the Syndicate tomorrow to give some directions concerning the [ground] which wants clearing from weeds. [Mr Murray] much wishes to have some [conversation] with you about the plan for the [Garden. He] wants to get to work at once. [He is very] pleased with the collection of plants in the garden[31] which he says is far more valuable than he had supposed. He particularly noticed a number of fine specimens of old plants which are now rarely to be found even in [botanical] gardens.[32]

Murray seems to have been an admirable choice for the demanding new post. He had established his reputation in the Liverpool Botanic Gardens at Edge Lane, where he was involved in the removal of plants from the old botanic garden to the Edge Lane site, which opened in 1836. In addition to having this practical experience, he was obviously very competent botanically, and both Henslow and Babington found him a pleasant young man with whom to work. The Botanic Garden still houses Murray's working plan of his Garden, complete with annotations[33] (Figure 29).

Murray set to work with a will, removing the herbaceous plants from the Old to the New Botanic Garden, and publishing in May 1848 a catalogue of the plants he had re-housed. The Preface is admirably clear as to its purpose:

> THIS CATALOGUE of the HARDY PERENNIALS, removed from the OLD to the NEW BOTANIC GARDEN, at CAMBRIDGE, has been printed for the purpose of inviting attention to the imperfect state of this portion of the Collection. It will be circulated in the hope of obtaining assistance from other Botanic Establishments. It is the earnest wish of the Curator to improve this Department of the University Collection as speedily as possible. He would feel obliged if the parties to whom copies of the Catalogue are sent would insert in the blank columns of one of them, and opposite their respective Genera, the names of all the Species they may be disposed to supply, and then return that copy to him. He will then collate the various copies thus returned, and inform the respective parties what Species he would wish each to forward him. This plan will prevent his receiving duplicates from different quarters. The returned Catalogues may be marked with an * against the names (in the printed list) of such Species as may be wanted by the parties offering to contribute; and he will endeavour to supply them as he has an opportunity of doing so.

The catalogue has 59 pages, each with a single column of c. 36 entries to the left of the page: a total of more than 2,000 species. We do not know what success Murray had in augmenting his collection from other 'Botanic

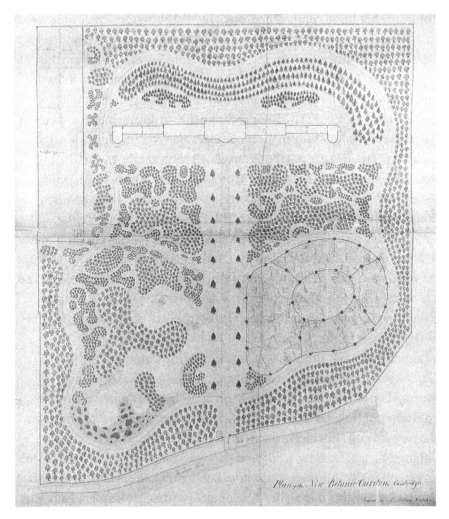

FIGURE 29 Murray's plan of the New Botanic Garden, 1846.

Establishments', but we can say that by 1850 his definitive new *Catalogue of the Hardy Plants in the Botanic Garden, Cambridge* contains nearly 5,500 species, most of which are herbaceous plants.[34]

Murray showed even more talent in acquiring and establishing sapling trees for the Arboretum, so that by 2 November 1846, when the New Garden was formally opened by the then Vice-Chancellor, Ralph Tatham, the outline of the Arboretum and the Garden proper could already be seen, occupying some 20 acres at the west side of the available ground.[35]

Murray's botanical competence can be judged by his Catalogue of 1850. In the Preface to this work, dated 15 April 1850, he tells us that his Catalogue lists all the 'Hardy Herbaceous Plants, Trees & Shrubs in the new Botanic Garden', and that 'the names generally adopted are those employed in de Candolle's *Prodromus* and Loudon's *Arboretum Britanicum* [*sic*]'. We can use this list with some confidence to document which trees in particular still flourishing in the modern Garden were in Murray's original plantings. The area of the Arboretum which we now call 'The Old Pinetum', still contains a number of original 'Murray trees'. The genus *Pinus* is particularly represented: indeed the Cambridge Garden is justly famous for the diversity of species of this familiar conifer, including several rarely seen in cultivation. The most valuable is a mature specimen of *P. gerardiana*, named as such in the Catalogue, which Murray must have acquired soon after its introduction from its native land in Afghanistan and the north-west Himalayan region. Its slender trunk with greyish-white flaking bark shows up impressively against the dark background of Cedars by the old rock garden (Figure 30).

Tragically, the very promising career of the first Curator of the New Garden was terminated by an accident: during the summer of 1850 Murray died of pneumonia, brought on by accidental immersion in the cold water of Hobson's Conduit, which runs from deep chalk springs on the Gog-Magog Hills along the west boundary of the Garden. It was Babington, not Henslow, who wrote Murray's obituary: a warm tribute to an exceptionally able man with whom Babington had obviously enjoyed day-to-day working during the development of the new Garden. Here is an extract:

> All who have seen what he has done in converting a corn field into a botani-
> cal garden containing one of the best collections of hardy trees, shrubs, and
> herbs in the kingdom (see the catalogue which he has recently drawn up
> and printed), can and must appreciate his eminence in his professional
> capacity; but it is only those who have become intimately acquainted with
> him, as I have done, who know his full value. Thoroughly understanding
> his business, both in its scientific and practical departments, he was
> respectful but firm in his well formed opinion towards his superiors, and a
> good master to those under him.[36]

During the academic year 1850–51, when the growth of the new Botanic Garden was much hampered by Murray's death, Babington felt keenly that

FIGURE 30 *Pinus gerardiana* in the Botanic Garden in 1999.

Henslow was not able to pull his weight, and relations between the two were somewhat strained. We have already quoted in Chapter 5 Babington's surprisingly outspoken comment to Balfour about how Botany suffered in Cambridge 'from a non-resident Professor'.[37] Henslow's commitments at Hitcham, ranging from the British Association meeting at Ipswich attended by Prince Albert, which Henslow as President of the Ipswich Museum was responsible for organising, to the marriage of Frances Henslow to Joseph Hooker, must have convinced Babington that Henslow had lost interest in Cambridge affairs. Indeed, as we now see it, Henslow's great enthusiasm for the Ipswich Museum at a time when his new Botanic Garden in Cambridge was languishing for effective leadership and promotion is somewhat difficult to understand.

The evidence given before the Graham Commission, set up in 1851 to 'inquire into the state, discipline, studies, etc. . . . of the University . . .', is also revealing. Henslow's submission officially as Professor of Botany is a minimal reply, whereas Babington's independent submission is much more informative and constructively critical.[38]

It fell to Murray's successor, James Stratton, to complete the removal of all the living plant material from the old Walkerian Garden. First, the transfer of the perennial herbaceous plants, begun by Murray, had to be completed, and then the construction of a glasshouse range and the transfer of all the tender plants. Work was in progress in the summer of 1854 when Henslow brought a large party of Hitcham parishioners by the newly-opened railway for a day out to see his University. We describe this remarkable visit in Chapter 12; here we can use extracts from the printed pamphlet prepared by Henslow for the visitors. They are timed to arrive at Cambridge railway station at 9.20 am:

> In about five minutes' walk from the Cambridge Station, we arrive at a back entrance to the NEW BOTANIC GARDEN. About 20 acres are laid out and planted. We pass the spot where the greenhouses and stoves are being erected, to contain plants from hot countries. This part of the garden is laid out for trees and shrubs; and we cross over to the opposite side, where herbaceous plants are grown. The different families are confined to separate beds, and are so grouped that those of the two great Classes of flowering plants, (viz. Dicotyledons and Monocotyledons) are ranged apart. A large space has been prepared for a pond to contain water-plants.

Later on in their schedule they visit briefly what remains of the Old Garden:

> Passing through the Old Botanic Garden, we find the Greenhouses and
> Hothouses still standing with a few plants in them. The Curator has kindly
> promised to attach the following numbers to the plants here named, so that
> they may be readily recognised as we pass by them:
>
> I. DICOTYLEDONS.
>
> 1. Coffee. 2. Tea. 3. Sensitive, which closes its leaves on being touched.
> 4. India-rubber Tree, which emits a milky juice on being pricked. 5. Cycas,
> from which a coarse sago is prepared.
>
> II. MONOCOTYLEDONS.
>
> 6. American Aloe. 7. Date Palm. 8. Dwarf Palm of Europe. 9. Banana.
> 10. Papyrus, from which the ancient Egyptians made their paper.
> 11. Sugar-cane. 12. Bamboo-cane. 13. Some Orchids, many of which are
> called Air-plants.
>
> III. ACOTYLEDONS.
>
> 14. Foreign Ferns.[39] [Figure 31]

With the construction in 1858–59 of the Lake and 'Island' – features that
essentially remain unchanged today – Henslow's New Botanic Garden was
complete, and the remaining, unused land, usually referred to officially as
'the field adjacent to the Botanic Garden' and occupying the eastern half of
the site, was not let off as allotments until well after Henslow's death.[40]

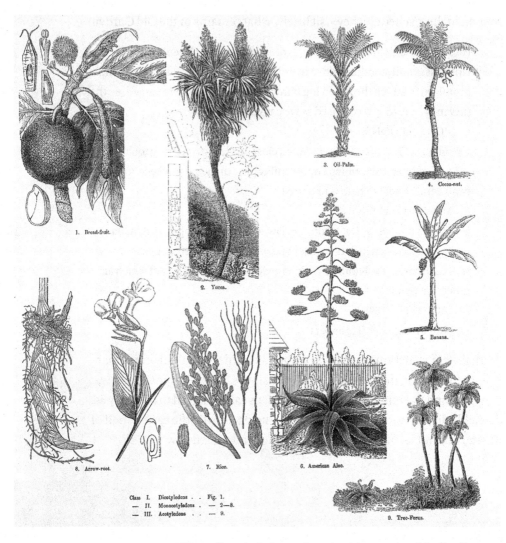

FIGURE 31 Plate of tropical plants, from Henslow's pamphlet for the parish visit to Cambridge in 1854.

9 A Liberal Churchman

An oft-quoted comment by Darwin on Henslow's religious belief, that he was 'so orthodox, that he told me one day, he should be grieved if a single word of the Thirty-nine Articles [of the Church of England] were altered', has all too often been used to suggest that Henslow's Christian belief was a dogmatically held traditional view that he had never questioned.[1] That such an assessment is quite wrong is, however, easily gleaned from reading Jenyns' *Memoir*. Jenyns, of all people, would be most anxious to present his deceased brother-in-law, for whom he had great respect and affection, in as favourable a light as possible, yet even he does not disguise the considerable difficulties the young Henslow had faced in thinking through the implications of his Christian faith. The first hint of early questioning is given by Jenyns:

> His parents had always been desirous that he should go into the Church, and he is said to have acquiesced in their wishes; yet he does not seem to have fully determined on this step till some years after he was of the proper age. It has been stated by a Unitarian connection of the family, that the reason why he did not take orders earlier, was from religious scruples; he could not reconcile the doctrine of the Trinity as held by the Church of England, with what he considered to be the true teaching of Scripture. If this be the case, which I do not remember to have heard him mention, and should hardly think likely, it is quite certain that his scruples were soon overcome, and that he never entertained a doubt of the truth of this doctrine in after-life.[2]

Jenyns is clearly reluctant to believe that his hero had entertained heretical views on the Trinity, although we might reasonably think today that, far from being a sign of weakness in Henslow, such questionings indicated a lively mind which from his childhood had seen the fascination of intellectual enquiry, and could not reasonably suppress or ignore the problems posed by the creeds and formulae of the Christian religion.

We have not been able to identify with certainty the 'Unitarian connection of the family', but it is surely significant that Henslow's grandmother, Ann Prentis, was of Presbyterian stock,[3] and that in the first half of the eighteenth century many Presbyterian Chapels, which were built in the years following the Act of Toleration in 1689, were 'Open Trust' places of worship whose purpose was solely 'for the worship of Almighty God'. In such assemblies, some of which eventually became Unitarian chapels and survive to the present day, belief in the doctrine of the Trinity was rare.[4] Nor is it the case that the Church of England itself was free from 'heresy'.

With this background, and with Henslow's enquiring mind, it seems more than likely that the doctrine of the Trinity had been questioned within the family circle, and that this and other questions of Christian belief were already exercising Henslow when he came up to Cambridge. Once in Cambridge, it is easy to see that these questions would continue to be explored. There is one person who was probably involved: an Anglican clergyman with strong Unitarian sympathies who became a firm friend of Henslow in Cambridge and later a regular visitor to the Rectory at Hitcham. He was Joseph Romilly (1791–1864), Fellow of Trinity and University Registrary. It seems more than likely that Romilly, five years older than Henslow, played some part in helping his young friend to think through his difficulties with Christian doctrine during his Cambridge years. Certainly the group of zealous reformers, mainly Trinity men centred around William Whewell, a group to which Henslow belonged, must have provided him with plenty of opportunity to test and form his views on the Church's doctrines, as indeed on many other matters of local and national politics.[5] Romilly's diaries, '41 note-books in a small but legible hand'[6] cover the period 1818 to 1864, and are full of delightful, sometimes even racy, accounts of people he met in his professional career in the University and his wide social life further afield.

We might also expect that Henslow would be familiar with the writings of Soame Jenyns (1703–87), a convert from deism who wrote much on the problems of 'revealed religion'; Henslow could hardly have married into the Jenyns family without at least being aware of the views of his wife's illustrious ancestor.[7]

In trying to find chapter and verse for the nature of any person's struggle to think through doctrinal difficulties, we are, it seems, unlikely to discover much in printed texts. If, as is the case with Henslow, no personal diary survives,[8] much must therefore remain conjectural or oblique. We have, however, the text of a single sermon delivered by Henslow during his Cambridge years, a sermon that he must have decided to have printed so that he could send copies to his friends. This is the sermon of which Jenyns tells us in the following passage:

> At one period of his life he gave much of his attention to the study of the prophetic books of the Sacred Scriptures, especially the book of Revelation; and he joined a small circle of friends in the University, who met together occasionally to discuss some of the more obscure passages in those books. It was whilst engaged in those studies that he preached in Great St. Mary's Church, Cambridge, before the University, a sermon 'On the first and second Resurrection', which caused much sensation, from the circumstance of his being to such a degree excited by the subject, that he burst into tears in the middle of his discourse, and for some time was unable to proceed.[9]

The title of the sermon, *On the First and Second Resurrection*, preached at Great St Mary's, the University Church, in Cambridge on 15 February 1829, is not immediately appealing, but a personal Preface, at ten pages nearly as long as the sermon itself, is quite revealing. The opening sentence inclines one to persevere:

> It will perhaps excite surprize that I should presume to publish a Sermon, in which there will be found no pretension to support my position, either by argument, or by the detail of the fruits of my own research.

So far as the printed sermon is concerned, we find it difficult to appreciate the niceties of the argument, in the absence of any knowledge of what Henslow had said on an earlier occasion when he preached in the University Church.[10] To quote his exact words, we do not know what were 'those views which I opened to you upon the last occasion on which I had the privilege of addressing you', and can only guess that he must have suffered some criticism from theologians and other colleagues for being too partisan concerning what were 'millenarianist' controversies within the Anglican Church. These eschatological arguments arose within evangelical circles, and centred round the reference in the twentieth chapter of Revelation, the last book of the Bible, to a thousand years, at the end of which would come the final battle of Armageddon, when Satan would be 'cast into the lake of fire' and finally destroyed. Wolffe (1995) can be consulted for further information on the controversy between 'premillennialists' and 'postmillennialists', whose interpretation of the Second Coming of Christ differed radically, the former preaching an imminent cataclysm, whilst the latter interpreted the 'thousand years' as a long period of 'the gradual triumph of improvement and religion'. It would seem likely that the liberal Henslow would have had more sympathy with the post-millennialists who were on the whole, as Wolffe explains, 'more content to participate fully in the existing structures of society'.[11]

Be that as it may, the Preface reveals to us three characteristic Henslovian attitudes, which we must now explore.

The first is as we have already seen in his scientific writings, that he was firmly in the camp of those who saw the wisdom of God revealed in the created world, believing that God has provided Man with his faculty of reason, which he should feel free to use to question everything, even the contents of the Bible. To quote further from the Preface:

> ... I think we may calculate upon a similar result from the further prosecution of religious research, to what we find take place in the pursuits of science. The deeper we advance, the clearer do our views become of the first principles. What once seemed to us 'hard to understand' presently becomes easy, until at length we are able to look back with surprize upon those minor difficulties which are now obstructing the progress of another.

A little further on we read: 'In critical and philological enquiries the Bible must be studied like any other book ...'

Henslow is revealed at the age of 32 to be quite sure that all human experience, including religious and artistic experience, is the legitimate subject of rational enquiry. There should be no 'banned areas'. In this respect he is not unusual in Georgian Cambridge, in which William Paley's *Evidences of Christianity*, together with Locke's *Essay Concerning Human Understanding*, were the two 'set books' that every undergraduate was required to study. Although some of the very widespread 'Enlightenment' questionings of the late eighteenth century were declining in Henslow's time, there were still many leading University and Church figures who were prepared to ask awkward questions about the central tenets of the Christian faith, and to look for evidence of divine design, as Henslow did, in the detail of the natural world. The opposition to 'natural religion' in the form of evangelical 'revealed religion' was, however, already a very powerful force in the early years of the century, and the tension between these two opposing views must have been very apparent to any young aspiring clergyman in Henslow's time. In particular, the appointment of Charles Simeon in 1783 at the age of 34 to be minister of Holy Trinity Church had supplied a central platform for a fiery exponent of a simple faith in the Bible as the Word of God and in Christ's death as an atonement for individual sin. Simeon 'exercised a profound influence both in Cambridge and far beyond. Macauley said of him (in 1844): As to Simeon, if you knew what his authority and influence were, and how they extended from Cambridge to the most remote corners of England, you would allow that this real sway was far greater than that of any primate.'[12]

This brings us to the second element in Henslow's character: tolerance of the right of people to express views he could not himself support. Again, such tolerance, applied to religious differences, is easy to illustrate from the Preface, as in the following passage:

> Let everyone then proceed in such investigations as his conscience may direct. Only let us be careful to assume nothing in contradiction to the express declarations of the New Testament, but rather let us take its precepts as the true key to the right understanding of the older prophets. Above all, let us be careful not to suffer any root of bitterness to prejudice our minds against our brethren merely because they may not agree with us in certain points of detail.

Henslow illustrates this tolerance of the theological positions of others in a letter to Hooker (senior). Written on 16 October 1829, obviously in reply to one from Hooker about young men coming up to St John's, Henslow says: 'I will do all in my power to make your young friends happy during their stay in Cambridge.' Later in the letter he is replying to a concern apparently expressed by Hooker about the influence of Simeon on undergraduates:

> Simeon is certainly at the head of the Evangelical party, but I should hope your friends would neither think the better nor the worse of him for that reason. We certainly mistake the privileges of our calling whenever we take offence, for whatever circumstances good or evil that affect us, and in religious matters especially I think the name of party should be avoided. Let them therefore stand aloof from high Church & low Church – respect the foibles of the weak & be not obsequious to the seeming vigor of the strong – & they will be happy in this world, and fearless for the future in doing the duties of the day.[13]

It was Henslow's problem, and one that has been familiar to 'liberals' in the late twentieth as in the nineteenth century, that he 'suffered no root of

bitterness' to prejudice his mind against any 'brethren', but faced intoler-
ance and downright bitter opposition in his turn when he stood up for his
principles.

Adopting Vidler's analysis of the Church of England in the early nineteenth century, which draws attention to three 'main groups or tendencies
with marks that distinguish them from the ordinary ecclesiastical conven-
tions', we can distinguish 'the Evangelicals, a more theologically-minded
run of High Churchmen, and a miscellaneous group that may be conveniently designated "Liberal".'[14] Charles Simeon was an Evangelical:
Henslow was clearly and consistently a Liberal. To complete the picture,
what should we say of the 'more theologically-minded . . . High churchmen'?

The term 'High Churchman' – or rather 'High Church' – can be traced back
to the seventeenth century, and was in common use through the eighteenth.
In Cambridge in the early decades of the nineteenth century an 'intellectual
party' was able to accommodate a wide range of tastes and theological positions. One Cambridge divine whom Henslow must have known well was
Hugh James Rose (1795–1838), a contemporary at Trinity College, and for
three years Dean of Bocking and Rector of Hadleigh (though before
Henslow took the adjacent parish of Hitcham). Rose's old High Church
position came more to the fore as his controversy with John Henry Newman
became more bitter and open in the last years of Rose's life.[15]

The third element in Henslow's religious belief apparent in the Preface is
his insistence that the essentials of the Christian faith are accessible to
everyone, whatever their station or education, with the corollary that those
fortunate enough to have received any gift of learning had a duty to use
their talents to the glory of God and the good of their fellow-men. Thus,
again from the Preface:

> I believe the 'simplicity that is in Christ' to be as much the property of the
> unlettered as of the learned, the method of comprehending it being the
> work of the Spirit, and not of the understanding.

This strong conviction must have played an important part in his eventual
decision to move to Hitcham and devote his talents to a much wider audience than the University could provide. This 'wider audience' was, of
course, apparent to him whilst still resident in Cambridge, where, as we
have seen, he involved himself in local politics where he saw a clear
Christian duty to do so.

In Chapter 6 we gave some detail of the relationship between Henslow the teacher and Darwin the pupil, but restricted this largely to academic and social aspects. We must now look more closely at Henslow's influence on Darwin so far as religious belief was concerned. From their early acquaintance it is obvious that they must have discussed aspects of the Christian faith in which they had been reared. Not only was Paley's 'Evidences' a required 'set book' for the Senate House examination – and we should recall that Darwin tells us that he enjoyed studying Paley! – but we know that in his second year Darwin was seriously considering 'reading divinity' with Henslow after graduating.[16] Further, we know that by November 1830 Darwin is curious about rumours he has heard about his teacher's religious views. Writing to his cousin Fox he says: 'I have heard men say that Henslowe [*sic*] has some curious religious opinions; I never perceived anything of it. Have you?'[17] It seems reasonable to relate this to the notorious sermon in Great St Mary's the previous year and its aftermath in provoking discussion and rumour; the implication is that Darwin looks to Henslow to present a rational balanced view of religion as every other subject and is rather worried by this 'emotional Henslow' he had not himself encountered! As we have seen, the experience in Great St Mary's seems to have been a turning-point for Henslow away from any taste for fiery, emotional preaching to the tolerant, practical Christianity he so effectively revealed in his Hitcham ministry.

What are we to make of Darwin's much-quoted comment on Henslow with which we began the chapter? Does this, as many seem to think, call into question any claim that Henslow was a 'Liberal Christian'?[18] In Henslow's time every man being ordained had to swear allegiance to the Thirty-nine Articles as set out in the Church of England's *Book of Common Prayer* and assent was the normal, customary procedure. In any case, was not Darwin recalling in later years, when he himself had lost his own faith, a remark of Henslow probably taken out of context? That Henslow would seriously give general assent to the Articles after he had settled his own doubts on the doctrine of the Trinity seems entirely reasonable: like most of his own peers he did not question the role of the Sovereign as Supreme Governor of the Church of England, nor his or her God-given right to lay down rules for the religious and secular life of his (her) subjects . . . and this is surely the purpose of the Articles.

Much of what is written on the controversy between science and religion presents an over-simplified picture. Perhaps this is inevitable, but, to understand the Darwin–Henslow relationship, we do need a more balanced view. It cannot be too strongly emphasised that the earliest and most serious area of controversy between 'science' and 'religion' was in the realm of geology, and that most serious thinkers in the early decades of the nineteenth century, whether they were Churchmen or not, had abandoned any attempt to cling to Genesis as a *historical* account of the Creation or Archbishop Ussher's 4004 BC as the first day of Creation. We now see Darwin 'as entering the debate not in 1859, with publication of the *Origin of Species*, but earlier, in the 1820s and 1830s, when he was working as a geologist'.[19] Darwin's teachers of geology in Cambridge were, as we have seen, Henslow and Sedgwick. Both were ordained Anglican priests, but neither of them was clinging to any literal reading of Genesis. As we saw in Chapter 6, when the young Darwin came under their influence, it is true that they were both holding some 'catastrophist' view of geological history and were opposed to Lyell's 'uniformitarian' view; but they were both reconciled to a long, complex geological history which was not easy to equate with Genesis except in a general, figurative way. Indeed, Henslow's only publication on the subject of 'The Flood', which appeared in 1823, puts forward 'a hypothesis of a non-miraculous cause for the deluge, one that would employ the ordinary means of nature and also coincide with the account given in Genesis'.[20] This short submission, entitled 'On the Deluge', is well worth reading. Henslow felt moved to send it in to the Editor of *Annals of Philosophy* on 15 October 1823, after reading an article (in the *Quarterly Review*) reviewing Buckland's *Reliquiae Diluvianae*.[21] Henslow gives us first the reviewer's own opinion that 'the various theories which have been adopted to account for the phenomena of the Deluge' are all unacceptable, and tells us he therefore 'seems to think that we ought to ascribe the whole to the miraculous interposition of Providence, "<u>excluding</u> the operation of ordinary nature" from our consideration'. Henslow disagrees, saying 'I see no reason for supposing that he [God] did not employ the ordinary means of nature as the instruments of his operations'. He then elaborates a theory which he attributes to Mr Greenough (President of the Geological Society which Henslow had recently joined), which postulates a catastrophic strike by a comet, explaining that 'it appears universally conjectured by the most accurate observers, that they are in great part, if not wholly, composed of aqueous vapour'.[22] There is a curiously modern ring about all this. Comets and asteroids are very much 'in the news', and every child now 'knows' that the extinction of the dinosaurs was caused by a catastrophic impact of a heavenly body with the Earth! The young Darwin was obviously not discussing religious belief with a fundamentalist who took as literal truth everything in the Bible, and looked for miraculous explanations of Biblical stories.

Once the 'Beagle' voyage was under way, Darwin became captivated by Lyell's interpretation of geological phenomena, and in particular, adopted Lyell's view of superficial gravel deposits as being caused by submarine deposition, not by catastrophic 'flood' events. Although he continued to use the term 'diluvian', reflecting what he heard when he was in residence at Cambridge, he had largely discarded the term by the time he was publishing his papers after the voyage. This is consistent with Sedgwick's own views about 'superficial gravel' deposits, as expressed in his Presidential Address to the Geological Society in 1831:

> It was indeed a most unwarranted conclusion, when we assumed the contemporaneity of all the superficial gravel on the earth. We saw the clearest traces of diluvial action, and we had, in our sacred histories, the record of a general deluge. On this double testimony it was that we gave a unity to a vast succession of phenomena, not one of which we perfectly comprehended, and under the name diluvian, classed them all together.[23]

We should now turn to the other subject on which Henslow and Darwin must have had much discussion, namely the mutability of species of both plants and animals. Some of the relevant evidence is only just emerging, but we now know enough to see that Henslow, who was clearly fascinated both by the variability of plant species and by the problems of hybridisation, had by no means a closed mind to the central idea in Darwin's picture of evolution. In one of his first published papers on a botanical subject, Henslow writes *On the specific identity of the Primrose, Oxlip, Cowslip and Polyanthus*,[24] thus joining the long tradition of Cambridge botanists who have been fascinated by the relationship in gardens and in the wild between our native *Primula* species (Figure 32). (This paper precedes the remarkable study of the hybrid *Digitalis*, already referred to in Chapter 6.) Although Henslow does not mention Ray in his paper, we might recall that his illustrious predecessor had given a paper to the Royal Society in 1674 *On the Specific Differences in Plants* and had indeed studied the British *Primula* species as relevant to such questions. There is a curious parallel between the

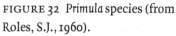

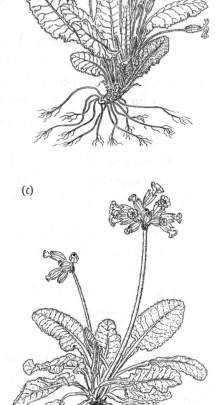

FIGURE 32 *Primula* species (from Roles, S.J., 1960).

(a) *P. elatior* Oxlip

(b) *P. vulgaris* Primrose

(c) *P. veris* Cowslip

opening sentences in these two papers. Here is how John Ray begins his paper:

> Having observed that most herbarists, mistaking many accidents for notes of specific distinction, which indeed are not, have unnecessarily multiplied beings, contrary to that well-known philosophic precept; I think it may not be unuseful, in order to the determining of the number of species more certainly and agreeably to nature, to enumerate such accidents and then give my reasons why I judge them not sufficient to infer a specific difference.[25]

More than a century and a half later, Henslow is setting out the same
problem, in much the same way, beginning with the following passage:

> Our knowledge of vegetable physiology has not been hitherto sufficiently
> advanced, to furnish us with any precise rule for distinguishing the exact
> limits between which any given species of plant may vary. Hence the most
> accurate observers often differ in their opinions, whether two or more indi-
> viduals should be considered as mere varieties of the same, or be raised to
> the rank of separate species. Indeed, the more accurate our powers of dis-
> crimination become, the more inclined we seem to be to multiply species.

It is very tempting to think that Henslow had looked up what Ray had written
before writing his paper, and indulged in a little innocent plagiarism! However
that might be, one conclusion can certainly be drawn, namely that the ancient
Universities had contributed remarkably little to the growth of modern biological science throughout the whole of the eighteenth century. Like Ray,
Henslow wanted a science based on careful, properly recorded observation
and experiment: as Henslow puts it in the *Primula* paper, 'It should now seem
that nothing but the multiplied results of direct and accurate experiment can
be allowed to form the basis of our speculations in this, any more than every
other department of science.' Henslow's special interest in the problem had
been aroused by finding in April 1826 'at a place called Westhoe, a few miles
from Cambridge, a peculiar variety of *Primula*, which I scarcely knew whether
to call the oxlip or the cowslip'. Stimulated by the great variation he saw in this
population, he also began to observe and study *Primula* plants in his own
garden,[26] and these small studies must have been familiar to Darwin early in
his Cambridge career. Typically, Henslow's paper concludes that further well-
documented observation and experiment are needed before any decision can
be made about whether to treat the *Primula* 'varieties' as parts of a single
species, as Linnaeus had done, or as separate species.[27] Not content with the
Primula studies, Henslow was also making hybrids between garden *Potentilla*
species on a considerable scale in 1830. In a short paper describing one of
these artificial hybrids he stresses how important he thinks such studies are:

> Notwithstanding the opposite theoretical position taken by some botan-
> ists, we believe, doubtlessly, that hybrid plants sometimes become estab-
> lished, and hold a permanent place in the vegetable kingdom; it is therefore
> but reasonable to notice them; and it is far better that their origin be regis-
> tered whilst it is known, in lieu of remaining to become the subject of future
> conjecture and error.

Henslow is here foreshadowing the rise of biosystematics (experimental taxonomy) more than a century later.[28] Nowhere in these papers does he mention 'special creation', or indicate that he felt any theological hestitation about the course he is taking.

Much more important to the development of Darwin's ideas, however, are the references in the *Primula* paper to the work of the Revd William Herbert, who had become something of an authority on the specific distinction of garden plants. Herbert, Rector of Spofforth, Yorkshire, published a number of papers in the *Transactions of the Horticultural Society* in the 1820s and 1830s, culminating in 1837 in his book on the Amaryllidaceae. His view that 'in fact, there is no real or natural line of difference between species and permanent or descendible variety' must have contributed to Darwin's eventual decision that both genera and species are artificial constructs of the mind. Henslow was the means by which Darwin, when he returned from the 'Beagle' voyage, was enabled to benefit from Herbert's knowledge. Typically Darwin had prepared a series of 'questions for Mr. Herbert' after reading his book, and had sent them via Henslow, presumably because he knew that if Henslow submitted them to a colleague whose work he had known for several years, they were more likely to receive careful consideration than if a mere young and as yet unknown enquirer sent them out of the blue. The stratagem worked, and Henslow received a long and careful letter from Herbert, dated 5 April 1839, which he duly sent on to Darwin, with an almost curt note appended at the bottom:

> My dear Darwin,
>   I have just got back from Milton. I find this letter wh. concerns your queres. – Ld. F's gardener declares that he does not believe in Hybrid Ferns – He raises 100s & sees 1000's of seedlings & never met with any thing of the sort –
>   Mrs H. is at Ely –
>   Kind regards to Mrs Darwin. Yrs ever sincerely J.S Henslow[29]

One can sympathise with the overworked Henslow, who must have had many matters on his mind. It is nice of him to add what he can, answering the only question that Herbert had declined to comment on, namely whether any ferns or mosses were known to hybridise! Darwin had earlier, in a letter of 3 November 1838, tried to stimulate Henslow into action about his Galapagos plants, pressing him to write 'pretty soon' and suggesting that the sort of answer he wants from Henslow 'will not take up more than an hour'. 'All I want', he says, 'is to know whether in casting your eye over my plants, how many cases there are of (near) species of the same genus, … one species coming from one island, & the other from a second island.' Such facts, he says, will 'support my case of the birds & Tortoises …' Henslow's reply, quoted by Darwin in the Addenda to his *Journal of Researches*, pp. 628–9, informed him that there were 'several instances of distinct species of the same genus, sent from one island only: that is, whilst the genus is common to two or three islands, the species are often different in the different islands. In some cases the species seem to run very close to each other, but are, I believe, distinct.'[30]

Darwin made much of Herbert's work, and it is mentioned in several places in *The Origin of Species*. Again we see Henslow facilitating Darwin's enquiries, and pointing him in a useful direction. Ironically, the one area in which Henslow might be said to have failed to continue to help his former pupil – namely the identification and publication of the Galapagos plants – is quite free, in itself, from any theological difficulty. The study, naming and classification of new plants was an acceptable activity for any Christian believer, for it revealed the extraordinary richness and variety in God's created world.

Before we leave Henslow's investigations of hybridisation, we should note one of the clearest statements made by him concerning the phenomena in plants in a short paper describing an artificial hybrid which Henslow names *Lophospermum erubescente-scandens*. This paper was published in 1841, as one of Henslow's contributions to Maund's journal, *The Botanist*, which we mentioned in Chapter 7. The hybrid in question had been made artificially by the Curator of the Bury St Edmunds Botanic Garden, Henry Turner, by pollinating flowers of *Lophospermum scandens* with pollen from *L. erubescens*. Not only does Henslow supply a very careful description of this hybrid plant, together with a tabular comparison of the hybrid and the parent species, to accompany the coloured plate, but he also takes the opportunity to explain his own attitude to hybrid plants:

> We are entirely opposed in sentiment to those persons who *regret* the intro-
> duction of hybrid plants, because it is troublesome to assign to them a place
> in our systematic arrangements. We would request the systematist to
> remember that the botanist has a higher object than merely describing and
> arranging specific forms. Such a branch of our science must ever be looked

upon as a means to an end. The ultimate aim of true science is to ascertain the laws by which nature is governed; and the more we multiply our experiments, and the more care we take in noting the results, the more likely are we to arrive at definite notions of those laws. At present no one knows with certainty what are the true limits to the variations in form which any one species may assume; and it is impossible to foresee whether multiplied observations on hybridizing may not lead us to some law of vegetation by which a botanist may be able to pre-determine the possible limits of every species, as accurately as a mineralogist can now define the limits within which all those forms of crystallization must necessarily lie, which belong to any particular simple mineral. We would therefore suggest to those persons who have opportunities of studying hybrids, that they should record their observations under some such form as the following table, where we may see at a glance to what extent the hybrid resembles, or differs from, both parents. If we were in possession of some hundred comparisons of this kind (and the more minute the better), we might then, possibly, be able to detect some general law by which the production of hybrids and the limitation of species is governed – but until greater pains has been bestowed upon such enquiries than have hitherto been taken, we can hardly expect much progress to be made in solving this mysterious question.

Ever the practical scientist, Henslow then proceeds to discuss what should be done about hybrid nomenclature and, although his solution, exemplified in his name for this particular horticultural artificial hybrid, has not found favour – presumably in part because it is too cumbersome – he is very clear that naming and recording (where known) the parentage of species-hybrids is very desirable:

> In the present instance, we doubt not, its proper appellation will be adopted, now that it is figured and published. The system of compounding the specific names of parent plants, between which hybrids have arisen, was first proposed in Maund's Botanic Garden, and applied to a plant raised by the author himself. Under No. 385 of that work [the *Potentilla* paper already mentioned] it is observed, 'Authors have not agreed on the most convenient mode of naming hybrid or mule plants. Some have thought that names may be completely arbitrary; some name them after the person with whom they originated; whilst others would altogether excommunicate such productions from botanical nomenclature.'

The passage continues with the extract quoted above. The modern procedure, now enshrined in the International Codes of Botanical Nomenclature
and the Nomenclature of Cultivated Plants, would undoubtedly have satisfied Henslow's requirements.[31]

Once Darwin had established his own position in the scientific world,
which could be said to date from 1845, when his friendship with the
younger Hooker began to be very evident, the correspondence with
Henslow takes on a very different flavour, and there is little evidence that
Henslow's opinion or advice was of much importance to him, whether on
matters of religious belief or scientific questions. Henslow's own views on
species and the broader evolutionary questions are not revealed much in
surviving publications or letters, but seem to change little in his Hitcham
years. A letter of his to Whewell in August 1847 contains an interesting
comment. Whewell must have been asking both him and Sedgwick for
their views on 'transmutation of species', and Henslow says: 'No botanist
so far as I am aware gives any credit to the [? tale] – How the results have
been obtained (in the 2 or 3 cases out of many that have been tried) is a
mystery – but possibly some carelessness or neglect or even trickery may
have been practised.'[32]

The final stages of the Darwin–Henslow relationship could be said to date
from the publication of the *Origin* in November 1859. Henslow, together
with Sedgwick and a number of others,[33] received an advance complimen-
tary copy, and a letter from Darwin dated 11 November reveals how much
Darwin hoped for a considered appraisal from Henslow:

> My dear Henslow
>
> I have told Murray to send a copy of my Book on species to you, my dear
> old master in natural history. I fear, however, that you will not approve of
> your pupil in this case. The book in its present state does not show the
> amount of labour which I have bestowed on the subject.
>
> If you have time to read it carefully & would take the trouble to point out
> what parts seem weakest to you & what best, it wd. be a most material aid to
> me in writing my bigger book, which I hope to commence in a few months.
> You know also how highly I value your judgement. But I am not so unrea-
> sonable as to wish or expect you to write detailed & lengthy criticisms, but
> merely a few general remarks pointing out weakest parts –
>
> If you are in *ever so slight degree* staggered (which I hardly expect) on the
> immutability of species, then I am convinced with further reflexion you will

become more & more staggered, for this has been the process through which my mind has gone. –

My dear Henslow | Yours affectly & | gratefully | C. Darwin[34]

Apparently Henslow did not 'find time' to read the *Origin* and reply in writing to Darwin, but he visited the Darwins at Down from the 14th to 16th February, 1860 and, as we see below, he used his visit to tell Charles what he thought about his book.[35] Jenyns supplies us with most information on Henslow's views, referring in the *Memoir* to what seems to be the only published reference made by Henslow, namely a letter to *Macmillan's Magazine*. It is significant that this letter is called forth by the claim that Henslow 'supported Darwin's views': in it he says 'though I have always expressed the greatest respect for my friend's opinions, I have told himself that I cannot assent to his speculations without seeing stronger proofs than he has yet produced.'[36] Jenyns had written two letters to Henslow towards the end of 1859, the first, dated 21 November, giving his preliminary reaction, and the second, on New Year's Eve, which is worth quoting in detail:

> I have just finished Darwin's book, & should very much like to know what you think of it when you have got through it yourself. I am not at all indisposed to accept his theory in part, tho' perhaps he would say – if in part, how can you refuse to go the whole way with me upon the same reasoning – I am no stickler for the multitude of so-called species created or adopted by so many naturalists of the present day, & can imagine the genera, & perhaps families, when perfectly natural, to have originated in a single stock; – but when I think of a whale or an elephant by the side of a little mouse & am told they have a common parentage I stand aghast at the boldness of the assumption . . . The existence of man too seems to me the great difficulty of all.[37]

It is this letter that Henslow sent on to Darwin. Henslow's inability to give Darwin his own reaction in writing is rather pathetic, but we know that his own views were similar to those of Jenyns, as Darwin tells his friend Lyell in a letter written on 15 February during Henslow's visit:

> Henslow is staying here: I have had some talk with him: he is in much the same state as Bunbury & will go a very little way with us, but brings up no real argument against going further. It is really curious (& perhaps is an argument in our favour) how different opposers view the subject. Henslow

used to rest his opposition on imperfection of Geolog: Record, but he now
thinks nothing of this, & says I have got well out of it: . . . As L. Jenyns has a
really philosophical mind, & as you say you like to see everything, I send an
old letter of his. In a later letter to Henslow which I have seen, he is more
candid than any opposer I have heard of; for he says though he *cannot* go as
far as I do, yet he can give no good reason why he should not. – It is funny
how each man draws his own imaginary line at which to halt. – It reminds
me so vividly what I was told about you when I first commenced geology, to
believe a little but on no account to believe all. [38]

In spite of his inability to convert his old teacher, Darwin as always greatly
appreciated Henslow's willingness to talk to him, and expressed as much
to Joseph Hooker, to whom he wrote: 'I much enjoyed the day Henslow
spent here.'[39]

Henslow defended Darwin's reputation as an honest scientist against
what he felt were somewhat intemperate attacks by Sedgwick (and even
more by William Clark, Professor of Anatomy at Cambridge), on the occasion of Sedgwick's paper to the Cambridge Philosophical Society on 7 May
1860. The paper was entitled: 'On the succession of organic forms during
long geological periods: and on certain theories which profess to account
for the origin of new species.' We have Henslow's own account of why he
was moved to speak and even criticise his friend Sedgwick: in a letter to
Joseph Hooker dated 10 May 1860 we read:

> Sedgwicks address last Monday was temperate enough for his usual mode
> of attack, but strong enough to cast a slur upon all who substitute hypothe-
> ses for strict induction, & as he expressed himself in regard to some of C.Ds
> suggestions as *revolting* to his own sense of wrong and right & as Dr Clark
> who followed him, spoke so unnecessarily severely against Darwin's views;
> I got up, as Sedgwick had alluded to me, and stuck up for Darwin as well as I
> could, refusing to allow that he was guided by any but truthful motives, and
> declaring that he himself believed he was exalting & not debasing our views
> of a Creator, in attributing to him a power of imposing laws on the Organic
> World by which to do his work, as effectually as his laws imposed upon the
> inorganic had done it in the Mineral Kingdom –
>
> I believe I succeeded in diminishing, if not entirely removing, the chances
> of Darwin's being prejudged by many who take their cue in such cases
> according to views of those they suppose may know something of the
> matter. Yesterday at my lectures I alluded to the subject, & showed how

frequently naturalists were at fault in regarding as *species*, forms which had (in some cases) been shown to be varieties, and how legitimately Darwin had deduced his *inferences* from positive experiment . . .

   I do not disguise my own opinion that Darwin has pressed his hypothesis too far but at the same time I assert my belief that his Book is (as Owen described it to me) the 'Book of the Day'. I suspect the passages I marked in the Edinburgh Review for the illumination of Sedgwick have produced an impression upon him to a certain extent. When I had had my say, Sedgwick got up to explain, in a very few words, his good opinion of Darwin . . . [40]

Hooker had sent on this letter to Darwin at Henslow's request, and it called forth an effusive letter of thanks that begins: 'I have been greatly interested by your letter to Hooker; & I must thank you from my heart for so generously defending me as far as you could against my powerful attackers . . .' Interestingly, the letter carries a long P.S. about the dimorphism of *Primula* flowers, and Darwin follows it a few days later with a long letter largely concerned with *Primula* dimorphism. [41] Here is a subject that Henslow undoubtedly introduced to his young pupil, as we have seen; soon after Henslow's death, Darwin presented his studies in a paper, and the material was eventually used in his book entitled *The Different Forms of Flowers on Plants of the Same Species* (1877).

   Henslow's final public involvement in the controversy over Darwin's ideas was at the famous Oxford meeting of the British Association which took place at the end of June 1860. Considering the extraordinary and long-lasting interest in the Oxford debate, when Bishop Samuel Wilberforce ('Soapy Sam' to his opponents) clashed with Thomas Henry Huxley over the descent of man from the animals, it is surprising that there is still to this day uncertainty about the exact turn of events. [42] One thing is certain, however: Henslow took the chair for the meeting at which the famous exchange of invective occurred. Henslow was there in his capacity as President of the Natural History section, but there is some evidence that he had not planned to take the chair, but did so only because Owen failed to show up. Darwin, of course, was 'too ill' to attend, and Hooker reported to him at length from Oxford on 2nd July. This letter commends Henslow's admirable tact in his difficult role where intemperate speakers had to be controlled: 'I must tell you that Henslow as President would have none speak but those who had *arguments* to use, and 4 persons had been burked [*sic*] by the audience & President for mere declamation.' [43]

1. Sir John Henslow,
Chief Surveyor to the Navy,
1784 - 1806.

2. Still life of molluscs painted by Henslow when a boy.

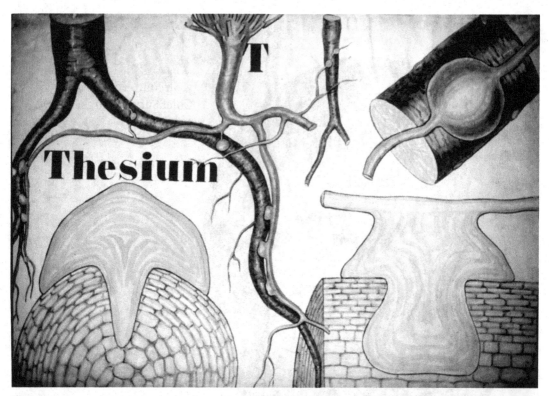

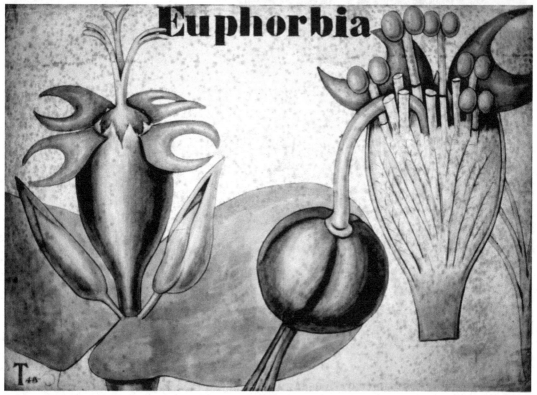

3. Two of Henslow's original teaching drawings.

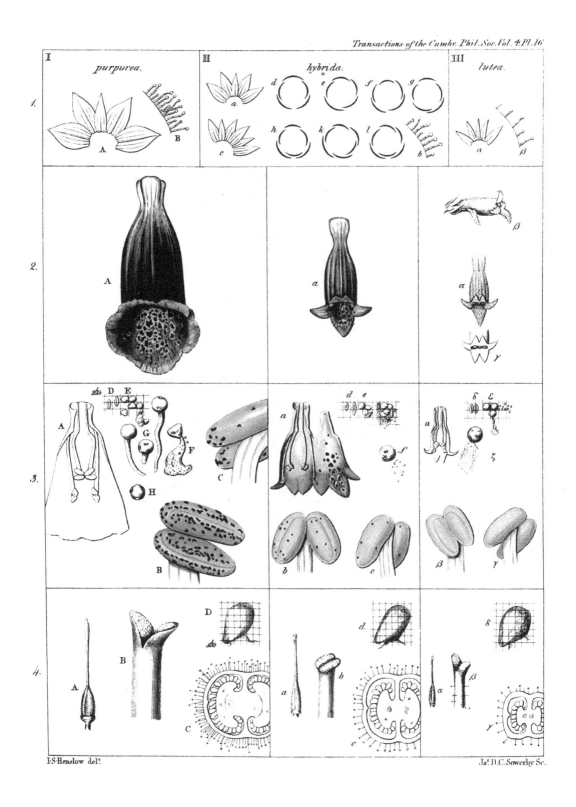

I·S·Henslow del.ᵗ   Jaˢ D.C.Sowerby Sc.

4. Illustration from Henslow's paper on a hybrid Digitalis, 1831.

5. The young Charles Darwin, pianted by George Richmond.

PLATE 70

6. Audubon's illustration of Henslow's Sparrow.

7. Day-flowering cactus. Original water colour for illustration no. 12 in Maud's *The Botanist*.

8. Sketch of Hitcham Rectory from family scrapbook.

9. Design for Ploughing Match Prize from family sketch book.

10. Silver cup presented to Henslow by the Hitcham farmers in 1854.

11. All Saint's Church, Hitcham. Henslow's grave is west of the tower.

12. Site of Hitcham Great Wood, 1999. World's End Wood in distance.

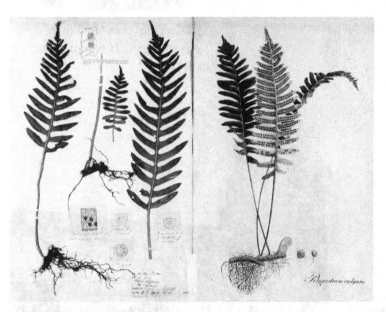

13. Polypodium vulgare: a Henslow sheet from the Cambridge University Herbarium, 1860.

14. The Henslow Walk in the Cambridge Botanic Garden. Howard Rice, 1996.

Jenyns' assessment of Henslow's reaction to Darwin's book is summarised in the following passage from the *Memoir*.

> In this instance, as in so many other instances, he showed his philosophic caution. He declined going a step further than he could well see his way; at the same time that he would not take upon himself to say that such and such things could not be, where there were no sufficient data for establishing the truth one way or the other. In previous letters to myself, he had told me he thought there were in Darwin's book too many suppositions, too many things assumed, which might or might not be true. Moreover, the further the question was followed up towards its source, the more it was beset with difficulties which were never likely to be solved. In fact, he said, he considered an inquiry into the origin of species about as hopeless as an inquiry into the origin of evil.
>
> With reference to the religious aspect of this question, he stated, during his last illness, that he thought it an objection to the hypothesis of all animal and vegetable forms having been evolved in succession, through countless ages, from one primitive germ, or from a few such germs, that it did not allow for the interposition of the Almighty. 'God,' he observed, 'did not set the creation going like a clock, wound-up to go by itself, but from time to time interposes and directs things as he sees fit.'

As we enter the twenty-first century, such doubts and misgivings about a totally 'Godless' view of natural selection as the driving force of evolution continue to make themselves heard, and are likely to do so indefinitely.[44]

Part III
Hitcham

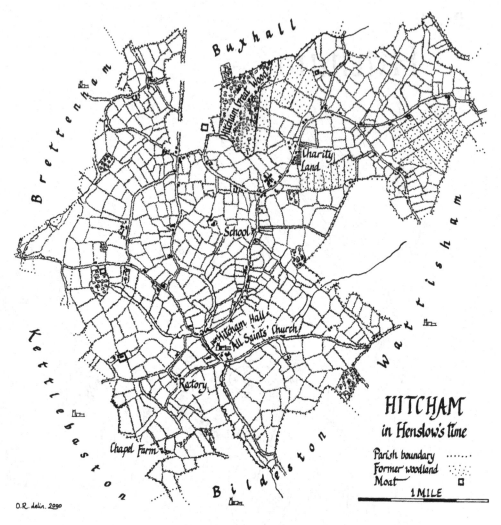

FIGURE 33 Map of Hitcham in Henslow's time (Oliver Rackham).

10 Early years as Rector of Hitcham

The industrial revolution and the rise of modern manufacturing cities with their rapidly-growing populations had meant that the period into which Henslow was born at the end of the eighteenth century was one of unprecedented social change. The Henslow family, raised in comfortable affluence and largely shielded from the cruder manifestations of the brash new urban society, must nevertheless have been well aware of new forces and dangerous new ideas. Undoubtedly the most obvious danger to members of the comfortable professional class was posed by the frightening events of the French Revolution and its aftermath in Napoleon and the Napoleonic Wars. Historians generally agree that the retention of a coherent social fabric was seen by the vast majority of educated Englishmen in the post-Napoleonic period as being of overriding importance. Such a consensus, often largely unexpressed, must explain the widespread antipathy to any 'rabble-rousing' idealists, whether they were Dissenting preachers or atheist revolutionaries. 'Law and order' was a strong card to play politically, and the majority of conscientious priests must have firmly believed that it was their Christian duty to uphold the established order, even when they may have secretly sympathised with the law-breakers.

Whilst this picture seems to be generally accepted, we should be wary of explaining too much in its terms. In the first place, the sheer neglect and corruption of areas of both Church and State that seem to have been features of the late eighteenth century might well have engendered a reaction as soon as the Napoleonic Wars were over. The young Henslow felt it to be his civic duty to take a significant part in fighting corruption in local election politics in Cambridge, and it clearly remained a strong conviction throughout his life that all 'right-minded citizens' should be prepared to play their part in purifying the body politic to the best of their ability. Towards the end of Henslow's life, of course, this very strong feeling of duty became the hallmark of Victorian respectability; at its best we see it as an impressive belief in material progress, whilst we reject or ridicule its areas of hypocrisy and arrogance. A second important *caveat* to the view that 'law and order' was the primary driving force in nineteenth-century social and political life arises from the peculiar position of the Church of

England *vis-à-vis* the other denominations. As Knight has very cogently argued, the period through which Henslow lived saw the Anglican Church move 'from a uniquely privileged relationship with the State, in which it was closely bound up with the political and legal system, to being one denomination, albeit still the most powerful one and still formally and legally Established, among several in a society in which it appeared that half of those who professed any sort of religious allegiance expressed a preference for a non-Anglican variety'.[1] In some ways, this very practical change might be seen as running counter to the conservative 'law-and-order' tendency, for it removed from the parson, for example, any automatic right to be concerned with local issues of law-breaking, and reshaped the average Anglican parish politics.

This was the social scene that Henslow knew when he began his Hitcham ministry. He faced a daunting task. For the first two years he attempted to combine continued residence in Cambridge with his duties at Hitcham. This involved spending most weekends in his Suffolk parish, but the state of that parish gradually persuaded him that he must give priority to his new appointment and leave Cambridge. It seems likely that a contributory factor to this decision may have been the evidence that his botanical lectures were less popular than they had been in the 'Darwin days'. As we saw in Chapter 6, the decline had set in by 1835, after the end of the monopoly granted to Oxford and Cambridge in training the future Fellows of the Royal College of Physicians, which removed a significant 'captive' audience of medical students from Henslow's course. One gets the impression that Henslow found the attempt to continue his Cambridge career, whilst acting as Rector of Hitcham, increasingly frustrating.

We may be tempted to say that he should never have tried to combine a University teaching career with reviving the social and religious life of a Suffolk village. If we do, however, we ought to remember that many academics showed no such conscientious scruple; it was commonly assumed that a 'fat living' rewarded by College, Ecclesiastical or political patronage did not require the Rector to live in, or even to visit the parish concerned, but simply to pay one or more Curates to conduct services there on Sundays. Since stipends of the order of £40 or £50 a year were commonly paid to Curates for such services, a £1000 benefice could easily be run from a distance.

The Established Church, which throughout the eighteenth century had tolerated (or even, according to its opponents, encouraged) pluralism and indolent or absentee Rectors and Vicars, was at last undergoing change, and Henslow's career spans the most important of these changes. Most of us derive at least some picture of the state of the clergy from the works of Jane Austen, whose novels were already widely read by Henslow's time, and are perhaps even more popular today than they ever were. Here is Miss Crawford in *Mansfield Park* (1814), musing on the position of the beneficed clergy after a visit to the Rushworths:

> Oh! no doubt he is very sincere in preferring an income ready made, to the trouble of working for one; and has the best intentions of doing nothing all the rest of his days but eat, drink, and grow fat. It is indolence, Mr. Bertram, indeed. Indolence and love of ease – a want of a laudable ambition, of taste for good company, or of inclination to take the trouble of being agreeable, which make men clergymen. A clergyman has nothing to do but to be slovenly and selfish – read the newspaper, watch the weather, and quarrel with his wife. His curate does all the work, and the business of his own life is to dine.[2]

Where Jane Austen entertains us with her delicious, gentle irony, another very different writer, William Cobbett, is prepared to turn his fierce invective against the Anglican Church and its parsons. In his *Rural Rides* (1830) Cobbett provides us with a vivid picture of the state of Hanoverian England and its countryside seen through the eyes of a 'champion of the people', and loses no opportunity in showing his hostility, not to religion as such, but to the bulk of the clergy, and in particular the tithe system. Here is a typical Cobbett extract, referring to the parish of Botley in Hampshire where he lived from 1805 to 1820: 'Tithes do not mean *religion*. Religion means a *reverence for God*. And what has this to do with tithes? Why cannot you reverence God, without Baker and his wife and children eating up a tenth part of the corn and milk and eggs and lambs and pigs and calves that are produced in Botley parish?' (author's italics)[3]

To some extent, and all too slowly for 'agitators' like William Cobbett, the Governments of the time, in a series of Acts of Parliament between 1813 and 1836, tried to correct the acknowledged excesses in the Anglican Church *inter alia* by requiring clergy to reside in their parishes and to pay their curates adequately. This gradually had some effect, reducing the number of parishes with non-resident incumbents from about half in 1827 to below 10% by 1850.[4]

Henslow was, as we have seen, a politically active man, who converted

from the Tories to the Whigs following Lord Palmerston. It is quite inconceivable that he would not have felt the pressure to reside in Hitcham as
Rector. Furthermore, the previous Rector of Hitcham, John Staverton
Mathews, had himself resided in the parish, a fact that must have weighed
heavily with the newly-appointed Henslow. A further considerable attraction must surely have been the Rectory provided for the incumbent, with its
well-planned living quarters, large garden and view down to the Church
some half-a-mile walk across the fields. This commodious Rectory was
built by Mathews' predecessor, Henry Close, in 1790, but Mathews
employed the London architect James Spiller to rebuild it substantially in
its present form in 1814. It is now a private house, called Hitcham House.[5]
(Plate 8)

All the evidence we have supports the bleak picture painted by Jenyns of
the state of Hitcham parish when Henslow took over in 1837 from the previous incumbent. Here was, in Jenyns' words,

> a people sunk almost to the lowest depths of moral & physical debasement.
> Ignorance, crime, & vice appear to have been rife, even to a degree beyond
> what was too generally prevalent at that time among the Suffolk peasantry.
> The worst characters were addicted to poaching, sheep-stealing, drunken-
> ness, and all kinds of immorality . . . The Church was all but empty upon
> the Sunday, and but little respect paid even to the marriage and baptismal
> rites.[6]

It seems to be the case that the previous incumbent, who was Rector from
1810 to 1837, was

> one of these easy-going men, of which there was formerly a large sample in
> the Church, who contented themselves for the most part with doing their
> Sunday duty, leaving the people to themselves during the week and making
> no exertion to promote, in other ways than what the law obliged them to,
> either their temporal or their spiritual welfare.[7]

Another reason why the contrast between academia and the Suffolk
parish was much greater was revealed by the Bishop of Ely in commending
Henslow to the post: 'the duty of squire, magistrate and rector must all fall
upon the latter'. There was no surviving squire in Hitcham Hall from
whom Henslow might have received some moral and practical encouragement, if only by the regular attendance of the family of the gentleman at the
big house at the Church services.[8] At the time of Henslow's arrival,

Hitcham Hall Farm was let to one John Harper, a local farmer and, in about 1843, on the sale of the whole Mapletoft Estate in Hitcham, Harper was able to buy the Hall and all its land. From 1840 to 1854 Harper was the Rector's Warden; as we shall see in Chapter 11 he was by no means cooperative about all the Rector's plans, and in particular opposed the allotment scheme.[9]

The Sunday services conducted in Hitcham Church by Henslow's predecessor apparently attracted at the time of Henslow's appointment congregations 'insufficient to fill one pew.'[10] Whether Henslow expected anything better than this is not recorded. What is clear is that he was not disheartened, and obviously decided that there were more important practical things he could do for his parishioners to win them over. Jenyns does not mince his words in explaining this:

> But the new rector was not a man to flinch or go back at the sight of difficulties. What he purposed, he was determined to accomplish, if possible. His aim was to recall his flock from sin and idleness to habits of soberness, honesty, and industry; to give them a love of independence, and to teach them to respect themselves, as well as those who were placed above them in life.
>
> Moral and religious lessons would in the first instance have been utterly lost upon them. It would have been throwing pearls before swine. He wisely began with trying the expedient of winning them over by kindness & conciliation. He did what he could to amuse them. He got up a cricket club, and encouraged other manly games of a like character ...[11]

It must have been an increasing realisation that he could not properly carry out this practical aim during the first two years when he was still resident in Cambridge that forced the decision finally to move into the Rectory with his wife and family, a move that took place in the autumn of 1839. There they were to remain and bring up their five children in what was a very comfortable home. George, the youngest child, was only four years old when they moved, and the two eldest daughters, Frances, aged 13, and Louisa, aged 12, must soon have been helping their increasingly invalid mother to run the Rectory. Frances, as we know, was to marry the younger Hooker, and it was to be the unmarried second child, Louisa, who took over the main domestic duties. As befitted their station in life, the Henslows hired local domestic staff. The census returns for 10 June 1841 record a total of six domestics in the household, four of whom are women ranging in age from

18 to 40: the two men are William Laflin, 36, described as the gardener, and John Leeks, aged 20. William Laflin (or 'Lafling') is the only one of these domestics still in service at the time of the 1851 census, when he is said to have 'a wife and 4 children'. Although described as 'gardener', he obviously acted also as coachman, for we are told by Romilly that he was met at Bury St Edmunds, on the occasion of his visit to the Henslows on 26 March 1842 and 'driven over to Hitcham by Henslow's man Lafflin [a third spelling!] thro. a series of hailstorms with a very cold wind . . .' Romilly, who does not mind revealing in his diary that he enjoys feminine company, is also much taken with the Henslows' governess, whom he describes on the occasion of his first stay at the Rectory (25 March 1840): 'much pleased with the governess (Miss Brind) a delicate pretty cheerful young lady of 22 (looking not more than 17) who dresses peculiarly neatly'. We hear no more of Miss Brind, who is not listed in the 1841 census return, except for a note of Romilly's during his second visit to stay with the Henslows, when he describes what they did on Sunday 6 September 1840:

> Fine day – Mrs Henslow not strong enough to go to church in the morng: she was wheeled in a garden chair in the afternoon. I read prayers both times: in the morng Henslow gave a short lecture on our Lord's early miracles from Mark (ch 1) – in the afternoon he preached from 2 Sam 24 on the sacrifice at the threshing floor of Araunah, & attacked his congregation for not coming to church. – We were followed home by the girls of the Sunday School to whom Mrs H gave a singing lesson on the lawn – Henslow & I then walked until dinner – Leonard as well as Fanny, Louisa & Miss Brind dined with us. We had a fire this evening as well as Sat evng At 9½ Henslow read half a sermon of Hare's loud to us & the servants: he always divides them into two.[12]

Henslow's 'Christmas letter' to his friend Hooker dated 17 December 1839 is full of interest, and worth quoting *in extenso*:

> . . . I found it too inconvenient & expensive to keep up two houses & after mature deliberations have given up my residence in Cambridge and now reside here continually excepting for the month in Spring when I have to deliver my lectures – I am not 40 miles from Cambridge & have already communication enough, but miss the public library, & can have only a part of my herbarium about me. However I am in a very retired spot & have much more leisure for my pursuit of botany, notwithstanding my parish avocations,

which I no way neglect, than I had in Cambridge – I have been occupied
every morning lately in studying our fossil [plants] & if you can send me any
fossil plants from your neighbourhood, or any ferns which are not named
(or even un-named) I shall be very thankful for them – I have undertaken to
assist Mr Hutton in continuing the Fossil Flora of Britain. I make it a rule to
work for myself i.e. botanically every morning till one o'clock except Friday,
& I have undisturbed evenings for similar occupations – I am in rather an
out of the way district amid great ignorance & none but farmers in my
parish. – Through the summer I have attempted to give the people a little
interest in intellectual pursuits by delivering a series of lectures on a variety
of subjects under a Marquee on my lawn – & now that I have determined to
reside among them continually I have settled to lecture once a month at this
time of year to a few whom I invite to my house – & under my Marquee in
summer as before –

To give an interest to this scheme I exhibit as many specimens as I can, so
consequently am collecting all sorts of objects of Science & Art for a
general museum which I call my village Adelaide Gallery – Animals, Plants,
Minerals, Models, &c &c – Prints & so forth. Nothing comes amiss & I have
been most unblushing in my application to all friends for assistance – Any
specimens which illustrate a manufacture from the raw material to the finished article are extremely acceptable to me – My last lecture was "Quartz &
Glass" & I got from a Glass manufactory in London not only all the materials for glass making, but materials for their [ ] – specimens of the tools for
glass-cutting & if ever I come to Glasgow again I shall get you to introduce
me to any of the manufactories there & beg hard for permission to walk off
with illustrative specimens – Now that I am away from the library I find my
general collection of plates very serviceable, & if you have not grown weary
of helping me in this way I shall be very thankful for any that may be
inclined to add to your former liberal gifts – I am not in a good botanical
district here, the only plant I had never met with before is the Equisetum
hyemale which I can produce in abundance if you want any [Figure 34]. I
have spoken so much about my scientific pursuits that I am afraid you will
think I am a careless country clergyman – which however is not the case –
For by a correct use of my time, I am not aware that I neglect any portion of
my duty – visiting my school at least 2 or 3 times a week, & discharging the
regular duties of the parish I hope satisfactorily – Indeed I believe that many
clergymen who have no other occupation than parish duty are often less
happy or contented than those who have no spare time on their hands &

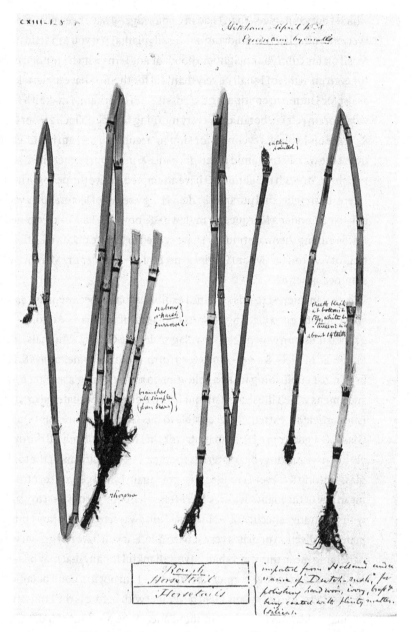

FIGURE 34 Henslow's annotated specimen of *Equisetum hyemale*.

never find a day too long – I hope Lady Hooker & your family are well – This place agrees well with mine, & Mrs Henslow has not felt so well for the last 2 or 3 years as she finds herself at present – My children are all grown beyond your recollection of them – & I am sadly afraid that the expression of any hope that you may intend to renew your acquaintance with them would be quite in vain – but I know you do visit Norfolk, & if some N wind would only blow you a little southwards I cannot venture to say how rejoiced I should be.[13]

It is evident from this account by Henslow himself that he was already using the Rectory grounds to entertain and educate 'the People' of his parish, in particular by holding events in a marquee on the lawn during the previous summer. This idea, as we shall see, developed into a series of summer shows and exhibitions which flourished until Henslow's death (Figure 35). Further evidence of the Rector's activity even before his move to Hitcham is provided by the date of establishment of the annual plough-ing matches, given as 1838.[14] (Plate 9)

An obvious priority seen by the new Rector even before his permanent res-idence at Hitcham was the provision of an adequate village school. There was a small 'Dame's school' already operating, but even Henslow, who was unwilling to make derogatory remarks about anyone, was moved to state in the Parish Accounts for 1841 that the woman running the school had 'no knowledge of the art of teaching'. Henslow opened a subscription list for a new school, and by 1841 was able to open the Hitcham Parish School in a new building on the opposite side of the road from the Dame School. Since 1830, a Government grant had been available to support village schools, and the National Society for Promoting the Education of the Poor in the Principles of the Established Church actively encouraged the formation of what we would nowadays call 'Church Schools'. The size of any Government grant was affected by the number of subscriptions from parishioners, self-help being suitably rewarded, so it was important for Henslow to exhort his parishioners and other local well-wishers to perform their Christian duty to educate the poor.

In spite of all his efforts, only 16 parishioners were subscribing to the Village School when Henslow engaged a teacher to begin work in the new building in 1841, and it is obvious that the Rector himself was paying the main part of the costs. Again, the absence of any 'landed gentry' from his parish made his task more difficult; in many country parishes the efforts by clergy to found new village schools were strongly supported by the resi-dents of the 'big house', partly, no doubt, because a local supply of literate servants was clearly to their advantage.[15] No such advantage accrued to the farmers of Hitcham, who saw the spread of education as a dangerous ten-dency, rendering the pool of agricultural labourers more and more discon-tent with their lot. The tension was especially great in Hitcham Parish,

HORTICULTURAL SHOW.

PROGRAMME FOR SEP. 14th, 1859.

8 to 10. A. M.   Specimens for competition received at the Rectory.

12.   Marquee Museum ready for inspection.

N. B.   Lecturets as opportunities may occur.   Among novelties observe

Animal. Skulls of Albatross, Condor, Toucan ;   Whales'-Food. Vertical Wasp-comb from Mexico.

Vegetable.   Grater and Sieve for Manihot, with Cassava, from Brazil. Palmite Brushes from S. Africa. Chinese Sugar-grass grown at Hitcham. Skeleton leaves from China.

Mineral. Iron Pyrites (from Essex, by H. Fisher,) changing to Green Vitriol.   Lava from Monah Rowah, Owyhee.

Miscellaneous. Jamaica Rat-trap. A Glass pebble found at Ringshall.

Stereoscopic Photographs on a stand near the Museum.

1. P. M.   Show-booth ready for inspection.

2¼.   Prizes to Village Botanists for

(1) Wild-fruit Posies.   The species not to exceed 20, and to be named and classed.

(2) Dried Grass Posies.   The species to be named and classed.

(3) Herbarium Specimens prepared since last year.   Any one who has been in the First Class, whether resident or not in Hitcham, may compete for this prize.

School Report.

3.   Allotment Report ; and prizes awarded for

(1) Superior Allotment Culture.

(2) Hatcher Sweepstakes.

(3) Specimens exhibited at this Day's Show.

(4) The Cottage Barometer, (by Dr. Lindley,) and a Metal Teapot, (by G. Knights, Esq.,) to the two Members of the Society who shall have obtained the greatest and next greatest number of marks for all prizes during the year.

Thanks to the Judges, Donors of Prizes, and to all others aiding and abetting our proceedings.

4.   No specimens to be removed from the Show-booth before 4 P. M. Short memories are then to remind their owners to return all Check and Prize Tickets to one of the Stewards.   Short days necessitate sharp movements, or we shall not have time by

5.   To take our tea orderly and sociably.   Rules for " ticketless babies " (0 to 2 years) and for " ticketed ditto " (2 to 4 years) continue as heretofore.

6.   God save the Queen. Good night. May every one present at the Show rise to-morrow with a conscience void of offence towards God and Man, in respect of any thing said or done during a day devoted to innocent and rational recreation.

FIGURE 35 Programme of the Hitcham Horticultural Show, 1859.

where most of the law-breakers were guilty of arson and crimes of violence. Converting the farmers of Hitcham to adopt a more liberal attitude to his education and self-help schemes for the rural poor continued to be an uphill struggle for Henslow throughout his life in the Rectory, and we return to this theme in the next chapter.

It should not be assumed that all parents supported Henslow's efforts to establish a village school. In an age when child labour was everywhere accepted, even though the movement to regulate and discourage the employment of children was beginning to make itself heard in and out of Parliament, the right of parents to supplement the family's meagre earnings through hiring out their children for casual work was quite strongly defended. Glyde, writing a history of Suffolk, describes the kind of casual farm labour on which children were employed in Hitcham and the surrounding Suffolk countryside in Henslow's time:

> Boys sometimes come at 2d a day: little things that can hardly walk come with their fathers. More usually nine- and ten-year old boys would be weeding, corn dropping, pulling turnips, 'crow-keeping', 'stonepicking' and carting for 1s a week. Girls were largely employed 'stonepicking' and 'keeping birds'. The older girls would absent themselves from school for 6–8 months of the year to earn 6d a day.[16]

This problem of irregular attendance of the older children persisted in country areas throughout Henslow's life; indeed, it could be said never to have completely disappeared amongst working-class families in spite of enormous social change in Britain.

The teacher appointed by Henslow to run his new school in 1841 was Jane Stow (or Stowe – both spellings exist). Aged 24, she lived with her mother and brother in Bildeston, and walked to Hitcham school and back every weekday, some four miles in all. We do not know what, if any, training she had. What we do know about her arises from Henslow's concern in 1850 to defend her against 'certain slanderous reports', a concern which led him to circulate one of his privately-printed 'Addresses'. This long, carefully-argued note is a typical Henslow product, combining real concern for the unfortunate young lady with quite scrupulous fairness. Thus he begins by conceding that she has 'not been so regular in her morning attendance as her duties seem to require', and explains that she lives 'in another parish, at considerable distance from the school, with an aged mother to look after, and not herself possessing a strong constitution'. His main concern, however, is the slanderous talk, which is 'quite distinct from the complaints to which I have just referred', and he now directs his strong attack here:

> We all know how readily the most absurd reports are manufactured by the evil disposed, and how greedily they are swallowed by many who, living in glass houses themselves, ought to be more cautious how they encourage others to throw stones. When once a slander has been either hastily or artfully spread, it acquires so much of plausibility that even the well disposed become staggered, and are inclined to imagine there must be some foundation for it.[17]

We have no way of telling how such admonition was received by those to whom it is formally addressed, namely 'subscribers to the Parish School' and 'the parents who send their children there'. The former category, though rather few, would at least be able to read it, but what of the parents themselves? As we shall see in the next chapter, Henslow became notorious in some circles for the verbosity of his writings, and we cannot help feeling that his arguments might have stood a better chance of acceptance had they been more concisely expressed.

Another area of social concern confronted by the new Rector was the wretched daily life-style of many of his parishioners. Many cottages were old, clay-walled, badly thatched and often totally undrained, with all the family sleeping together in a single room. Even where the husband was employed, the poor wages were hardly enough to feed the whole family, and their diet was barely adequate. Again, quoting Glyde:

> . . . their diet consists principally of bread and potatoes. There are, however, some who when their families are grown up, by putting their earnings together, occasionally get a piece of meat at their supper time and their Sunday dinner.[18]

Henslow lost no time in forming self-help clubs, amongst them a coal club to provide some winter fuel for cooking and heating, a children's clothing club and a 'wife's Society'. Early in 1841, he had printed and circulated the first of a series of formal reports on the success or otherwise of all the local Clubs and Societies, and from this report we learn that the Medical Club is the most flourishing, with 186 paying members – 'married persons pay thirteenpence for themselves and families, and single persons above sixteen years old pay sixpence halfpenny (excepting those in service who pay ninepence,) quarterly'.[19] This particular Club was, however, already established at Hitcham when Henslow arrived, as we learn from a long 'letter' he wrote to the local Press in 1844, addressed to the 'The Country Practitioners of Medicine in Suffolk': here is an extract:

> Now I think we may consider a good medical Club in a country village as a very fair substitute for such town advantages. In the present position of our agricultural labourers it affords all who are not in the very lowest stage of poverty an opportunity of avoiding the humiliating necessity which the better spirits among them still acutely feel, of being obliged to have recourse to parish relief. Through these Clubs the poor are provided, at very trifling cost to themselves, with advice and medicine. I feel that I may more readily direct public attention to the character of the Medical Club established in this village, because I had nothing whatever to do with its formation. I found it in activity when I came to reside in Suffolk, and no alteration of any importance has been made in its rules since then. It contains about 200 subscribers out of a population of 1060 souls. In fact, there are very few of our poor who do not readily avail themselves of its privileges; unless it be those who happen to belong to some benefit club through which they are entitled to receive medical assistance.[20]

It seems that none of his new clubs were as popular as the Medical Club, though the Children's Clothing Club (22 children enrolled), the Cricket Club (23 members) and the Ploughing Match Society (22 members) seemed to be modestly successful. At the end of the year, in a printed circular letter dated 25 November 1841, Henslow appealed to the 'Landowners and Occupiers of the Parish' to respond more generously, and 'favour [him] by Christmas Day with an account of whatever annual subscriptions they may be disposed to contribute in future'. To encourage those not already doing their duty, he attached to this appeal letter a detailed list of all subscribers and potential subscribers, from which it is very apparent that Henslow himself (and 'Mrs J.S. Henslow' who is listed separately) is far and away the most generous subscriber. Some absentee landlords, including 'King's College Cambridge' and the 'Canterbury Dean and Chapter' pay nothing![21]

This type of action was by no means unique to Hitcham: it was rapidly becoming fashionable among liberal Christians to advocate such schemes, whether they were priests or other influential members of the community, and it fitted the Whig philosophy prevalent in the country. Biblical texts could easily be found to justify helping the poor to help themselves, and it seems in Henslow's case that the farmers of Hitcham were at least initially not opposed to some of these self-help clubs. They agreed to deliver the coals free to coal club members – although even this meagre level of support was apparently withdrawn when they opposed Henslow's attempts to introduce an allotment scheme. We shall read more about this in the next chapter.

What was Henslow's attitude to the Sunday services at Hitcham Church? Was it simply a matter of discharging his duties? As explained by Russell, discussing the average Parish Church services in the early nineteenth century,

> The clergyman's duties were those of reading the service and preaching the sermon, which came at the end of the service, unless it was a Sacrament Sunday. It was inevitable that the educated clergyman, who took pains to express himself as a scholar, in carefully composed literary exercises, spoke only to the educated members of his congregation . . . Clearly, Hogarth's cartoon of the sleeping congregation recorded an experience with which many could identify, and a large part of the appeal of non-conformity lay in the more energetic and colloquial preaching of ministers whose social and educational situation more closely approximated that of the congregation.[22]

The ever-loyal Jenyns tells us what he thought of Henslow's sermons at Hitcham, in a long passage in the *Memoir*. The following extract will suffice:

> His Hitcham sermons were all more or less practical. They were composed rapidly, and written offhand. Yet there was nothing in them to give the idea of their having been written in over-haste, or without sufficient thought. They were alike sound in doctrine and reasoning; the language clear and well-chosen . . . If deficient in anything, Professor Henslow's sermons were deficient in warmth and animation as regards the way in which he treated his subject. Truths, doctrinally important, and highly necessary to be enforced, were often too abstractedly considered; little being said to lead his hearers to make a personal application to themselves of what they heard. Perhaps his sermons would have been more attractive if more had been yielded to the circumstances and conditions of his hearers. They were addressed to a congregation for the most part but scantily educated, and yet they required in some measure an educated mind, – not so much to

comprehend them, – as to be impressed by the lessons they conveyed. They were directed more to the understanding than to the heart. The common people, it is well known, like to have their feelings excited by the preacher, and unless this is done to a certain extent, though they may listen attentively to the sermon while being delivered, they seldom reflect much on it afterwards.

Jenyns continues:

For this reason, it may be doubted whether his public ministrations in the church had all the effect upon the poorer part of his flock which his private ministrations had, and the truly Christian example he displayed in his every-day life. In like manner, his preaching was not after the best models of pulpit eloquence. His voice was loud and sonorous – at times over-loud – with strong emphasis upon particular words and passages, where there did not seem to be occasion for the stress he laid upon them. Preaching, certainly, was not natural to him in the way that lecturing was, in which he excelled. His manner of reading the prayers and lessons was much like that of his preaching.[23]

What is concerning Jenyns is that Henslow had been criticised by various friends and colleagues for giving prominence in his ministry to 'good works' and neglecting the 'saving of souls'. If by the 'saving of souls' is meant emotional preaching on Sundays, then Henslow was clearly guilty, and there is evidence that his congregation was often seduced by dissident preachers delivering a message of individual salvation in language the ordinary people could understand. There was (and indeed still is) a Baptist Chapel in Wattisham, in an adjacent parish to Hitcham and equally accessible to many Hitcham families, and in Henslow's time there was a satellite meeting house in Hitcham itself. With a population in Wattisham of only about 200 in the middle years of the nineteenth century, congregations of more than 500 were regular at the Sunday services![24]

Characteristically, Henslow printed for private circulation only a very small and carefully-selected number of his Hitcham sermons. These are nearly all on 'good works' themes, mostly of local concern, and were obviously intended to remind all his parishioners of their Christian duty to succour the poor and needy. Typical 'outside' subjects are the Irish famine and an appeal for relief for Christians subjected to persecution in the Holy Land.[25]

Guests at the Rectory at Hitcham, we can assume, were not infrequent during this period. It was a hospitable place with a lively family of young children. Babington's Diary records that he stayed with the Henslows for three days in June 1842, arriving on the 8th and departing for Ipswich and the Suffolk coast on the 11th. Highlights were a visit with Henslow to Hadleigh, where Babington appreciated 'a remarkably handsome church & parsonage', and on the following day what we presume was a conducted tour of Henslow's favourite botanical site, though Babington records only the following: 'Found *Myosotis sylvatica* in plenty in Hitcham Wood, with peculiarly large flowers.'[26]

We have no record of any other visit by Babington, but as we have already seen, Romilly was a frequent visitor, providing us in his Diaries with fragments of information about the pattern of life with the Henslows on Sunday in Hitcham, and occasionally telling us about their attendance at Church, where he usually helped by reading the prayers and even occasionally preaching himself. From these references we glean that the normal pattern was a morning and an afternoon service, presumably following the morning and evening prayer (Matins and Evensong) of the Book of Common Prayer. It would be good to know what part music played in these services, but Romilly is silent on the subject. All we know is that the Church had had an organ installed about 1800, well before Henslow's time.[27] Henslow himself seems to have had little or no musical taste, unlike his wife and daughters who clearly enjoyed both singing and playing. Romilly greatly enjoyed such social occasions with the Henslows, as the following account of a dinner party illustrates. The date is 4 June 1850:

> The dinner party today consisted of Mr & Mrs Calvert & their youngest son Charles, Kirby of Clare H(AB 1817) & wife, Mr & Mrs Knox (I think he is an Oxonian; – Mrs K is his 2d wife, is about half his age & plays the piano most exquisitely); a Mr Raymond (a very young man, a neighbrghd Curate) & our 5 selves . . . The dinner went off exceedingly well – Mr Gedge came in the evening – he sang 2 or 3 songs (one a laughing song) rather pleasingly. Mr Raymond joined Louisa & Anny (I think) in a song . . . a very feeble performance of all 3 – Louisa & Anny played a duett on the harp & pianoforte tolerably; – & Mrs Knox played 2 long & very difficult pieces without book in the most finished & masterly style – After she was gone Mrs Henslow pronounced that that manner of playing did not at all do for the country: – she of course spoke as a mother who had no pleasure in hearing a performance

wch utterly annihilated that of her own children. Mrs Kirby only distinguished herself by snapping the string of her bracelet & letting the sham
pearls run about the carpet – several were trampled on & broken: but as she
had never opened her lips before, so neither did she now: her husband said
very truly it was of no consequence – Company left at 10.[28]

Chronologically, an 1850 dinner party takes us too far, and well into the
period covered in the next chapter, but such social gatherings were a
feature of the Henslow family life in Hitcham from the earliest days,
and the passage makes a suitable transition to the account to come.

11 The Rector

For the first two years of his holding the living, Henslow, still resident in
Cambridge, operated like so many of his fellow-clergy by employing a
curate who performed most of the duties. Henslow's induction took place
on 30 June 1837, only ten days after the accession of Queen Victoria, and he
answered briefly as best he could the questionnaire that preceded the
Visitation in that year carried out by the Bishop of Ely. In this he states that
the 'officiating minister' is the 'Curate' who is paid a stipend of £150 and
that his own residence in the Parish is 'occasional'. The Curate was Richard
Broome Pinniger, graduate of Pembroke College, Oxford in 1825, who had
been Henslow's last Curate in Cholsey; he left when Henslow came into
residence, to become Rector of Whichford, Warwickshire. Henslow writes
at the end of the 1837 questionnaire: 'I hope to answer these questions
more satisfactorily on a future occasion.'[1]

We are very fortunate to have the Hitcham Vestry Minute Book, a large,
well-bound volume obviously especially bought by Henslow himself so
that he could open a new book for his first Vestry Meeting[2] (Figure 36).
The first Minute is quite short, and reads as follows:

> March 14th 1840
> Rev J. S. Henslow Chairman
>> At a Vestry meeting held this day it was proposed by Rob Mapletoft Esqr.
> that the occupiers pay 6d per Acre for surveying & mapping of the Parish,
> and the Landlords 3d per Acre in addition to the sum charged for the two
> copies. Also that the Landowners pay 75£ for the appointment of the Tythes
> and the occupiers 25£ for the valuation for the Rate being together the
> Amount of Mr. Sheldrake's contract for both. At the same time it was pro-
> posed as an amendment by Mr. Pilgrim that the occupiers pay 3d and the
> Landlords 6d per Acre upon the above charges. At the same meeting Mr.
> James Peart and Mr. Ralph Hitchcock were chosen to serve the Office of
> Assessors for the ensuing year. Adjourned to the 25th of April at 12 oclock.

Henslow's first Vestry Meeting, as here recorded, dealt with 'the surveying
and mapping of the Parish'. Consequent to the passing of the Tithe
Commutation Act in 1836, each Parish was required to prepare a detailed

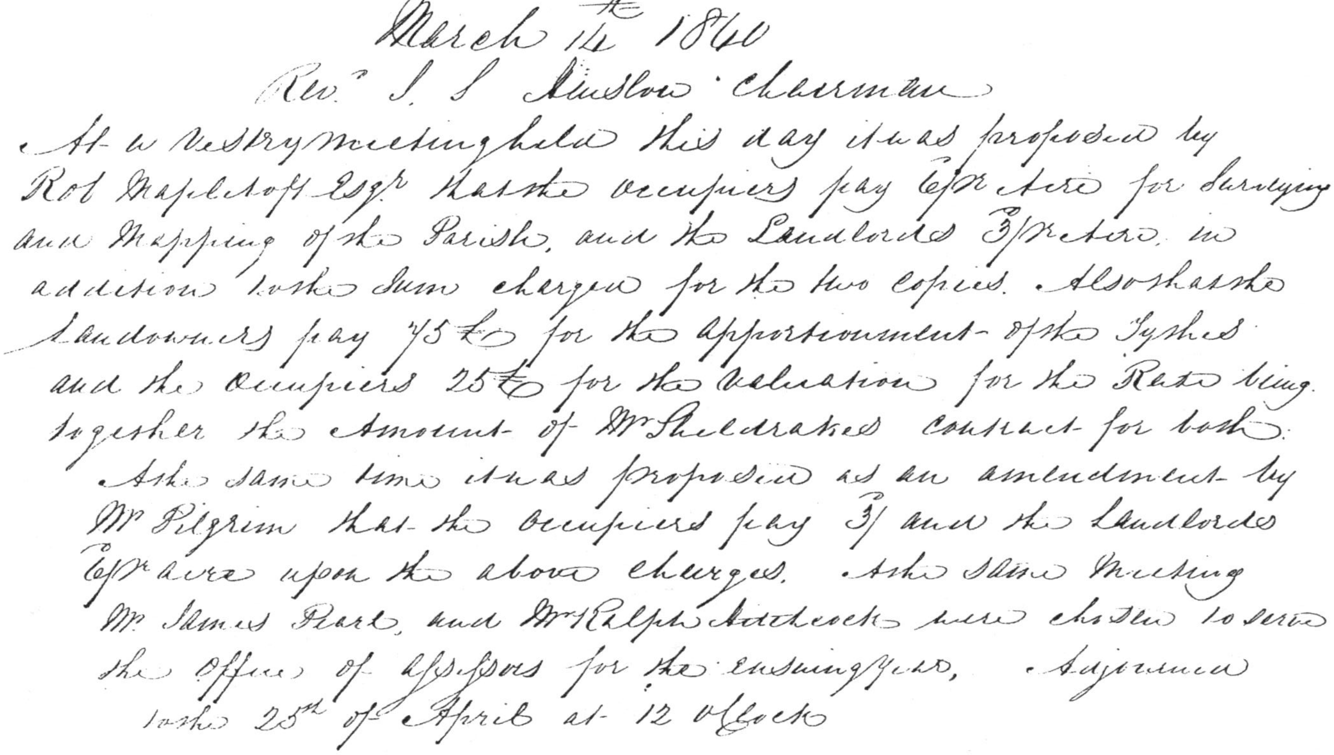

March 1st 1841

Revd J. S. Henslow Chairman

At a Vestry meeting held this day it was proposed by Rad Bayldoff Esqr that the receiver pay officers for surveying and repairing of the Parish, and the Landlords Expences in addition to the sum charged for the two expired. Allowances laid down by 75 £ for the apportionment of the Tythes and the receiver 25 £ for the Valuation for the Rate now together the amount of Mr Allchards contract for both.

This same time then proposed as an amendment by Mr Pilgrim that the receiver pay if and the Landlords expences upon the above charged. And this Vestry appoint Mr James Hart, and Mr Ralph Catchesett new Wardens now chosen to use the office of Wardens for the ensuing year. Adjourned to the 20th of April at 12 o'clock.

FIGURE 36 First page of Hitcham Vestry Meeting Book recording Henslow's first annual meeting.

survey and map so that the appropriate tithe payment (no longer in kind) could be collected from each owner and occupier of the land. The great Tithe Map was complete in 1839. It showed that, of the 4,117 acres in Hitcham Parish accountable for tithe, only 714 acres were owner-occupied. The rest of the land was owned by at least 25 absentee landlords including King's College, Cambridge (25 acres) and the Dean and Chapter of Canterbury (20 acres).

The haggling about money recorded in this Vestry Meeting was common at the time, when the Vestry fixed Parish rates and taxes. The landed gentry (represented by Robert Mapletoft, the Lord of the Manor) proposed that the 'occupiers' or tenant-farmers pay twice as much as the owners of the land for the survey and mapping of the Parish, and the occupiers (represented by John Pilgrim, of Chapel Farm) tabled an amendment to the opposite effect! Adjourning the meeting to the 25th April seems not to have improved matters, for the (very terse) minute of that date reads only: 'At the Vestry meeting held this day pursuant to public notice, in consequence of the necessary documents not being signed by the Commissioners, this meeting is adjourned to the 20th day of June at 12 oclock.'

There is no entry for 20th June, so we are left guessing who won: indeed the next minute is dated 25 March 1841 and records the election of officers 'for the ensuing year'. This is the annual general meeting legally required to be held on or about Lady Day, and in the minutes of these annual meetings are recorded the names of all the officers for the ensuing year. Henslow's first annual meeting, dated 25 March 1840, records as Churchwardens Mr. John Harper as 'chosen by Rev. J. S. Henslow' and Mr. Rob Luckey 'on the part of the Parish'. As we saw in Chapter 10, Harper was the tenant of Hitcham Hall owned by Mapletoft, and bought the Hall and farmland from the Mapletoft Estates when they were sold in 1843. In Henslow's estimation Harper was the nearest thing he had to a 'squire', and it would be natural that the new Rector would choose him as his Churchwarden.

No minutes or papers seem to have survived which tell us how Mathews had run – or attempted to run – the Vestry meetings of his incumbency but, as we have seen in Chapter 10, Jenyns thought that the Rector had been powerless to control the meetings at which the poor rate and other local taxes were fixed: 'parish relief was not infrequently levied by bands of forty or fifty able-bodied labourers, who intimidated the previous rector into instant compliance with their demands'.[3]

Troublesome though these tensions between landlords and tenants might be in the Vestry Meetings, they were probably far less demanding to Henslow than the constant problems of 'the lower classes'. There is plenty of evidence that crime and lawlessness had reached a peak in the early 1830s in the Suffolk countryside in the period when the writings of Cobbett and the activities of 'Captain Swing'[4] were everywhere powerful manifestations of unrest among the unemployed labourers, and had then died down somewhat. But Henslow's residence in Hitcham more or less coincides with a new, even more alarming rise of rural crime as the 'hungry forties' came on, a burst of crime which Henslow as local Magistrate must soon have had to try to deal with and understand. The Suffolk historian Glyde, whose work provides a mass of relevant statistics, analysed the kinds of crime in the county, and found that arson in particular had shown the most spectacular rise. Thus, committals for arson for the years 1838–42 in the whole county were only six, but for the next five-year period, 1843–47, numbered 83, and for the subsequent period 1848–52 dropped somewhat, but remained high at 61. In fact, to quote Glyde:

> The year 1844 was one of intense alarm to the owners and occupiers of farms . . . Fires were of almost nightly occurrence during the long evenings. The occupiers of land lived in a state of nervous excitement, looking about their premises every night before retiring to rest, apprehensive of their crops being destroyed by the match of the prowling incendiary.[5]

Hitcham and the surrounding parishes were among the worst affected areas. Here is Henslow himself, in a letter to Whewell dated 1 March 1844: 'We have almost nightly fires about the neighbourhood and as I was lecturing at Hadleigh on Wednesday, a cry of fire interrupted and spoilt all . . .'[6]

The strain of being both parson and magistrate is exemplified by the trial of three of Henslow's own parishioners, Thomas Quinton, Henry Knock and Thomas Powell, who came before the Bury Magistrates (Robert Mapletoft and John Henry Heigham) on 15–17 March 1842 on a charge of sheep-stealing. A fourth man, Robert Powell, who turned Queen's evidence, gave to the Superintendent of Police at Stowmarket a detailed account of what happened:

> I went to Mr Hatton's barn-yard with the three prisoners; we took two sheep away from the yard to the second field from the barn; we killed the animals there, skinned them, took the entrails out, cut the heads off, and

left the skins in the pond, we then carried the mutton home to my house, where we each had a quarter of each sheep.

There was a second charge involving one sheep stolen from another farm in the nearby parish of Chelsworth. The three men were found guilty on both charges, and sentenced to transportation for 15 years.[7]

On the 20th March, Henslow preached a sermon which has come down to us *verbatim*, because he had the text printed 'for private distribution only'.[8] He must have felt that he would like to use this distressing affair to explain how he as a parish priest reacted to convicted criminals amongst his flock. Reading this 24-page sermon would soon convince anyone that the term 'liberal' is a relative one: an attitude that could have been described as liberal in 1842 would hardly qualify as such 150 years later! He chooses as his text that familiar passage from St Paul's letter to the Ephesians (6:14): 'Wherefore, take unto you the whole armour of God, that ye may be able to withstand in the evil day, and, having done all, to stand', and begins, in sombre tones:

> The events of the past week afford us matter for serious reflection. The offended laws of our country have been fearfully avenged. Those whom we are bound in all Christian charity to consider as neighbours, friends, and brethren, have been proved unworthy of the holy calling of Christians and have been condemned to a long and terrible exile . . . I have no desire to dwell upon the particular circumstances which have led to these wretched results. It would be too painful to some of you, and much too disagreeable to myself, to treat of the private character of individuals from the pulpit . . .

He quickly turns his attention to the causes of 'much of the evil that prevails around us', and tells his flock there are two main ones, namely 'the neglect of all public religious duty' (i.e. people do not come to Church) and 'attend-ing "unworthily" to the public duties of religion' (i.e. coming to Church but ignoring Christian teaching in their daily lives). There follows a short but revealing passage addressed to those 'who are not members of the Established Church', which shows a degree of tolerance: 'if they [i.e. the Dissenters] see good reason (as they suppose) for rejecting our faith, and for despising my ministry, they must carry their reasons to the throne of God . . . But I quarrel with no man who differs from me. I would judge no man. Conscience is too sacred a property to be interfered or trifled with.' There is abundant evidence from many different parts of the country that members of 'Dissenting Chapels' were quite often prepared to attend services in their Parish Church. Indeed there are recorded cases where such attendance was normal, as in the village of Stretham, Cambridgeshire, where the Curate reported: 'It is customary for people to attend the church at those times when a sermon is preached and the Methodist Meeting house at other times'.[9] There were both Baptist and Primitive Methodist chapels in Hitcham in the 1840s.[10] That Henslow should explicitly allude to 'those of my parish . . . not members of the Established Church' – suggests – though we have no evidence – that one or more of the convicted sheep-stealers could have been from a Dissenting family. Henslow was, of course, as we have seen in Chapter 7, well aware of the problem of the civic rights of Dissenters from his early years in Cambridge. His great friend Richard Dawes had, indeed, sacrificed his chances to become Master of Downing by voting together with Sedgwick, Whewell and Henslow to admit Nonconformists to the University. Now Henslow found a new facet of the 'Dissenters problem' in his own parish.

Reverting to his Sermon, we find no hint in it that these criminals in his own 'flock' might have been unemployed married men who, seeing their wives and children at near starvation levels, took the law into their own hands. Of course the Rector would have known them and their families personally, and may have judged that there were no extenuating circumstances. We do not know, and therefore should not judge. But the climax of the Sermon is an exhortation to his parishioners to:

> Take the sword of the spirit, and learn to acquit yourselves like men. Determine for the future to discountenance all wickedness in others, as well as to forsake it yourselves. Be ready to rebuke it, be prepared to forbid it: and if nothing else will prevail, why then, be bold enough to expose it. But do not seek to expose it as some do – out of malice, or expecting favour, or looking for reward. Expose it, (if needs be) as a part of your duty to God and to man. You may be thus the blessed instrument of saving some, perhaps many, from the corrupting influence of evil example, from evil counsel, from being lured to crime. We are not called, as christians, to be always standing on the defensive only; but our calling is clearly offensive also. We are to attack, and by God's help to beat down Satan under our feet.

This reads like an appeal to the wives and mothers to be prepared to denounce, if necessary, their husbands and sons when they resorted to violent protest against their abject condition. It is not, to our ears, a liberal attitude. Yet we should perhaps remember that Henslow, like most village priests of his time, was primarily concerned not to alienate the only educated people in his parish, which he would most certainly have done had he shown any tendency to condone the sheep-stealing, as so many of the supporters of Cobbett and other 'rabble-raisers' would have done.

Glyde, the Suffolk historian from whose work we have already quoted, is very sure that the combination of priest and magistrate in one person is undesirable. This surprisingly outspoken passage follows closely on some praise of Henslow (and a fellow-cleric, the Revd Tighe Gregory) 'for the local improvement of the labouring class . . . by lectures, summer excursions, and innocent amusements':

> We believe that a clergyman's connection with the civil power, as magistrate or guardian of the poor, is detrimental to his spiritual influence. We know that the difficulty of finding duly-qualified laymen to serve as magistrates is considerable, and may be pleaded as an excuse for the appointment of clergymen to that office; but we feel that the pastor of a parish serves a higher Master than the highest of those who make the human laws, and we would not have them mixed up with the painful details of business which magistrates who do their duty to the public must constantly encounter. Besides, the clergyman is frequently needed to admonish or console the relatives of offenders; but these poor unlearned creatures cannot be expected to attend to the instruction of the man who has just sentenced their companion. Such a clergyman is frequently looked upon with fear instead of love, and the higher interests of the Gospel as well as the influence of the Church suffer. By some it has been contended, that to interest himself in the proper administration of Poor Law relief in his own neighbourhood, is an occupation for which a country clergyman is peculiarly fitted. We believe, however, that it is not desirable to place the pastor of a parish at the head of the parochial establishment for the support of the poor. We would have the clergyman be as little as possible the administrator of civil government, and have as little as possible to do with secular affairs. The harshness of legal authority is better placed elsewhere.[11]

This sums up very well the problem facing Henslow. How can you, as a clergyman, console the mother or wife of a man you or your fellow-magistrates have just had to sentence for a crime, especially when that sentence – transportation – would almost certainly mean that they would never see the man again?[12] Henslow must indeed have had an unusually stubborn streak in his nature. Once he had decided on a course of action, he pursued it.

We must now move on to consider, as the violent social unrest became worse to reach its climax in Suffolk in 1844, the kinds of efforts Henslow made to persuade his fellow-men to see what should be done socially and politically to educate, in the real sense of the word, the labouring class so that they would no longer resort to desperate, violent protest. These efforts seem to be directed not only at the immediate short-term relief of the most abject poverty, but also at longer-term measures to influence the politicians to legislate in such a way as to correct the most obvious failures in social policy. Behind both kinds of activity lay Henslow's steady, consistent belief in the value of formal school education for all people, on which, as we saw in Chapter 10, he took action at a very early stage in his ministry.

Action by Henslow to influence the public that really mattered included a use of the local press as a publicity vehicle. Two important developments took place around the time when Henslow arrived in Hitcham. The first was the reduction (and eventual abolition) of the Stamp Tax on newspapers, as a result of which, for example, a copy of *The Times* which had cost seven pence in 1830 could be bought for three pence by 1870. The second, which had an even more dramatic effect on written and printed communication, was the arrival of Rowland Hill's Penny Post in 1840. Ordinary letters which might have cost in the 1830s up to tenpence to send across England could now be sent for a flat rate of one penny. Like everyone else in the professional and business classes, Henslow availed himself of this marvellous new facility, and the Henslow manuscript archive contains some philatelically very attractive 'penny black covers' carrying the new adhesive stamps bearing the young Queen's head (Figure 37). Advances in the technology of paper-making and book publishing all contributed to the dramatic fall in the price of books and pamphlets, and a new mass market for both fiction and non-fiction opened up rapidly. An example cited by Mitch neatly illustrates the phenomenon: 'In 1820 a penny bought a broadside consisting of a few doggerel verses describing the trial of Dick Turpin. By 1850 a penny could buy a Gothic romance of fifty to one hundred pages written in reasonably fluent prose.'[13]

During the winter and spring of 1842–43, Henslow undertook what must have been a very strenuous campaign to win the support of the tenant

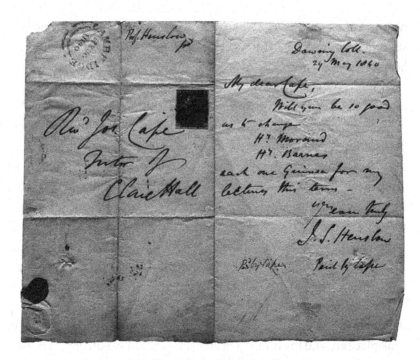

FIGURE 37 'Penny Black' letter from Henslow at Downing, May 1840.

farmers of the Hitcham neighbourhood to carry out improvements to their ways of working, primarily by interesting them in scientific information and experiment on, for example, the use and manufacture of manure and what we would now call 'artificial fertilisers'. By joining the Hadleigh Farmers Club and occasionally taking part in discussion meetings there, he succeeded in getting an invitation to the Anniversary Meeting of the Club on 16 December 1842 at which he delivered a lecture on the subject; he followed this successful occasion with a series of fifteen *Letters to the Farmers of Suffolk* published originally in *The Bury and Norwich Post*, and then as a separate pamphlet including the Anniversary Lecture. That this whole operation was carefully planned is apparent from Henslow's remarks in the preface to the pamphlet, remarks which seem to refer to the criticism from friends who thought his 'drollery' unsuited to his Professional dignity:

> I must be content to lose caste in the eyes of the more discreet if I have been unfortunate in selecting an improper 'time to laugh'; but I shall not fear to give an account of the few idle words in which I have indulged, *seeing they were intended to serve a special purpose,* [our italics] which might otherwise not be attained. In catering for the world's wide stage, it may not be amiss to

remember that our audience are not all called to sittings in the dress circle; and a little sprinkling of broad farce may serve better to point a moral for some parties, than more sober comedy could have done.

The 'special purpose' is clearly revealed at the end of the text of his Anniversary Lecture, reprinted in the pamphlet:

> Before I sit down I shall venture to say a few words upon another subject on which I feel myself much more qualified to give an opinion than upon how your crops should be managed. There is a description of culture which requires its special manure, and in which I conceive you are as deeply interested as in any which you carry on in the fields. You have the proper cultivation of your labourers to look at. This is not the place, nor is this a befitting occasion for me to appeal to you on any higher grounds than mere worldly policy, for recommending attention to their moral, intellectual, and social condition. One of the best manures which you can provide for the description of culture I now allude to, is to secure your labourers constant employment. I shall not enter upon the wide field which this question embraces; but I put it simply to you as a matter of worldly policy to do so. I am no prophet; but it needs no prophet to foreshow you what will certainly come to pass if your labourers are thrown out of employ. If profits are to depend in future upon increased produce, and not upon high prices, then there must be an increase of general intelligence among your labourers, to enable you to take advantage of improved methods of culture; and there must be increased labour, also, to carry them out. I recommend to your serious attention that glorious maxim of the wisest of earthly monarchs, 'There is that scattereth and yet increaseth; and there is that withholdeth more than is meet, but it tendeth to poverty' – Prov.xi.24.[14]

Henslow is here giving a clear warning to the farmer-employers that unless they adopt a more responsible attitude towards those whom they employ (or fail to employ), the present high level of unrest will get even worse. Hitcham, we should remember, was one of three Suffolk parishes with the highest proportion of its male adult population classified as 'agricultural labourer': at the 1841 census the figure is c. 64%.[15] Although Jenyns thought that 'it must have been some consolation to Professor Henslow to think that he had spoken out on this subject . . . and that he had himself done, and was still doing . . . all he could to amend the evil',[16] the year 1844 began, as Henslow clearly feared, with even worse unrest, and

his considerable energies were directed on to the wider political scene.
A spate of long letters, addresses and even sermons appeared under
Henslow's name in the local newspapers during the summer of this year,
achieving also publicity in the London *Times*. A substantial pamphlet,
written by Henslow and dated 3 July 1844, is entitled: *Suggestions towards an
Enquiry into the present condition of the Labouring Population of Suffolk*; the initial
paragraph alludes to the author's difficulties with *The Times* newspaper,
which seem to arise partly from the fact that Henslow was not initially
aware that one of his letters to the Editor had actually been published, and
then was unable to get copies of the correspondence in *The Times* columns
relevant to his concern. He writes, almost petulantly: 'I have desisted from
further enquiry, as I have too little leisure at my disposal to be at much
pains about them.'[17] Henslow is incensed by the fact that most of the
newspaper correspondents who are writing on the 'Incendiary Fires which
have prevailed in the County' are writing anonymously, a practice which he
himself as a matter of principle will not follow. He therefore turns to Sir
Henry Bunbury, an enlightened landowner of Great Barton near
Mildenhall, and re-publishes a letter of Bunbury's dated 14 June 1844
which had appeared in the local press.[18]

Bunbury's letter starts with a dramatic statement: 'The fearful progress of
these incendiary fires continues without abatement; yet nothing is done in
the county either to check the course or to trace the causes of these crimes.
Suffolk is acquiring a notoriety that is little to be envied.' This is followed
by a disarmingly naive statement: 'Everybody [sic] reads *The Times*, and
views the picture which it presents on our condition.'

The picture being presented by *The Times* was the result of the activities of a
Times correspondent, Thomas Campbell Foster, who made it his particular
concern to give accounts of the problems of rural life in various parts of
Britain and Ireland, and as a result had 'brought down on his head an angry
storm of criticism'. He visited Norfolk and Suffolk in 1844, and produced
detailed case-studies of the varying plights of labourers and their depen-
dants. Hitcham was one of the villages in which he operated: here is the
account, published in *The Times* in June 1844, of an example of a parishioner
of Henslow's, Henry Squirel, who might reasonably be thought of as a 'rep-
resentative of those in regular employment at Hitcham':

> Henry Squirel has a wife and nine children, four of them are off his hands
> and five entirely depending upon him. He has 8s. a week regular wages.

He can hardly tell how he manages to live. His family requires four stones of flour a week for bread, and that consumes the whole 8s. His wife and children sometimes earn a little, and got a job to do now and then; 'but a woman', said he, 'who have five children to attend to cannot do much'. He pays £3 rent for his cottage, and he earned the last harvest £2.15s. above his wages, and the children and his wife picked up a little corn in gleaning, and that helped them out. 'We can't hardly tell you', said he, 'how we get clothes; we pinch ourselves of bread for them.' They drank peppermint and 'bread crust tea' (toast and water) at their meals, and got a bit of cheese and butter when they could. His wife, a neat and orderly-looking woman, was laid up with a bad foot, but the cottage was clean and neat, as possible, and the rose tree looked smiling at his door. His cottage had two rooms and a closet for lumber on the ground floor, and one room upstairs with three beds for himself and wife, the five children, and two big lads, off his hands, who supported themselves, but slept at home. I ascertained that this was one of the best labourers in the parish, and he is also a man of superior intelligence to those around him.

A much worse case reported by Foster was that of William Death, with nine children, who had been forced into the workhouse for a period and was now surviving very precariously on 'irregular work and charity'.[19]

It is very clear why Henslow chooses to begin his pamphlet with Bunbury's letter: in him he finds an influential supporter who is pleased to be quoted and can express succinctly Henslow's own thoughts. This is how Bunbury explains 'the circumstances which . . . have engendered discontent among the labourers . . . The great grievances are

> the inadequacy of wages compared with the work performed; the frequent dismissal of the labourers from employment because the weather is too wet, or the weather is too dry, or on similar pleas; and that, generally speaking, the ill-paid, or discharged labourer, has nothing to fall back upon as a resource. To these circumstances, which have been in full operation for some time past, must be added the deep jealousy and dislike which the labouring class bear to the New Poor Law. It is useless to palter with the question; there is the detestation; be it reasonable or not reasonable, there it is.

Glyde, whom we have already quoted, was equally sure that throughout Suffolk in general the harsh working of the new Poor Law was to be held largely responsible for the widespread unrest of 1844, although he appreciates the weaknesses and disadvantages of the old system. 'The old Poor Law system and the low rate of wages which labourers receive, have induced a habit of relying on the poor's rate in all cases of emergency; which has undermined prudence, prevented them from feeling the degradation of pauperism, and rendered them insensible to the honest spirit of independence.'[20]

It was, of course, to make a fundamental change to this old system that the Whig government passed the Poor Law Amendment Act in 1834, under which most of the responsibility of the local community to the poor was removed from the Church Vestry meeting and given to new, more centralised powers. As Glyde puts it: 'The new Poor Law Act was, among the wealthy and influential classes one of the most popular measures that was ever passed by the Legislature; but among the poor, the feeling towards it amounted to hatred.'[21]

Of course, Henslow had no experience of the old, crude and chaotic system under which his predecessor had suffered; but he was soon aware, like most of the educated liberal opinion throughout the country, of the inhumanities of some of the workhouses to which paupers and their wives and children could be separately assigned by magistrates. Charles Dickens' novel *Oliver Twist* had appeared in monthly parts in 1837–38 and achieved very wide publicity: even today the cruelty of Victorian workhouses is likely to be familiar to people through the continued popularity of this particular Victorian novel. Henslow must have had much personal evidence of the fear and hatred amongst his labouring parishioners of the 'Union House'. Here is an example quoted by Glyde from an official source, illustrating the fact that people might actually prefer prison to the workhouse:

> In the Visiting Justice's Book of the Ipswich County Gaol the following entry will be found.
>
> "February 4th, 1845. – The Visiting Magistrates feel it incumbent on them to notice the commitment of six out of seven women for the period of two months, from the Stradbroke Union, who have acknowledged that they prefer being in the County Gaol to remaining in the Union House, as they are employed in picking oakum, and the food and cleanliness in the Gaol is much superior to that which they experienced at the said Union House."[22]

Henslow's 1844 pamphlet is a very carefully-argued appeal to the politicians to set up an official enquiry into the conditions of the labouring population of Suffolk, and in particular into the working of the New Poor Law, the advantages and defects of which he assesses realistically:

> I was not residing in the county under the Old Poor Law, but all classes have represented to me, the debased and disorganized condition of the labouring population in this division, as having been extremely frightful; and the testimony is equally general to their decidedly improved habits and bearing, since the passing of the New Poor Law. If I were asked to sum up in few words the most prominent distinction between the Old and New Law, I should say that the Old Law placed the rate-payers too much at the mercy of the labourers, and that the New Law places the labourers too much at the mercy of the rate-payers. Under the Old Law some of the labourers grossly abused the advantages they possessed, – but whether the rate-payers have equally abused their present power, is a subject I should hope none of us would shrink from seeing fully investigated.[23]

More generally, the pamphlet appeals to patriotic Englishmen to ask why

> certain masses of the poor of England are not so happily circumstanced as they ought to be; or why they may not be so steadily advancing with those of some other nations, in their opportunities of enjoying the comforts of life. Surely England ought to be as much in advance of all other countries in the social prosperity of all classes of the community, as she is a-head of them in the arts of War necessary to the preservation of her liberties, or in the spread of commerce proportionate to the enterprise of her merchants.[24]

In the second half, the pamphlet reviews all the other causes of rural deprivation and discontent, and makes a strong case for the long-term advantages of popular education and of securing 'a sufficiency of harmless amusement and manly recreation to the labouring class'. Henslow finishes with an appeal to Christian morality: if we do all these things 'We shall be living in mutual respect for each other; the rich for the poor; the poor for the rich.' It is the dream of all reformers, but many find the frustrations and setbacks too great, and abandon the vision. Henslow, at least in his younger, more vigorous years, seems to be able to take them in his stride – even the schemes that fail dismally, like his attempts to get unemployed men to 'seek their living elsewhere.' Here is Henslow's own description:

I very lately summoned the unemployed labourers of this parish to meet me; and between twenty and thirty of them assembled. I then read to them letters from Newcastle, in which they were offered wages from three shillings to four shillings per day in the collieries. I engaged they should have all befitting arrangements made for them, both at starting from Ipswich, and again upon their arrival at Newcastle. They were even offered to be sent back again at the end of a month, free of expense, if they did not like their quarters. Not one of them would stir! A wholly uneducated and uninformed agricultural labourer is sometimes more like an animal, guided by mere instinct, than a man directed by reason. And yet there are persons still prejudiced against their intellectual improvement![25]

A modern commentator might reasonably side with the unemployed labourer who objected to being pushed around by a 'do-gooder' and told to leave his village! We might also question whether 'intellectual improvement' would solve this particular problem of mobility of labour which is still with us today. What about the quality of loyalty to a community and way of life? Ronald Blythe's classic study of a Suffolk village, *Akenfield*, as he found it after the Second World War, shows that, a century after Henslow, there were Suffolk farm labourers who had spent their whole lives on local farms, and hated and feared what they saw as disruptive modernity, whether in terms of mobility or the mechanisation of farm work.[26]

That Henslow was increasingly aware of adverse criticism, even downright hostility, to some of his ideas and plans is evident enough from both his letters at this period and his published articles. He is even prepared to admit that the powerful emotions which drove him to such an outburst of articles, sermons and letters to the Press around this time might have carried him too far, as in the following passage from one of the opening shots of his campaign to establish allotments for his parishioners to rent:

I believe I became so much excited by the importance of the object in which I had engaged, as to lose, in some respects, the sobriety of mind with which such enquiries should ever be conducted, at least this seems to have been the opinion of some persons to whose judgment I am willing to defer. Still I have no intention of desisting, so far as respects my own parish, from making further endeavours to palliate the evils under which our poor are labouring: and I have not yet lost any portion of that hope which I before expressed, that those evils may be and will be removed to a very great extent.[27]

He does not tell us which critics he has in mind, but it is a reasonable guess that the reports by Foster, *The Times* correspondent, brought not only the state of the Suffolk labourers to a much wider audience, but also brought Henslow some heavy criticism from some of his peers.

He is not, however, prepared to give in, and is in fact on the point of launching in Hitcham the scheme which produced the strongest outburst of opposition, namely the provision of allotments. In an important pamphlet dated 1845, with (to our ears) the quaint title: *An Address to Landlords on the advantages to be expected from the General Establishment of a Spade Tenancy from among the Labouring Classes,* he sets out his case.[28]

During the Napoleonic Wars, the allotment movement had been mainly promoted so that the 'labouring poor' might improve their meagre diet by growing potatoes and thereby cut down their expenditure on wheaten flour to make bread. There had been enlightened pioneer landlords such as Thomas Escourt of Long Newton, Wiltshire, as early as 1802 who had let allotments on a fourteen years' lease to labourers, specifying that at least one quarter was planted with potatoes and also that the tenants did not apply for poor law relief; but there was much local resistance. The movement grew sporadically and on the whole most successfully in the Midlands and the Eastern Counties, where wheat was the major crop, and many cottage gardens had been lost in the enclosures. Against this background, we can see Henslow again at a disadvantage in his parish with no resident powerful and enlightened landowner to take up the cause. His friendship with the Bunbury family at Mildenhall, who were model landowners in this respect, must have been very important to him at the time. Sir Henry Bunbury had more than 100 allotments on his estate in Mildenhall, and wrote a paper on *The Allotment System* in 1845.[29]

In spite of local and wider appeals, Henslow found the majority of tenant farmers implacably ranged against him. Not only were they themselves unwilling to sublet any small plot, say one acre, as allotments to the farm labourers, but they opposed the whole idea, on the grounds that their work-force would put their energy into digging their own ground and raising their own crops, and therefore work less conscientiously for their employers. In vain did Henslow plead that, since seasonal or total employment was so bad that most of the takers were not in fact able to earn a regular wage as agricultural labourers, the fear was totally misplaced. The Rector, of course, gave a practical lead and set aside from 'his own small tenure . . . a single half acre, which I purpose to let at Michaelmas between two labourers', and further explained his aim 'to procure about 40 acres in any part of the Parish which may be convenient for allotments, and shall feel deeply obliged to any Landlord who will allow me to rent this much . . . for a term of years'.[30]

By 1849, he had made no progress, and there were still only the two allotments in his Parish. Nothing daunted, he tried a different tack: some charity land was due to be re-let at Michaelmas, and he accordingly set out in his Easter Address of that year 'a scheme to use this "Town Lands Charity" land for allotments'. Interestingly, he uses as his model the scheme set up by his brother-in-law and biographer Leonard Jenyns for his own Parish of Swaffham Bulbeck, and prints the rules and regulations so that everyone can see it is above board.[31] Neither Henslow nor Jenyns explicitly states this: in Henslow's case he might have felt it better not to admit the connection, and in Jenyns' biography of Henslow the author hardly ever 'intrudes' upon the narrative. The plot succeeds: by the following Easter, when Henslow again reports to the Parish, he is able to state under *Allotments* as follows:

> Under this head I can thankfully and most satisfactorily report of the progress that has been made during the past year. In addition to the only two allotments hitherto in the Parish, there are now fifty others, of one quarter of an acre each, laid out in garden ground; and also a pasture of one acre, allotted to a father and son who keep cows. There are numerous applicants still on my list for allotments, and I trust the day is not far distant when every labourer in the parish, who is desirous of possessing one, may be thus assisted.[32]

It was a tough battle, not helped by the overt opposition of the Rector's own Churchwarden, John Harper; but persistence was rewarded. By 1849 the extremes of violent protest were receding into folk memory, and *The Times* columns no longer featured Suffolk. Indeed, events in Ireland, where the great potato famine caused a national calamity in the years 1846–48, had pushed the relatively minor civil unrest in England out of the national press.

'Potato murrain' or, as we now call it, potato blight, first known in North America from 1842, arrived in Continental Europe in Belgium and spread to France and England by 1845. Henslow's friendship with Miles Berkeley meant that he was naturally interested in the scientific study being undertaken by the latter into the cause of the disease, which had affected the crops of his allotment holders. In a paper published in January 1846, Berkeley set out clearly his view, subsequently proved correct, that 'the decay is the consequence of the presence of the mould, and not the mould of decay'. Interestingly, Lindley, who had dramatically announced in September 1845 the failure of the potato crop around Dublin, continued to hold the view that the mould (fungus) was the result, not the cause, of decay, and it was not until the work of the German plant pathologist Anton de Bary was published in 1863 that the role of *Phytophthora infestans* was clearly established.[33]

Characteristically, Henslow's reaction to the local 'potato murrain' was a severely practical one, namely to produce a three-page leaflet setting out in great detail 'the method . . . of preparing a wholesome flour from decaying potatoes'. Dated 10 October 1845, this was widely distributed: it gives several examples of satisfied users, including this gem: 'Mr. W. Baker, Carpenter at Hitcham, has long been in the habit of procuring Arrow Root from the Shops at 1s. 4d. p lb. for an invalid wife, who uses a good deal of it: but he now finds that he can prepare for himself, from his bad Potatoes, an article equal to what he has been accustomed to buy.[34] (Figure 38)

Throughout his whole career in Hitcham, Henslow showed a great and sustained interest in the improvement of agricultural practice, and turned his attention to a range of subjects such as the diseases of wheat and the local failure of the red clover crop.[35] Nowhere, however, did his enquiring mind succeed in his aim of scientific agricultural improvement more than in the matter of the 'phosphatic nodules' or coprolites, though the great development of that industry was to take place after the Professor's death. It was on a family holiday on the coast at Felixstowe in 1843 that he first discovered the hard 'phosphatic nodules' in a coastal deposit known as the 'Red Crag'. Henslow's son, George, describes how he was present with his father when he made the discovery. Just eight years old, George was, in his own words, 'initiated into the delights of the science of geology' on this holiday, and explains to us how his father was fascinated by the rich variety of fossil material that they unearthed from the eroding cliffs at Felixstowe.[36]

Although these nodules were already known to geologists, Henslow seems to have been the first scientist to show that on analysis they contained high proportions of calcium phosphate, and could be used in

THE POTATOE CROP.

SEVERAL applications have been made in Hadleigh, for copies of a statement that was put forth by the Hadleigh Farmers' Club at their monthly meeting, on Sept, 13th, respecting the Potatoe crop. As these applications were mostly made for the purpose of ascertaining the method, which was given in that statement, of preparing a wholesome flour from decaying Potatoes; it has been determined that a fresh statement should be put forth, embodying such additional information as the experience of the past three weeks has supplied towards improving and shortening the process recommended.

I. Thoroughly wash the potatoes. It is unnecessary to peel them, unless a very superior article is required.

II. Rasp them to a fine pulp with a common bread grater; such as may be procured for $5\frac{1}{2}d$. or may be made by punching a piece of tin with a nail. The more finely the pulp has been rasped, the more flour will be obtained.

III. Place some of the pulp (say about a quarter of a pail full at a time) into a pail. Fill the pail with water and stir up the whole well, in order to work the flour out of the pulp. The water will soon become thick from the quantity of flour it will hold in suspension.

IV. Pour the contents of the pail (before it has had time to settle,) upon a horse-hair or other sieve, placed over a large tub. The water will readily pass through the sieve with most of the flour in suspension, whilst the pulp remains behind, and may be emptied from the sieve into another pail, to undergo a second or a third washing, after all the pulp has been similarly treated. This plan is better than putting the pulp first on the sieve and then pouring water upon it. As some small portion of the pulp often finds its way through the sieve, it may be removed at once by allowing the water to fall upon a loose coarse cloth stretched over the tub; but in this case the water must be continually stirred whilst it is filtering through the cloth, otherwise the flour is apt to clot and fill up the passages.

V. The flour will rapidly subside in the tub, and in less than ten minutes there will be a compact layer formed at the bottom. But as the finest particles remain suspended in the water much longer than the rest during this first part of the process, it is advisable to let the whole stand for three or four hours, when the water may be poured off.

VI. The flour thus procured should be washed at least once or twice more, by filling the tub with fresh water, stirring it up well, and then allowing it to settle again, which it will now do very rapidly; and as soon as the water above it is clear, it may be poured off.

FIGURE 38 Henslow's Potato leaflet, 1845 (first page).

powdered form as 'fertiliser' on arable crops in place of bone meal. His first publication on the subject appeared in 1843, and was followed by four others. One of these papers, in the *Gardener's Chronicle* for 1848, contained the following passage:

> It occurred to me that possibly certain nodules of an anomalous character, abounding in the upper green sand in the neighbourhood of Cambridge, were in some respects allied to those in the crag, and would possibly be found to contain phosphate of lime. Upon directing the attention of Mr. Deck, of Cambridge, who is a practical chemist, engaged in making analyses

for agricultural purposes, to these nodules, he readily detected in them the presence of earthy phosphates, in porportion varying from 57 to 61 per cent. Whether these various nodules, thus abounding in phosphate of lime, can be made available for agricultural purposes, must depend upon the possibility of their being collected at a cheaper rate than an equal quantity of bones can be.[37]

By the time Jenyns wrote the *Memoir* (1862), this speculation about the possible agricultural value of the nodules had turned into a whole industry of coprolite-digging in and around Cambridge. Quoting the above passage, Jenyns continues:

> The stratum of green sand in which they are found, 'although never more than a foot thick, occurs near the surface over many square miles in the vicinity of Cambridge,' and the land is now in places 'honey-combed with pits' from which they are dug, finding employment for large numbers of labourers, and bringing in immense profits as well to the farmers as to the proprietors of the soil. The same operations are being carried on in some of the adjoining counties. Yet, great as was the benefit here conferred by Professor Henslow upon the farmer, no acknowledgement was ever made of his services. He never indeed, looked for any himself. It was not his habit to work for reward. He was satisfied with having made a useful discovery, and it was no sooner made than he 'at once gave it the widest circulation in the local papers, without reservation of any kind,' in the same liberal spirit in which he always acted for the advantage of others rather than of himself.[38]

Henslow's wider concern to promote knowledge of the natural sciences on a national scale is evidenced in his lifelong association with the new University of London, which received its Charter in November 1836. The attraction that Henslow felt towards this new university must have arisen from his frustration at the resistance to change in his own ancient unreformed university. To quote Green (1969), the new 'red-brick' universities

> came into existence to meet a genuine demand which Oxford and Cambridge could not or would not supply. They were not intended to rival but to supplement the older institutions. Oxford and Cambridge were essentially finishing schools for the children of the governing classes, for the clergy and the higher administrators; very few sons of the working or of the lower middle classes found their way there . . . They retained a strongly

Anglican bias, even after the abolition of the religious tests in 1871. The best of their studies were humanistic rather than scientific. Their teaching was not so much bad as increasingly out of step with the intellectual developments of the times. 'It seems,' Joseph Priestley had written in 1764, 'to be a defect in our present system of public education that a proper course of studies is not provided for gentlemen who are designed to fill the principal stations of *active life*, distinct from those which are adapted to the *learned professions*.'[39]

From the first nomination of examiners in 1838, Henslow acted in this capacity for 22 years until 1860, the year before he died. Joseph Hooker supplied at Jenyns' request a detailed assessment of Henslow's contribution to the new University, which is given *verbatim* in the Memoir:

He was the first to introduce into the botanical examination for degrees in London the system of *practical* examination. He insisted that a knowledge of physiological botany, technical terms, minute anatomy, &c., were not subjects by which a candidate's real knowledge could be tested, for the longest memory must here win the day. Still less did it test the observing or reasoning faculties of the men. Further, any amount of book-knowledge on these subjects was consistent with the most profound ignorance of the very rudiments of botany, and of the most elementary knowledge of the parts of plants, the relations of these parts, and the mutual affinities of plants. He therefore insisted in all his examinations that the men should dissect several specimens, describe their relations, uses, and significance, in a physiological and classificatory point of view; and thus prove that they had used their eyes, hands, and heads, as well as their books.

By this it must not be supposed that he overlooked pure physiology and minute anatomy. At every examination he gave questions in those branches, but their knowledge of them was tested by practice. In short he regarded the acquirement of an abstract knowledge of these branches as an accomplishment, which the tyro can only obtain by reading, and not of more educational value as a means of training the faculties.

He complained of the methods of teaching botany adopted in London particularly, where, especially in the purely medical schools, every branch of the subject is crowded into a three months' course, and the men leave the class with a most indistinct notion of the very rudiments of the science, and with minds positively impaired by the cramming they had undergone, instead of strengthened, as they should have been, by judicious use of specimens, and by confining the course to what men of average intellect can acquire *well* in the time. 'They are badly taught' was his constant complaint: 'Botany has done them more harm than good.'

He was further one of the few who thoroughly appreciated the diverse faculties of young men. Himself good at mathematics, classics, and the biological sciences, he yet clearly understood that with the majority of men a faculty for acquiring one or more of these branches of knowledge was deficient: in some, one was over-developed at the expense of the others; and in his appreciation of the merits of candidates he bore this in mind. Still he always held that a man of *no* powers of observation was quite an exception; and that the lowest development of that faculty was capable of such further development as to fit a man for any calling.[40]

Henslow never abandoned his hope that his own University would eventually see the light, and undertake a radical reform. His increasing preoccupation with the affairs of Hitcham after he moved there in 1839 had, however, prevented him taking more than a minor part in any Cambridge agitation for reform. He clearly supported Whewell in particular, the polymath Master of Trinity with whom Henslow associated in a lifelong friendship. Whewell was one of the principal agitators for reform which, though delayed and meeting much resistance, eventually took place in a series of steps between 1850 and 1860. In Henslow's 1846 'Address', from which we have quoted earlier in Chapters 7 and 8, appears the following:

> But still I must consider the claims of botany are not sufficiently appreciated among us. There are persons of great mathematical and classical attainments, who have very erroneous notions respecting the ultimate aim and object of this science. Many persons, both within and without the universities, suppose its objects limited to fixing names to a vast number of plants, and to describing and classing them under this or that particular 'system'. They are not aware that systematic botany is now considered to be no more than a necessary stepping-stone to far more important departments of this science, which treat of questions of the utmost interest to the progress of human knowledge in certain other sciences which have been more generally admitted to be essential to the well-being of mankind. For instance, the most abstruse speculations on animal physiology are to be checked, enlarged, and guided by the study of vegetable physiology. Without continued advances in this latter department of botany, the progress towards perfection in general physiology must be comparatively

slow and uncertain. And, here, I cannot refrain from making some allusion to the station which botany might be made to occupy in an improved scheme of liberal education. But as this claim has been so fully recognized by Dr. Whewell, in his late publication 'On Cambridge Studies,' I shall prefer reminding you of his opinion, to urging upon you whatever be my own view.[41]

The institution in 1851 of a separate Natural Sciences Tripos called forth from Henslow his pamphlet entitled *Questions on the Subject-matter of Sixteen Lectures in Botany, required for the Pass-Examination*, an explicit title to a guide for the 'poll-men' students taking this new examination. We have seen in Chapter 5 how, quite dramatically, the number of men attending Henslow's courses rises again from 1851, once some examination recognition of their study becomes available. Henslow's own son, George, was one of the successful candidates who obtained a First in the Natural Sciences Tripos in the interim period when his BA degree had to be awarded on his Mathematics and Divinity. George had entered Christ's College in 1854 and took his BA in 1858. It was not, however, until 1860, that all students 'who shall pass with credit the examination for the Natural Sciences Tripos . . . be entitled to admission to the degree of Bachelor of Arts'. Henslow, very fortunately, was able to examine the first Botany students in the fully-developed Natural Sciences Tripos that we know today; he did this in March 1861 only a few weeks before his death. Apparently, only three names appeared on the Tripos list, one in each of the three classes, and all were admitted to their degree accordingly – the first Cambridge scientist BAs! The man awarded a second-class was Robert Cary Barnard, who married Henslow's daughter Anne in 1859 (Figure 39).

In spite of his involvement with parish affairs, the Ipswich Museum and London University, Henslow found time in 1852 to negotiate the gift to Cambridge of the main part of a large private Herbarium of Charles Morgan Lemann (or Le Mann). Lemann was a Cambridge-educated botanist of private means who matriculated in Trinity in 1824 and obtained the medical degrees of MB in 1828 and MD in 1833. He was a friend of R.T. Lowe, who graduated in Christ's College in 1825; Lowe came under the influence of the young Henslow and was helped by him to obtain a travel scholarship (see Appendix 5). Lemann helped Lowe with his studies of the Flora of Madeira, and his herbarium contained many Madeira plants[42] (Figure 40). The preparation and transfer of Lemann's Herbarium to Cambridge was effected by George Bentham, who promised to undertake

FIGURE 39 Anne Henslow (later Barnard) with Floss.

the work as Lemann was dying early in 1852. A letter from Bentham to Henslow on 6 October 1852 sets out the details:

> The herbarium of the late Dr Lemann has just been made over to me by his brother and sister. I went to town last week to pack it up and have now unpacked it here and looked into it enough to judge generally of its condition and value.
>
> According to an understanding entered into with my friend last spring I now propose to select for myself (besides the Afghan collection which he had long previously given to me) a small number of species which I do not

FIGURE 40 Herbarium sheet from Madeira of *Ilex perado* in Herb. Lemann.

possess in my own herbarium and to have the greatest bulk glued down and named and arranged after my own herbarium adding from my own duplicates which I hope will prove an equivalent to what I shall keep of the herbarium; and to present the herbarium so formed in Dr Lemann's name to the University of Cambridge provided the University will not object to paying the necessary expenses . . .

    If you think you can undertake that the University will accept this gift from the late Dr Lemann on the above conditions, I will send you for your approval a sample of the paper, as I am anxious to commence operations as soon as possible.[43]

Henslow replies with commendable promptness on 11 October 1852:

I found your letter on my return home on Saturday – I had not been aware of Dr L's death. From what passed between the Vice Chancellor [Oakes], Dr Hooker and myself I can have no hesitation in saying the University will defray the necessary expenses – and that I can guarantee as far as the 150£ for the first necessary outlay.[44]

The acquisition of this large Herbarium, which contains more than 50,000 specimens, marks the beginning of the growth of the modern Herbarium collections in Cambridge. Henslow's part in this growth should not be forgotten. Although the re-housing of the Herbarium and Museum in a new and larger building on the Old Botanic Garden site did not take place until 1865 under Babington, it is to Henslow that the credit should go for the acquisition of Lemann's collection, sorted and named by the specialist Bentham, and well mounted and labelled under his supervision. Next to Lindley's Herbarium, that of Lemann is the largest single nineteenth-century acquisition.[45]

Undoubtedly these middle years of Henslow's life at Hitcham were very demanding, and correspondence with Whewell reveals something of this strain. Whewell had evidently asked his friend for help in writing the *History of Botany*, and Henslow replied in August 1846:

> I shall be very ready to attend to your wishes about your History & will think the matter over as to how I can assist you with suggestions. There has been a good deal done in some parts of Cryptogamic botany of late years by Berkeley & others – ... I fear (indeed I know) that I am not so well qualified as I ought to be to speak authoritively on these matters – but I dare say with a little enquiry on some points of those who are more enlightened I may be able to help.
>
> When I came to reside at Hitcham I had fancied that I should have more opportunity than ever to devote my time to Botany – but I well remember a remark of Willis 'you will get entangled in other interests & less able to pay attention to your Professorship' and so it has proved.
>
> It is now above 3 years and a half since I have been absent from Hitcham on a Sunday – & what with double duty to provide for – ordinary parish duty – justice business – and now and then the necessity of acting as Chaperon to my daughters (Mrs H. being ill) my time is pretty fully occupied – The girls kept me at a dance till 3 of AM only three nights ago.[46]

In our final chapter we shall see how the Rector's activities began to take a toll of his health and strength, but in spite of warnings he was unable – or unwilling – to cut down significantly, and achieve a more undemanding life-style.

12 The later years

FIGURE 41 Portrait of Henslow in later years.

One of the more unusual ways in which the Rector used his talents to educate his Hitcham parishioners was by the production of a 'Bioscope or Life-Dial' (Figure 42) on which one's progress through life could be recorded at each birthday. How much success he had in finding among his flock young parents who would conscientiously fill in the birth and baptism details for each child – and continue to keep it up to date – we do not know; but at least it indicates that Henslow himself was content to date the decline of life from the age of 50. In his own case, there is a good deal of truth in this. The awful social tensions of the early 1840s, which called forth

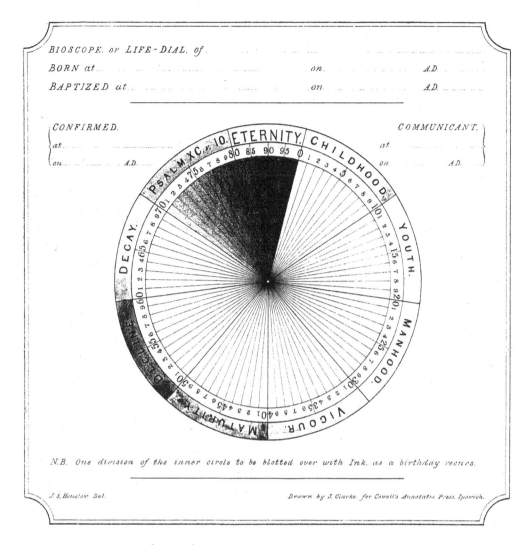

FIGURE 42 Henslow's 'Bioscope'.

such devoted response from him, must have taken their toll, and by his
fiftieth birthday in February 1846 he could see only limited success in many
of his plans, and certainly no prospect of much leisure to pursue natural
history for its own sake. Nor were the early deaths of all his younger brothers, not one of whom reached fifty, remotely consoling.

There were, however, some developments which must have given
Henslow great satisfaction in these later years. Among these was the
launch at a public meeting in Ipswich Town Hall on 30 November 1846 of a

new Museum 'to promote the study, and extend the knowledge, of natural history in all its branches, and of the arts and sciences generally'.[1] Henslow's old friend, William Kirby, the eminent entomologist, now in his eighty-sixth year, was elected President of the Museum, and Henslow himself, together with the Astronomer Royal, Professor G.B. Airy, were elected Honorary Members. George Ransome, a Quaker manufacturer and amateur ornithologist, grandson of Robert Ransome, founder of the Ipswich agricultural implement firm, became one of the two Secretaries of the new project, and a firm friendship grew up between Ransome and Henslow. George Ransome was fifteen years younger than Henslow, but they obviously shared their concerns for social improvement.

The donations and subscriptions for the Museum came in quite abundantly, and the new building, designed by the Ipswich architect Christopher Fleury, was officially opened on 15 December 1847 by the Bishop of Norwich, Dr Edward Stanley, himself an amateur ornithologist of some standing. Kirby, the aged President, was present at the opening, and William Buckland, the eminent geologist and teacher of Charles Lyell, was the chief speaker. A series of Museum Lectures, in the arrangement of which Henslow played a main part, opened on 9 March 1848 with an Introductory Lecture by Henslow himself. The account of this lecture, fortunately included by Jenyns in his *Memoir*, reveals that it was a reasoned plea for 'the advantages and great importance of science in general'. In this lecture Henslow 'regrets how much [science] has been undervalued, not only by the illiterate and ignorant, but, in some cases by those who have studied deeply other departments of human learning'. He makes a strong plea for the inclusion of natural history and other sciences in the curricula of all schools, and looks forward to 'an efficient system of national education' throughout the country. The address also faced the increasingly relevant question of, as Jenyns puts it, the 'supposed interference of science with religion', and is a spirited presentation of Henslow's tolerant liberal theology.

The Anniversary Meeting, held over the 12th and 13th December 1848, was a very successful occasion. Henslow had persuaded his old teacher, Adam Sedgwick, Woodwardian Professor of Geology in Cambridge, to give the invited lecture, which duly took place 'to a class of six or seven hundred people', in Ipswich on the evening of Tuesday the 12th. The lecture apparently took over two hours to deliver, and was ostensibly devoted to the anatomy and physiology of three of the gigantic extinct mammals of South America, on which Charles Darwin had written in his account of the voyage of the 'Beagle' published three years earlier. Clark & Hughes, Sedgwick's biographers, comment as follows on the lecture:

> It is difficult to understand how such a subject could have been made inter-esting to an audience not specially acquainted with paleontology – but we are told that they were 'alternately charmed with the vast learning of the lec-turer, and thrown off their gravity by the most grotesque illustrations.' No doubt Sedgwick did his best with his extinct mammals – he was always fond of paleontology; but, just as the most important part of a letter is not seldom to be found in the postscript, it is evident that he threw his whole strength into the second part of the lecture, where he dilated, with charac-teristic vigour and emphasis, on the lessons to be learnt from such discov-eries. Sedgwick's lectures – no matter what the subject-matter with which they began – usually ended with whatever was uppermost in his mind at the moment. When he appeared before his friends at Ipswich he was forging fresh thunderbolts against the author of *The Vestiges*; and in consequence we find a considerable space devoted to the argument from final causes. He concluded with a noble passage on the true method of reconciling science with revelation. 'If', he said, 'we believed, as we did believe, that the Author of Nature allowed us to call Him Father, and that He called us children, and He gave us the hope of a heavenly hereafter, let us also believe, at the same time, while obeying those instincts which he implanted in our bosoms, and reading the records of His will as written in the solid framework of the earth, or in the glittering phenomena of the sky, or in what the broad gener-alisations of science had manifested, that these two kinds of truth, embod-ied in physical history and revealed religion, so far from being conflicting, were entirely in unison and harmony, if we investigated the one, and read the other, in a right spirit.'[2]

In 1847 the Revd Edwin Sidney was presented by Edward Stanley, Bishop of Norwich, to the living of Little Cornard, a parish near Sudbury only about 10 miles from Hitcham. Sidney, who had been an undergraduate at St John's, Cambridge from 1817 to 1821, must have known Henslow from those days and, more particularly, was much influenced by Sedgwick. Little is known of Sidney, who left no papers, diaries or journals, but we do know that he became a Vice-President of the Ipswich Society when Henslow was

President, and that Henslow was present at the last lecture in a course of agricultural lectures given by Sidney in Little Cornard. He and Henslow had much in common: in particular they shared the strong belief that the exploration of natural science reveals the wisdom and goodness of the Creator, and should be used to educate the 'common people'.[3]

Henslow himself contributed several lectures to the Ipswich Museum programme, including one on 'Geology', given in November 1849, during which he used material of coral received from Charles Darwin. Kirby died, aged 90, on 4 July 1850, and Henslow succeeded him as President of the Ipswich Museum at the two-day Anniversary meeting held on 18th and 19th December of that year. The guest speaker on the evening of the first day was Edward Forbes, Professor of Botany at King's College, London, who addressed 'a numerous and highly select' audience on the 'Natural History of the European Seas'. Amongst the 'highly select' was Dr Nathaniel Wallich, a Danish botanist who had been Superintendent of the Calcutta Botanic Garden; Henslow's acquaintance with this expert on the Indian and Malayan floras dated back to early Cambridge days when Henslow was largely responsible for Wallich's election to the Philosophical Society in 1832 (see Appendix 5).

In his acceptance speech, Henslow makes touching reference to the influence Kirby had on him when a boy, and indeed later in his career:

> ... I have enjoyed his personal acquaintance more or less, for several years, and I may justly say that it is to his writings, and in some respects, to the slight personal knowledge I had of him, when I was quite a boy, long before I went to college, that I am indebted for the first degree of interest I ever took in natural history. I began, as many do, with entomology, and when a boy still at school, having the acquaintance of the celebrated Dr. Leach, who was a most intimate friend of my venerated predecessor, I had now and then the opportunity of meeting him at his house, and reading his works, and I became possessed of that most classical work of his, *Monographia Apium Angliae*. Under these auspices I began my career as a naturalist, and, of course, I looked up to him as the father of anything I knew upon the subject. Curiously enough, later in life, an incident occurred, which may not be known to many of you. Mr. Kirby was a botanist as well as an entomologist, and at one time of his life he had a desire to hold the botanical chair at Cambridge, about a quarter of a century ago, when I was a candidate for that office. He wrote me a very kind letter, saying that if he had been a little

younger he would have been my opponent on that occasion, but that he most handsomely withdrew any pretensions that he might have to the chair, considering I was a younger man, and likely, perhaps, to act with more energy. Whenever I have had the pleasure of his acquaintance it has always been to me a source of instruction, and I may say that I have venerated his memory.[4]

The main body of Henslow's address was devoted to 'a few practical remarks' about the acquisition, display and use of the Museum collections, with a special emphasis on the importance he attached to the commitment to contribute 'towards the free education of the working classes in the science of Natural History, by providing for them a good museum library, lectures and classes'. He recommended what might be called 'a typical arrangement of a certain portion of the Museum' using a 'number of common objects [to be] grouped in two or three cases, so as to illustrate the main groups of which the mineral, animal and vegetable kingdoms are composed'. This 'typical arrangement' was to be 'made the constant object of short lectures to be given in the room from week to week, throughout the year'. He gave notice that when the British Association came to Ipswich the following year, he would ask them to appoint a Committee to draw up a specimen 'typical arrangement' for each branch of natural history. Inevitably, when the Association did meet in Ipswich in 1849, it happily accepted Henslow's plan and told him to get on with setting up a Committee. He spent a good deal of time on this project in the later years.

Henslow shows here two intertwining sides of his character. With many Victorians he was an avid collector – his rooms in Cambridge had been a storehouse of natural history specimens admired by the young Jenyns, as we saw in Chapter 3 – but he was also intolerant of mere random collecting, and wanted museums to adopt a more organised policy of acquisition and display. There must be a purpose behind museum collections, and to Henslow that purpose was education for the masses. Admitting *hoi polloi* did not, however, go entirely smoothly: the free evenings on Wednesdays and Fridays excited wide interest, and 750 people crowded the Museum on Christmas Eve, 1848. How exactly 'a Committee member went round the room and explained every object of interest to the working classes' with such crowds we are not told, but the Curator, Dr William Clarke, a local naturalist, was not amused by the behaviour at times of the 'lower orders', and was moved to write to the newly-installed President an intemperate letter, of which the following is an extract. He is complaining about the

> ... demoralizing result from the indiscriminate nature of the admission, troops of lads and girls that are howling, shrieking, whistling with fingers in their mouths, and rushing up and down the stairs. On mornings following these events, I have picked up a quantity of orange peel, shells of pin-patches [sic], peas, sticks, etc., which are usually thrown by these visitors at each other, and at the glass. The dresses of many are frequently observed to be covered with dirt, mortar, soot and smelling so abominably that respectably dressed people could not approach them; they are the vagabonds of the town and neighbourhood.

Clarke resigned in December 1850 in protest against Henslow's policy: the new Curator, David Wooster, was apparently more tolerant, but sixteen voluntary stewards were appointed to attend on the free admission days 'to maintain order and decorum among the lower orders' [sic].[5]

After a visit by Ransome to Buckingham Palace in February 1851, Prince Albert visited the Museum on Friday 4 July during the Ipswich meeting of the British Association, and Henslow as President read an address of welcome to the Prince Consort (Figure 43). It seems that the Museum and the enthusiasm of its promoters made an enormous impression on Prince Albert, for Queen Victoria is on record as saying that he talked of scarcely anything else for several days after his return! Henslow already knew the Prince and admired greatly his concern for the reform of University education; Whewell and Sedgwick had been largely responsible for Prince Albert's successful candidature for the post of Chancellor of the University of Cambridge in 1847, and Henslow was always a supporter of reform to the Cambridge examination system which was hopelessly out of date. 1851 was, of course, the year of the Great Exhibition, which must have given a great impetus to all plans for the furtherance of state support for scientific education. Henslow arranged an elaborate three-day visit to London for twenty of his parishioners, which took in the Great Exhibition, the Zoo, London Polytechnic and Kew Gardens.

Financial difficulties, and the resignation of George Ransome in 1852, presented Henslow with unwelcome problems in the Museum administra-

FIGURE 43 The Prince Consort at the Ipswich Museum.

tion; he seems to have tackled these with his customary practical flair, using the recent Public Libraries Act (1850) to propose that the rate-payers of Ipswich should vote whether they were willing to take over the Museum. A two-thirds agreement was necessary. The poll was held on 4 February 1853 and to Henslow's relief the support was overwhelming: 709 for and only 69 against. The Museum was able to re-open under Ipswich Corporation management on 6 June 1853, and admission was now free for all. Henslow was re-elected President under the new administration, and found he could easily cooperate with George Knights, the new Curator. Indeed a firm friendship developed between Knights and Henslow, and throughout his few remaining years Henslow seems to have continued to find time to prepare a large range of museum material, including his local Herbarium, still preserved in the Museum at Ipswich. Jenyns, whose occasional visits to stay with his sister and brother-in-law in the Rectory afforded him ample opportunity to observe their life-style, gives us this sympathetic description of Henslow at work:

> He never seemed more thoroughly contented than when at work at his table, on a long winter's evening, with all his specimens and tools about him; now selecting his best specimen of each kind or examining it with a lens; or in some cases paring it with a knife in order to show the interior; – now cutting stiff card-board with strong scissors of a peculiar make he had on purpose; – now moulding soft clay into suitable form to serve as a stand for specimens that could not otherwise be so well inspected . . .
>
> In such work he could be quite absorbed with his family around him, taking little notice of what others were about. It was always evident that his mind was quite as much engaged as his hands. He would seem to be turning over in his thoughts for what uses each specimen might serve, the impression it was each likely to make on the spectator's eye, the instruction it was capable of affording . . .[6]

The construction of railways throughout East Anglia, which took place very rapidly in the decade from 1845, provided Henslow with a marvellous opportunity to entertain and educate his parishioners by organising a series of village excursions. His son George, in one of his series of *Reminiscences of a Scientific Suffolk Clergyman* (i.e. his father!), provides us with an ample account of these very successful activities, which covered the period 1848 to 1855.[7] As George explains:

> Hitcham being situated seven miles from Stowmarket and five from Hadleigh, the nearest railway stations, it is not surprising that many, or rather most, had not only never travelled by rail, but had not even seen a train before they joined these excursions. This novelty was, therefore, at first an important part of the trip, not to mention passing through the tunnel at Ipswich, which caused an unlooked for excitement, resulting in much whistling and shouting!

Through his friendship with George Ransome and the Ransome family in Ipswich, Henslow was able to arrange the first excursion 'to inspect the Iron Foundry of Messrs Ransome & May'. To make up his party, he invited all Hitcham members of a local self-help society, the 'Stoke and Melford Benefit Club', to join and bring a friend; and he extended the invitation to the Hitcham farmers also, to whom he offered the excursion as recompense for his having discontinued the annual Tithe Dinner which traditionally the Rector had provided. How many of the farmers actually took part in the Ipswich visit we do not know, but some conveyed the 'fifty villagers of both sexes (over fourteen years of age)' to Stowmarket railway station to begin their exciting railway trip. The railway company agreed to a 'greatly reduced fare' for the day trip, and a good time was obviously had by all. The party 'reached home again by 11 pm'.

Obviously elated by the success of this first village excursion, Henslow was prepared to run a much larger trip the following year. This duly took place on Wednesday 25 July 1849. From Henslow's own account we learn that they 'were joined on this occasion by several of the farmers to whom the good conduct of the party on the previous year had been reported. Our party now had reached to 180 including members of my own family, and all my servants.'

For this excursion, the party enjoyed a trip down the Orwell Estuary in the steam-boat 'River Queen' to Harwich. The Ipswich local paper wrote up the occasion, and allowed itself a little criticism of a packed programme:

> After strolling through Harwich, examining the fort, etc., which was kindly exhibited by Mr. Sharpe, and partaking of luncheon at Dovercourt, the signal for return was given, and the steamer arrived at the Griffin Strait just one minute before the time for departure for the train. Doubtless it will be a day long to be remembered by the good people of Hitcham – an event for many years to be chronicled amongst the recollection of 'Our Village'.
>
> The utmost good feeling and order was manifest throughout the day's proceedings and every heart apparently beat happily. Perhaps there was too much of the 'move on' spirit of the age in so extended an excursion, and a few hours quiet lounge on the cliffs of Harwich might have been so much additional pleasure; but on the whole it was one of those delightful occurrences, which, whilst scattering happiness abroad, tends also to bind society together in the bonds of unity.[8]

For the following year's excursion, Henslow had decided to draw up a list of 'rules', which were duly printed and circulated in the parish – another characteristic Henslovian document, leaving, one might think, nothing to chance:

HITCHAM EXCURSION TO IPSWICH AND LANDGUARD FORT, ON TUESDAY, 30th JULY, 1850.

———————

I. According to my expressed determination last year, I am making arrangements for a village excursion to Ipswich and Landguard Fort; having received assurances that we shall be welcome at both these places.

II. The successful issue of such an excursion (provided no untoward accident should occur) will mainly depend upon a general attention to a few rules, which I here propose to those who wish to accompany me.

III. Every one is to be in good humour, accommodating towards all, and especially attentive to the ladies of the party. If the weather should prove unpropitious, every one is to make the best of it, and not to complain more than he can possibly help.

IV. Those who are invited on the present occasion are the occupiers of farms in Hitcham, the resident members of the Stoke and Melford Club, and those who attend the adult class on Saturdays. Every one who joins the party may also name a friend, residing in Hitcham, for whose good conduct he will be responsible.

V. The tickets to be issued will be limited to 200, at 1s 6d. each. All applications to be made at the Rectory on or before Friday the 26th. Should any tickets remain unapplied for after that by the parties invited, those who have already received tickets may apply for more for other friends. By this arrangement I hope to accommodate all, or nearly all, who may be wishing to take part in the excursion.

VI. The party are to assemble on the platform at Stowmarket by half-past eight.

VII. As the object of the party is not to be mere eating and drinking, but wholesome recreation for body and mind, the refreshments provided will consist of bread, cheese, butter, cake, with lemonade, and one or two pints of beer for those who may apply for an order to that effect when they receive their tickets. These orders for beer are not to be transferred to other parties; and if not needed are to be returned.

VIII Should any one be prevented at the last moment from joining the party, 1s of the money paid will be returned; and the remaining 6d will be appropriated towards the expenses that will have been incurred, upon the supposition that such person would have been of the party.

IX. Every one must contrive for himself how to get from Hitcham to Stowmarket, and back again. If he is not able to persuade any one to give him a lift, he must consent (as others have done before) to wear out a little shoe-leather.

J. S. HENSLOW.[9]

Here is George Henslow's memory of this excursion, which provided most of the participants with their first sight of the open sea. George was a youth of 15 at the time, and is recalling the event in his later years:

> A few hours, and the vessel is at rest; but no sea is yet to be seen. The party disembark and climb a bank of sand and shingle, when there lies before them the big ocean, far surpassing their most sanguine expectations; which were, I know, in some cases, limited to that of a good-sized horse-pond!
>
> Immediately a rush is made for the shore; pebbles, seaweed, bottles of salt water are carried off as mementoes. But this was not all; for the officer in command at Landguard Fort, understanding the nature of the party, kindly volunteered to exercise the artillery in a variety of field movements, drill and gunnery practice, for the amusement of the villagers; who were then conducted through the fort; while several joined the privates at cricket and football.
>
> A curious circumstance occurred: while a private was just going to fire a gun, one of the Hitcham girls ran up to him and called him by name. It turned out that he, as a Hitcham lad, had left home two years before, and nothing was known of his whereabouts, till his sister recognised him.
>
> Thus the hours flew by till the call for tea was made, and all, seated upon the grass near the fort, partook of the usual fare, augmented by a huge mass of Wenham Lake ice, which Mr. G. Ransome contributed, to the utter astonishment of the rustic party at seeing and handling such a substance on a broiling day of July. After a hearty farewell to the soldiers, who had thoroughly fraternised with the Hitchamites, the band struck up 'God Save the Queen,' and 'The River Queen' once more bore them, but this time homewards.[7]

Other excursions, apparently equally popular, visited Norwich, and the series culminated in a visit to Cambridge on 27 July 1854, made possible by the completion of the railway connection to Bury St Edmunds from Ipswich and Stowmarket earlier that year. No doubt Henslow had been waiting for this opportunity, for he prepared a very detailed eleven-page printed and illustrated pamphlet which all the participants received in advance. Some 250 people took part in the Cambridge excursion, and it was particularly gratifying to the Rector that a good many farmers and their wives joined the party. As George Henslow puts it: '. . . the trip to Cambridge proved too strong a temptation, and they came in a body to ask to join. They, of course, were welcomed. So delighted were they that they were no longer able to withhold their admiration for the Professor and united in presenting him with a silver cup.' (Plate 10).

The planned visit, set out on the first page of the booklet, is typical of Henslow. It has clearly been timed in advance, and represents a very tight schedule. We do not know how far the actual excursion carried out the whole programme, but can only assume that the Professor, setting a cracking pace himself, carried it through successfully. The key element in the day's programme was the generous hospitality of Downing, Henslow's 'adopted College'. Tea in Downing Hall at 10.15 am, and a dinner back there at 2 pm, for which, to quote the pamphlet, 'we shall find the Vice-Chancellor and his Lady[10] have kindly added to our ordinary frugal fare by a present of a Barrel of Beer, and Plum-puddings enough for the whole party'. Finally, the party was to return there before returning home: 'Quite ready for a cup of tea before starting, we adjourn once more to the hospitable Dining Hall, and resting till 6 o'clock allow plenty of time to make our way to the Station by 6.25'.

During this trip, Henslow is scheduled to demonstrate, from what is now Regent Street, 'the house in which I last lived when I resided in Cambridge', and also in St John's College 'the rooms in which I kept (as we say in Cambridge) during the time I was a student'. George Henslow, in his account of the Cambridge trip, points out, however, that his father, though mentioning 'Corpus College', omits to mention the notorious inscription 'Henslow Common Informer', which must have been very visible in 1854!

We have already mentioned in Chapter 8 the description Henslow gives of the Old and the New Botanic Gardens in this pamphlet. His devotion to museums also finds a place in the schedule, there being long descriptions of what his party should admire in both the Fitzwilliam Museum (a relatively new building, begun in 1837) and the Woodwardian Museum of Geology which Sedgwick had built up, and which was then housed in the Old Schools, together with the 'Public Library', which became the present University Library.

It is inconceivable that the 250 members of the party, however obediently they followed instructions to proceed in an orderly manner, could have received more than a very fleeting glimpse of most of the described objects. In the account of what the Woodwardian Museum had to offer, Henslow cannot resist a small general lecture on the significance of fossils, mixed with a warning not to touch anything!:

On the ground-floor, below the new gallery, is a very valuable collection of fossils and mineral specimens, in what is called the Woodwardian Museum. Here are carefully arranged the remains of many thousand species of plants and animals which belonged to the earliest periods of the Earth's history, of which no other record has been vouchsafed us than that which these objects can afford when duly studied by the light of that reason which God has endowed man. We must be cautious never to touch any articles in a Museum, lest some damage should ensue. Here is the entire skeleton of a large Elk found beneath the bogs of Ireland; also the very large bones of certain gigantic Mammals which belonged to earlier periods, and those of large Reptiles older still. The shells of Sea Mollusks, from their hard nature and earthy composition, are more abundantly preserved than any other kind of fossil, and those from our Suffolk Crag are some of the most perfect, belonging to a later period than most others. Those from earlier periods are generally remodelled in stony materials which have gradually replaced the natural substance of the shell. Numerous remains of Ferns and other plants are embedded in the slaty rocks which lie between the beds of coal. This most important inflammable mineral has resulted entirely from the decomposition, and other chemical alterations which have taken place, during a long lapse of ages, in the substance of such plants. Though some of these plants greatly resemble the Ferns now found on the earth, yet none of those belonging to our Coal-fields are really of the same species with these.

It must have been an enormously heartening experience for Henslow, as we can tell from a letter he wrote to his friend Thomas Martin (whom we meet later in the chapter). The letter is dated 29 July 1854:

> I send you a programme of an excursion which came off yesterday – the best (I think) & most satisfactory we have yet had, though I have no cause to complain of former years – we were 250, & could easily have more than doubled the number, if I had thought fit – from my own & neighbouring parishes but I only admitted about 30 from the latter – I did not meet with a single unpleasant circumstance & was glad to see an unusually large attendance of Farmers – They all seemed perfectly delighted with the days work – & I feel a little weary, but not at all otherwise the worse for a hard day, including the ascent of Kings College Chapel with about ¾ of the party – ! We overflowed Downing Hall & filled the Combination room into the bargain . . .[11]

And in a similar letter to Knights on 31 July he even embellishes the story by increasing the size of the party: 'Our Cambridge trip turned out a perfect success . . . We had a large body of farm men, and for the first time the Baptist dissenters of their party joined us. We sat down 283 [sic] to dinner.'[12]

One more excursion took place the following year, but this seems to have been the last. George Henslow tells us that it was greatly increased charges by the railway companies that forced his father to discontinue the series, but we might think that other factors, including overwork and overstrain, must have contributed to the decision.

Of all Henslow's educational innovations in Hitcham, none aroused more general interest outside the parish than his lessons on botany given to the village school-children.[13] These special classes were introduced by the Rector in 1852, after he had become confident that the parish school he had founded in 1841, and which he had regularly and carefully nurtured, was educating the village children well in 'the three Rs' – reading, writing and arithmetic. A voluntary botany class was offered on Monday afternoons to some of the older children, and the Rector himself conducted these classes. That learning botany was to be no casual affair is evident from the 'printed scheme' which Henslow prepared (Figure 44): no child 'wishing to learn Botany' would even be admitted to the class until they had learned to spell correctly a selection of technical terms including 'Monocotyledons', 'Gymnospermous', and 'Thalamifloral'! In spite of this considerable hurdle, the class proved very popular and eventually had to be limited in size 'solely by the room available'.

Henslow was undoubtedly encouraged in his desire to teach a scientific subject to his parish school-children by the acceptance by members of the Government, following the Great Exhibition at Crystal Palace in 1851, of the need to promote scientific education in general. By 1853 Government Departments of Arts and Science had been set up under the Board of Trade, and a serious attempt was being made to introduce, not just literacy, but scientific and technical training into the new, rapidly-developing State education system. Henslow certainly shared in the aims of the 'Society for the Diffusion of Useful Knowledge', with its increasingly popular Utilitarian philosophy associated with the names of Jeremy Bentham and John Stuart Mill, and made a point of giving wherever it was appropriate, in his Monday afternoon classes, references to economic and other uses of wild plants and their cultivated relatives. In particular, he stressed their food

VILLAGE-SCHOOL BOTANY.

Children wishing to learn Botany will be placed in the 3rd Class when they shall have learnt to spell correctly the following words.

Class.	Division.	Section.
(I. Exercise.)	*(II. Ex.)*	*(IV. Ex.)*

1. Dicotyledons
  - 1. Angiospermous
    - 1. Thalamifloral
    - 2. Calycifloral
    - 3. Corollifloral
    - *(V. Ex.)*
    - 4. Incomplete
  - 2. Gymnospermous

(III. Ex.)

2. Monocotyledons
  - 1. Petaloid
    - 1. Superior
    - 2. Inferior
  - 2. Glumaceous

3. Acotyledons.

Children in the 3rd Class who have learnt how to fill in the first column of the Floral Schedule, and to spell correctly the following words, will be raised to the Second Class.

Pistils & Carpels } of Ovary (with Ovules), Style, and Stigma.

Stamens, of Filament and Anther (with Pollen).

Corolla, of Petals
Calyx, of Sepals } or Perianth, of Leaves.

Children of the 2nd Class who have learnt how to fill in the second column of the Floral Schedule, and to spell correctly the following words will be raised to the 1st Class.

C. Mono—di—&c. to poly—phyllous,—sepalous,—petalous,—gynous.
V. Mon—di—&c. to poly—androus,—adelphous.
Di—, tetra—dynamous...............Syngenesious.

V.	C.	V.	C.	V.	C.
0. An—	A—	5. Pent—		10. Dec—	
1. Mon—	o—	6. Hex—		11. Endec—	
2. Di—	—	7. Hept—	a—	12. Doded—	a—
3. Tri—	—	8. Oct —		20. Icos—	
4. Tetr—	a—	9. Enne—		∞. Poly—	

Children of the 1st Class will learn to fill in the 3rd column of the Floral Schedule and to spell correctly the following words.

Hypogynous   Epipetalous
Perigynous   Gynandrous.
Epigynous

Monday Botanical Lessons at 3 P. M. at the School, to include—
1st. Inspection of a few species, consecutively, in the order on the plant-list. Any thing of interest in their structure or properties will then be noticed.
2nd. Hard word exercises. Two or three words named one Monday are to be correctly spelt the next Monday.
3rd. Specimens examined, and the parts of the flower laid in regular order upon the dissecting boards. The Floral Schedule to be traced upon the slates, and filled up as far as possible. Marks to be allowed according to the following scale.

	No.	Cohesion, Proportion.		Adhesion (Insertion)		Classification.	
P̲ C	1 3	a,—mono—,&c. gynous	2	Superior or Inferior	2	Class Division	1 2
St. f.— a.—	1 1 1	a,—mon—,&c. androus { mon—,&c. adelphous Di,—tetra,—dynamous Syngenesious	2 3 3 2	Hypo—&c. gynous Epipetalous Gynandrous	4 4 3	Section Order Genus Species	3 4 3 2
{ C. P. { C. S. or P. L.	1 1 1	a—,mono—,&c. petalous a—,mono—,&c. sepalous a—,mono—,&c. phyllous	2 2 2	Hypo—&c. gynous Inferior or Superior	4 2		

4th. Questions respecting Root; Stems and Buds; Leaf and Stipules; Inflorescence and Bracts; Flower and Ovules; Fruit; Seed and Embryo.

Regulations respecting Botanical Prizes and Excursions.

Prizes awarded according to the joint number of marks obtained at Monday Lessons, from Schedule Labels filled in at home, and for species first found in flower during the season.
Botanical Excursions attended only by those who obtain a sufficient number of marks at Monday Lessons. Two private Excursions during the summer, within the precincts of the Parish, open to Children in each of the three classes. Other Excursions within the parish are open only to those of the second and first classes. An Excursion to a distance from the Parish for those of the first class only who obtain the requisite marks.
The first class may attend (at the proper season) at the Rectory on Sundays after Divine Service in the afternoon, of Natural History, in the Animal, Vegetable, and Mineral Kingdoms .... as may tend to improve our means of better appreciating ....

FIGURE 44 Village-School Botany: Henslow's pamphlet.

value, and used this unashamedly to win his audience. As Russell-Gebbett explains, he

> deliberately highlighted any gastronomic uses for the Order under study and illustrated them by a *bonne bouche*. He knew this was a sure way of imprinting the group on a child's mind. Thus we read that at the close of the lecture he collected any plant-list schedules that had been prepared over the week and anyone who produced the minimum required of him received a reward in the form of 'some little gastronomic illustration (such as rose lozenges, to represent Rosanths, olives, the order Oleanths etc.) of the economic properties of the orders of the day'. The presentation was undoubtedly a popular conclusion to the lesson and Henslow made considerable efforts to acquire appropriate material. His long-suffering colleague Knights of Ipswich was required to go shopping. Henslow wrote, 7th September 1855,
>
> 'I have promised my botanical school children some angelica as a botanical sweetmeat; would you be so kind as to procure me about 2/6 worth and either send it to me or bring it with you on the 19th?'[14]

The terms 'Rosanths' and 'Oleanths' in this passage represent an aspect of Henslow's village botany that came in for a good deal of comment and, indeed, some light-hearted adverse criticism. His own son-in-law, Joseph Hooker, was impressed by Henslow's aim at teaching botanical science, and agreed that, *where the technical term is necessary to convey definite information*, the fact that it is a long and 'difficult' word (e.g. 'perigynous') should not prevent the children from being expected to learn it. He cautions, however, that the draft Henslow has sent him contains too many difficult words: in a letter dated 12 December 1854, he writes:

> My own impression is, that it would be better to make the demonstration of the Bean first, simple, clear and to the point, giving no words except the simplest. I object to 'axis', 'relative', 'modification' etc., when super-added to the necessary and unavoidable technicalities; each of these, though familiar to us, being a subject of thought to the 'village school' before understood. Fanny [Hooker's wife, and Henslow's daughter] has been looking over parts of it, and quite agrees with me that the words underlined in pencil will be so many stumbling-blocks to village school children and even higher class ones. In short the whole is not only too scientific but in too scientific language.[15]

Darwin himself had views on the problem of teaching botanical terms, and expressed these to Henslow in a letter of 11 December 1851:

> Now that my children are growing up & I think of educational processes,
> I often reflect over your inimitably (as it appears to me) good plan of teach-
> ing correct, concise language & accurate observation, namely by making
> your pupils describe leaves &c. I never profited myself by this, but very often
> I have wished I had. Has it ever occurred to you, (I have often wished for
> something of the kind) that a most useful volume might be published, with
> woodcut outlines, & on separate pages well-weighed, concise descriptions
> in Saxon, & not scientific English. What a habit it would give to youths of
> thinking of the meaning of words, & what powers of expressing them-
> selves! Compare such habits with that of making wretched Latin verses.
> I did not intend to write so much; but it is an old wish of mine, that you or
> someone would undertake such a task.
> My dear Henslow   Yours most truly   C. Darwin[16]

The desire to liberate the new popular education from the stranglehold of the classical languages that Darwin expresses here represents another facet of the problem of teaching technical terms in science. Both Joseph Hooker and Charles Darwin, representing a younger generation than Henslow, clearly held strong views, but they did not quite see eye to eye on the solution. It remains true to the present day that controversies about the desirability of using classically-based technical terms in botany when addressing different kinds of audiences are quite unresolved, as anyone teaching botany will know. It is, however, worth remarking that Henslow's own compromise with respect to the names of plant families ('natural orders' to Henslow), which was to use an English plural ending on the clas-sical scientific name (e.g. 'Rosanths' for Rosaceae, the Rose family), seems to have died a natural death.

In the Darwin–Henslow correspondence of this period, we have several allusions by Darwin to what he calls 'your little girls' – that is the village botany class – who are to be employed collecting wild seed for the great scientist. Here is a fascinating example, in a letter to Henslow dated 2 July 1855:

> My dear Henslow,
> – Now it has occurred to me that it would be an interesting way of testing
> the probability of sea-transportal of seeds, to make a list of all the European

plants found in the Azores, – a very oceanic archipelago – collect seeds and try if they would stand a pretty long immersion. – Do you think the most able of your little girls would like to collect for me a packet of seeds of such plants as grow near Hitcham, I paying, say 3d for each packet; it would put a few shillings into their pockets and would be an ENORMOUS advantage to me, for I grudge the time to collect the seeds, more especially, as *I have to learn the plants!* The experiment seems worth trying; what do you think? Should you object to offering this reward or payment to your little girls? You would have to select the most conscientious ones, that I might not get wrong seeds . . .

My dear old Master,
Yours affectionately,
C. Darwin.

P.S. Perhaps 3d would be hardly enough; and if the number does not turn out very great it shall be 6d a packet.[17]

Apparently the exercise went ahead, and a list of 22 species common to both Hitcham and the Azores was agreed between Henslow and Darwin. Although we do not know exactly what 'the little girls' were able on this occasion to collect for Henslow to send, we do know that Darwin was very impressed by the material supplied, for we find in a letter to Henslow dated 28 July the following ecstatic tribute: 'What wonderful, really wonderful little girls yours are in the Botanical line.'[18]

Some of the 'wonderful little girls' went on to become pupil-teachers who received their first training at Henslow's school, which was recognised in 1856 as a 'suitable establishment for educating pupil-teachers'. Russell-Gebbett gives details of the success of Hitcham's first pupil-teacher:

The achievements of Hitcham's first pupil-teacher, Harriet Sewell, certainly help to substantiate that all-round and effective education had been provided by the school. In 1861, the *Report of the Committee of Council on Education* notes under the list of candidates awarded Queen's Scholarships (about eight hundred training college places were available to young women in this scheme), 'Females . . . Sewell, H., Hitcham.' She was twentieth in her year. Harriet Sewell did not remain in school teaching all her life and after a short while became governess to the Bryce family with whom she travelled widely . . . 'Aunt Harriet's scrap book', still treasured at her Hitcham home, contains a captivating log of her journeys written in an effective and educated prose style conveying colourful description and thoughtful reflection. Her education, received entirely at this little village school, took her even further professionally and it is recorded in the scrapbook that the latter part of her life was spent as superintendent of a 'Home for ladies whose means had dwindled'. This Victorian lady, revered by her relatives to this day, was strong in character and a keen, tenacious worker. Her achievements were certainly due to her own persistence and ability but the opportunities to develop her talents and encouragement to do so were first provided by the professor and his school.[19]

One facet of this story which remains obscure is the proportion of boys to girls in Henslow's voluntary class. That there should be a preponderance of girls is obvious enough, for what applied to attendance at the village school in general must have applied even more to the botany class, namely that most parents could see a clear advantage in education for their daughters, many of whom would go 'into service' in the houses of the rich, but no corresponding advantage for the boys. Darwin was not right, however, in designating them *all* as Henslow's 'little girls'. We know of one boy who attended the Botany class, because his name appears on a 'schedule' issued by Henslow to pupils 'entering the third class'.[20] Walter Barton was the second son of Robert Barton (described as 'labourer') and his wife Susan; born in 1848, he would have been 11 or 12 on 20 June 1860, the date of the schedule. One Barton family remains in Hitcham to the present day, descended from Walter Barton's brother Joseph.[21]

Although this pioneering venture in teaching science in schools seems to have received general support outside the parish, both from official schools inspectors and eventually from the highest levels of the state and monarchy, there was an undercurrent of adverse criticism at the more parochial level. George Henslow is at pains to counter such criticism in the eighth of his Reminiscences, entitled 'The Minister'. In a footnote, he tells us that

> A clergyman once said to me at a clerical meeting: 'Hitcham never knew what Christianity was till your father died.' I thought silence was the best reply; but it seemed to me that the speaker was somewhat wanting in Christian spirit himself.[22]

As an example of more balanced criticism, we may cite the answer given by the Churchwardens to one of the series of questions they were asked to answer at the Bishop's Visitation in 1854. Concerning 'the Minister's' performance, John Harper and Robert Ennals wrote:

The Rector is resident in the parish. Divine service is regularly performed at proper hours and the sacrament is duly performed also the duty is performed strictly in baptisms marriages and burials but we must think the sick cannot be so well attended as they might as the Rector is so frequently absent from his parish. The children are instructed in the catechism and prepared for confirmation.[23]

It seems that there were two strands of adverse criticism, to some extent intertwined. The Churchwardens' case, which must have been that of other parish residents, was that their resident Rector spent too much time on pursuits outside the parish (such as, for example, the Ipswich Museum) and neglected his pastoral duties. The other critical strand was somewhat anti-scientific: science was seen as potentially dangerous and 'ungodly', and Henslow's liberal theology was seen by some as misleading his flock, to whom he should be preaching a straightforward message of individual salvation through repentance.

Part of the botanical training that Henslow's pupils received was in the preparation and mounting of good herbarium specimens, often collected on special excursions within the parish. Henslow kept a list of all the plants he and his pupils found within Hitcham Parish, which amounted to more than 400 species, and members of his Botany class were encouraged to collect, press and label as many of these as they could find. Most of these are still preserved in the Ipswich Museum, and cover the years 1838 to 1857. Alec Bull, an East Anglian botanist, lived at Hitcham a century after Henslow, and has provided us with a very interesting comparison: although his list contains almost as many species as Henslow, the losses of native species and the spread of 'new' casuals and plants of garden origin mean that there has been a fundamental change.[24]

Many of Henslow's records were from the ancient 'Hitcham Wood', which survived in reduced size (c. 100 acres) in Henslow's time, but was cleared and converted to arable land about the time of his death.[25] (Figures 33 and 45, Plate 12)

In 1857 was published Henslow's *Series of Nine Botanical Diagrams*. These diagrams were based on 'rough sketches' prepared by Henslow, with the help of his daughter Anne Barnard, for the artist Walter Fitch, and the publication was undertaken by the Committee of Council on Education of the Government Department of Science and Art. They are very early examples of printed wall-charts (Wandtafeln) mainly of German origin that were

FIGURE 45 Hitcham Great Wood: aerial photograph (1940).

much used in teaching botany in schools and universities in late Victorian times.[26] Technically, the Henslow–Fitch Diagrams represent an innovation; they were too large (40″ × 30″, i.e. roughly 100 × 75 cm.) to be printed on a 'litho' stone, and Fitch drew them directly on to a large zinc plate.[27] (Figure 46)

During most of his life at Hitcham, Henslow pursued, together with several colleagues who were also members of a special British Association Committee, experiments on the dormancy and longevity of seeds. Henslow was very sceptical of the claims concerning the so-called 'mummy wheat' samples from Egyptian tombs, insisting that experiments involving germination trials need to be conducted with scrupulous care. A paper *On the Supposed Germination of Mummy Wheat* was one of his last botanical writings.[28]

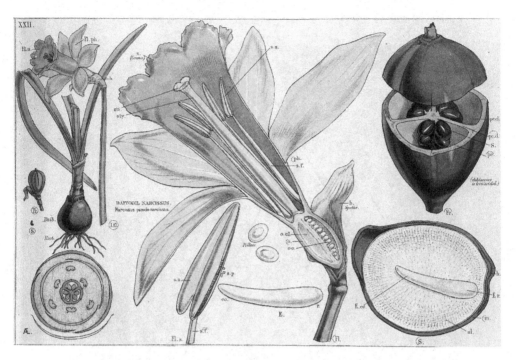

FIGURE 46 One of the Botanical Diagrams published in 1857.

Henslow's last botanical publication, which appeared in 1860, less than a year before his death, is a joint *Flora of Suffolk* for which his fellow-author is Edmund Skepper, a local botanist in Bury St Edmunds. The Preface to this work, signed by Henslow at 'Hitcham, 4th June 1860' reveals that Skepper has done all the work of assembling the records, and seeing the book through the press. Henslow says frankly: 'my part in the work has been that of a consulting, but, otherwise, sleeping partner', and goes on to use the occasion as another opportunity to give publicity to his botanical classes – 'Whoever may be desirous of seeing the plan we have adopted at Hitcham, will find the children at their botanical exercises every Monday at three o'clock.'

The title-page states, correctly, that it is 'a catalogue of the plants (indigenous or naturalized) found in a wild state in the County of Suffolk'. 'Hitcham' is given frequently as a locality, but there is little habitat comment. One interesting feature is that *all* plants are included, even bryophytes, lichens and algae.[29]

In his later years, Henslow corresponded with his friend Richard Dawes, by now Dean of Hereford, the Downing connection of which we wrote in

Chapter 7. From Dawes' letters which have survived,[30] two main subjects
are Henslow's concern to find a school-teacher in 1853, and his hope that
Dawes can find a suitable curacy for his son Leonard (1856).

In addition to such friendships made in Cambridge and, through his
botanical interests, in a much wider field, Henslow found a quite different
friend in Thomas Martin, a medical practitioner in Reigate, Surrey, with
whom he exchanged a long correspondence between 1845 and 1858.[31]
Martin was born in 1779 and, though a much older man than Henslow, outlived him, dying in 1867 not far short of his 90th birthday. We do not know
how they became acquainted, but a letter from Henslow dated 30 April
1845 suggests that Martin might have taken the initiative to write to
Henslow sending him details of one of the voluntary clubs he had helped to
set up in Reigate. In his reply, Henslow says:

> I am much obliged by your communication   The interchange of our ideas
> on such subjects cannot but have its uses where we understand each others
> motives   Your field is much larger than mine as you will see by the
> enclosed account (which I print annually) for the use of the parish as I think
> it tends to inspire an interest among some who would not otherwise be
> brought to understand the truth of its being 'more blessed to give than to
> receive'.

From this beginning the correspondence flourishes. Henslow does not
restrict his news to his activities to help the Hitcham parishioners. Thus on
25 August 1853 he writes, about his 'botanical girls', quite light-heartedly
as follows:

> You would have enjoyed seeing them on Friday exploring the margins of
> a large pond in Brettenham Park where we spent the afternoon & added
> 2 species to our Flora, winding up the day with Bread & Cheese, jam and
> cake, Lemonade & Raspberry Vinegar [sic] & bidding goodbye to the soli-
> tary woods & [glen] with a few songs & hymns.

In the following year, he unburdens himself about his health, as we shall
see below, and in a later letter, acknowledging papers Martin had sent him
about his Reigate clubs, Henslow produces this further light-hearted piece
about village reactions to Cambridge:

> I have separately asked 3 of the 4 little Botanists of my School whom I took
> to Cambridge, what they liked best – & each replied they could not tell, as

everything was so interesting. On asking what would you wish to see again if I said I would take you to see only <u>one</u> of the objects you saw – they each immediately said the 'Botanic Garden' & in a way that convinced me that this was their <u>preference</u>, acquainted as they are with the grouping of our wild flowers, they readily understood the merits of our classifying all plants into natural groups, & could appreciate what they saw in the Garden more than they could of the architecture of the buildings however striking. I have been interested in hearing which different parties preferred. The Pork butcher exceedingly admired the buildings but his chief preference was the anatomical museum, & the wax models of subjects under dissection! Several mistook these for real subjects, & one asked if one was wax, or a newly-killed . . .'

A much more serious matter is aired in Henslow's letter dated 18 February 1857:

We are sadly afflicted in the parish with a low typhus – & I have just returned from burying a stout hale man of 40 who died on Sunday leaving a widow & 2 children dangerously ill – I hardly think that one lad of 14 will recover but the other, a nice little girl of about 10, seems to have turned the corner after 3 or 4 weeks illness & is smiling again. The widow was an old servant of ours & lost two children last year from measles and whooping cough. Luckily her eldest daughter is in Cambridge with her Grandfather or she also would most probably have taken the fever – It has been brought this year into the parish from Stowmarket . . .

By the end of 1853, the strains of over-exertion forced Henslow unwillingly to seek medical advice. There is clearly evidence of this in a letter to William Hooker on 21 November, and in a letter to Whewell on Christmas Eve, 1853, we read that 'the doctors advised me to take a little relaxation – so I have been to see Darwin . . . I am told my chest pains are neuralgic and entirely due to over-exertions and without any ugly organic symptoms – I must in consequence take a few irons out of the fire – go to bed earlier and attend to all the etcs.' But it is, quite understandably, to his medical friend Thomas Martin that Henslow gives most information. Writing on 8 February 1854,

I have had rather too many Irons in the fire for a man now 58, & for the first time in my life, my constitution has given me warning that I am getting old. About two months ago a pain in my chest made my medical attendant apprehensive that I might have some organic affection of the heart, but he and Dr Billing (who examined me) now think it arose from dyspepsia. I can now walk about freely again. Instead of going to bed at 1, I get there as soon after 10 as I can – live a little more freely – & clothe myself more warmly – & I trust with a little less work than heretofore I shall before long find myself re-established – indeed I have only slight traces of the 'Affection' left.

Jenyns seems to have been unaware of this early medical warning – or perhaps his estimate of when Henslow first complained of chest pains and took medical advice was just inaccurate, for he tell us that

> for much of the latter part of Professor Henslow's life his mind seemed always on the stretch while his outdoor exercise was hardly sufficient to preserve health under such circumstances. It was *about five years before his death* [our italics] that he first complained of an affection of the chest, for which he took advice, and which obliged him for a time to live by rule, and to avoid over-exertion.[32]

There seems to be good evidence that Henslow felt himself to be over-worked and overstrained in 1855: his father had died in 1854, his mother was seriously ill and Henslow and Harriet moved over to Bildeston to look after her. She died in January 1856. A letter to Knights about arrangements for the Ipswich Museum lectures dated 1855 certainly reveals a man under great strain:

> I write in a great hurry having done nothing for the last 3 days (and almost nights) but get ready a set of specimens for a lecture and demonstration at Stamford. I have now finished the job and packed up. I must seriously protest against the lecture programme [Here follows a mass of detail about lecture arrangement, and the letter finishes]: The Stamford job has set me 3 days behind, and made me half ill, and now I am sent for to go to Lavenham to inspect a set of minerals for sale – and then something else will turn up, as generally does once and a half per day that I can't find time really for what I consent to do and must not take more upon me.

Russell-Gebbett, who quotes the letter *in extenso*, comments that poor Knights had some difficulties over Henslow's suggested re-arrangement of the lectures. Henslow in his agitation had given the incorrect date![33]

All was, of course, by no means 'doom and gloom'. On 15 July 1851 his eldest daughter Frances married Joseph Hooker, son of Sir William, in

Hitcham Church, and their first child, William Henslow Hooker, was born on 24 January 1853, followed by a daughter, Harriet Ann, on 23 June 1854. These two grandchildren, both born at Hitcham, must have been a great joy to the Henslows. After his health warning in the winter of 1853, Henslow relinquished for the next few years his regular Church duties to his eldest son Leonard who performed his first Curacy at Hitcham.

Henslow's museum enthusiasms led to his preparing a set of wax models illustrating the structure of fruits for the Paris Exhibition of 1855. This followed a request by Lyon Playfair, Secretary of the new Government Department of Science and Art, that he should send some exhibit of his models used in his botanical teaching. The venture got off to an unfortunate start, when the printed Catalogue of the Exhibition produced something of a 'howler', as Henslow pointed out to Knights in a letter of 3 March 1855:

> In the printed catalogue they have assigned my specimens to a position among surgical instruments under the heading 'Carpological Apparatus'. Dr. Playfair says he will see that this error is corrected, and that they shall be placed among objects intended for educational purposes . . . I am also designated as Professor Henslow of Hadleigh! So easily are blunders committed.[34]

But worse was to come. It was a hot summer in Paris, and apparently the models, which had managed to travel to the Exhibition intact, began to melt and deform in the heat. All was not lost, however. One Corporal Key, presumably a minor English official at the Exhibition, succeeded in rescuing and restoring the exhibit, and a mollified Henslow paid a visit to the Exhibition later in the summer together with a number of scientific colleagues. Henslow received a medal for the 'Carpological Illustrations', and eventually some refund of expenses for preparing two sets, one for the Exhibition and a set for the Education Museum in London. The visit to Paris, accompanied by three of his children, George, Louisa and Anne, seems to have been, after all, quite a happy and successful event, only marred by a stormy passage, as he describes in a letter to Sir William dated 17 October 1855. They stayed at 'the Hotel de Londres on 3 rue Bonaparte', and Henslow continues:

> If you get the room I occupied I think you will find yourself perfectly comfortable – so far as I saw of the others, they were all clean & comfortable

enough – The only drawback in the hotel is the usual wretched style of non-water closet – It was clean but small & offensive . . . & I generally migrated to the Palais Royale after breakfast for a quarter of an hour . . . We always dined at the Caffee [sic] Caron a short way off at a corner in Rue St Peres, (which seems parallel to Rue Bonaparte). Our dinner varying (for four) from 17 to 26 francs, generally we made them 21 francs (say 5 pence each per day) . . . We enjoyed ourselves very much at all times except when paying our respects to Mons Neptune who was somewhat uncivil especially on our return & had induced Boreas to delay us an hour or so beyond the usual period for passing. 7½ hrs passage in stormy weather (from Dieppe to Newhaven) is no joke to non-sailors. We have kept ourselves within bounds of the money set apart by each for this trip & I have found the facilities for getting to Paris so much improved since there 28 years ago that I feel quite inclined for another attempt when another opportunity may offer itself. I went to the Horticultural Show on last day – & saw the model of the germinating embryo of the cocoanut – It certainly is not equal to that my sister had prepared . . .[35]

A large extension of Henslow's museum interests developed from his friendship with William (later Sir William) Hooker, eleven years his senior, a friendship that had begun as early as 1827 when Henslow consulted Hooker, then Professor of Botany in Glasgow, about the plans for the projected New Botanic Garden in Cambridge. This initial contact developed into a firm personal friendship, as evidenced by the exchange of letters about some of Hooker's students coming up to Cambridge, and the possible influence of Charles Simeon, as we saw in Chapter 9.

The Hooker–Henslow correspondence, fortunately preserved, reveals a more or less continuous exchange of ideas and a developing personal friendship between the two families. The final years in Glasgow were depressing for William Hooker, but his fortunes turned dramatically in 1840 when Lord John Russell, heading the Whig administration, called him to London to offer him the post of first Director of Kew Gardens, a post he took up early in the following year. The story of the Hookers, father and son, successive Directors of Kew throughout the rest of the nineteenth century, during which Kew became the leading institute for botanical taxonomy in the world, is, of course, well-documented elsewhere. Our concern is with Henslow's friendship, and how this developed to mutual advantage. One link had developed early, when Elizabeth, fourth child of

William Hooker, went to the same school as Frances Henslow, where they became firm friends. This led to visits to Hitcham by Elizabeth, who like many of her contemporaries suffered from poorly-defined bouts of ill-health and was constantly being sent 'for her health' to relatives and friends who lived in places where the air was judged to be purer than in London. In this way, Joseph Hooker came to know Frances Henslow, a charming young lady eight years his junior, and the acquaintance blossomed at Oxford during the meeting of the British Association there in 1847, to which Henslow took Frances, and Joseph took Elizabeth. They became engaged, to the mutual satisfaction of both families, though Joseph's prolonged botanical stay in India meant that the marriage was postponed until after his return in 1851. As Mea Allan tells us:

> Three days before their marriage they sat happily side by side at the British Association meeting at Ipswich on the 12th of July 1851. Frances was just the wife for him. The daughter of such a father, she was both by birth and training able to help in his work and share all his scientific aims and enthusiasms. To Lady Hooker 'dear Frances' was 'an affectionate and amiable girl, who seems to take to us with a hearty good will'.[36]

Throughout the early 1850s Henslow and Sir William (he was knighted in 1836) corresponded largely about the material Henslow was supplying to Hooker's new Museum of Economic Botany which opened to the public in 1847. As an example of the enthusiasm Henslow showed in supplying both materials and ideas, we select just two, intentionally diverse, cases. The first, an exhibit of a local craft using plant material, is referred to by Henslow in his letter to Hooker dated 14 October 1853: 'I have got the wonderful Hitcham rush brooms for you – they are the simplest of the simple – & the woman thought herself overpaid when I gave her 6d for 4 of them.'[37]

The second example is quite different. It arises from the publicity given to the arrival in Britain of the first Wellingtonia trees (*Sequoiadendron giganteum*) in the early 1850s. These famous giant trees from California, which the Victorians wished to name after the Iron Duke, naturally caused a sensation, and Hooker at Kew was much involved in receiving reports and specimens of wood and bark from mature trees, including a trunk and sections showing annual rings. Henslow was soon informed, as a series of letters in August and September 1856 reveals, and the correspondence culminates in a letter from Henslow dated 15 September in which Hooker is encouraged to construct at Kew

(a)

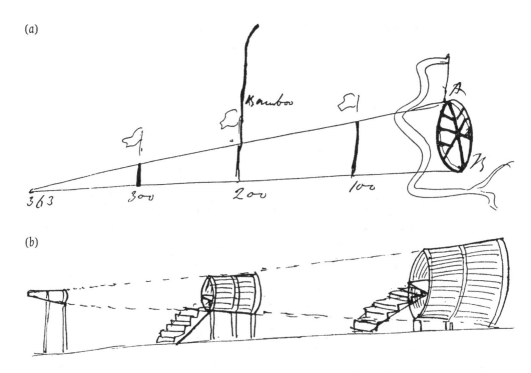

(b)

FIGURE 47a & b Henslow's sketches for *Wellingtonia* exhibit at Kew.

> ... an interesting model ... to show the dimensions of the Wellingtonia.
> I judge from the slight attempt I have made on my lawn, which you could
> much improve upon & make more instructive and attractive ... I have
> merely had the carpenter construct a circle with planks, 31 feet in diameter,
> to be raised to represent the section of the trunk, & had poles at intervals to
> mark the diameter at distances of 100 feet.

(Here follows a sketch: Figure 47a.) Not content with this, Henslow's
imagination runs riot, and he continues:

> In a more perfect scheme you might select a better (& better executed)
> series of trees & objects & place them into a room with a modelled portion
> of the Wellingtonia as follows: [Figure 47b]. If 3 or more sections were
> thus framed & <u>raised</u>, so that the diameter was horizontal, a perfect idea
> of the size would be conveyed. It might be constructed somewhere in the
> shrubbery, & fitted up as a room with instructive representation of trees.[38]

After this *tour de force*, Henslow signs off abruptly, showing every sign of
exhaustion!

We have no record, unfortunately, of how Sir William took this suggested addition to his empire at Kew. Perhaps it would have been better to draw a veil over Henslow's grandiose design, but that would have been a pity: surely it is impressive that he should continue to enjoy even great impractical visions so near the end of his life. Unfortunately, the tone and handwriting of the letter favours a rather more sombre interpretation: is this not more likely to be evidence of a talented man whose mind is beginning to lose some of its control?

The death of his wife Harriet on 13 November 1857 after a long and varying period of ill-health must have made the winter of 1857–8 a particularly difficult one for Henslow, although his eldest daughter Louisa, now thirty years old, was obviously competent to run the Rectory household – and indeed because of her mother's frequent illnesses must already have been in charge for some years.

We have good reason to think that Henslow's collecting for teaching and museum display must have continued to provide for him an absorbing relaxation even after his health warnings. Plate 13 shows a remarkable teaching specimen, fortunately preserved in the Cambridge University Herbarium, of the fern *Polypodium vulgare*, collected by Henslow, or one of his botanical pupils, on 20 March 1860, and mounted together with a colour plate and pencil drawings of sporangia and other parts.

In Henslow's last two years he was much involved in an archaeological controversy centred on the flint implements found in quantity at an old brick-pit site near the village of Hoxne, near Diss, Norfolk. Jenyns devotes nearly six pages of the *Memoir* to this affair, which he calls 'celts in the drift.'[39] Looking back on this controversy, which concerned many scientists from John Frere in 1800, whose original paper in *Archaeologia* was ignored for over 50 years,[40] one has to conclude that Henslow's contribution was a relatively minor one, and that his reluctance to accept the evidence that the flint implements or 'celts'[41] were many thousands of years old and therefore strong evidence for the antiquity of man was part of his general concern about the problem of human evolution. Henslow visited the Hoxne site in November 1859, and contributed a series of letters to the *Athenaeum*, in which he originally disputed the conclusions of Prestwich and Evans, but gradually modified his objections. The Hoxne 'celts' were the subject of Henslow's last public lecture, given in Ipswich on 13 February 1861, following a visit he paid to the famous gravel-pits at Amiens and Abbeville in Northern France in the autumn of 1860. From Jenyns we learn that he

> is said to have been preparing to lay his final conclusions on the celts question before the Cambridge Philosophical Society at the time of his being taken ill; and Dr [Joseph] Hooker 'believes that he had, at last, convinced himself that these implements belong to a period long antecedent to that usually attributed to man's existence on the earth, though by no means so distant as some geologists suppose.'

For a detailed, modern assessment of the controversy, Van Riper 1993, especially Chapters 5 and 6, is strongly recommended.

Henslow's terminal illness began in the middle of March 1861, after he had caught cold on a visit to 'friends in the South of England'. He reached home on Saturday 23rd, and was ill with chest pains and breathing difficulties. By Thursday 28th he was so obviously worse that his local doctor summoned Joseph Hooker immediately; Joseph arrived on the following day, Good Friday, bringing with him a physician from London, Dr Walshe, who said he had 'bronchitis, congestion of the lungs and liver, and enlargement of the heart', and held out no hope of recovery. Henslow lingered on for several weeks, and died on 16 May. In the weeks before his death, his eldest son, Leonard, and his son-in-law, Joseph Hooker, were almost continually with him, and in particular shared the night-time vigils. Joseph found that his father-in-law was always more talkative at night, and reported that he often recalled the events of his life and discussed the successes and failures with him. Leonard kept a diary of his father's conversations during his final illness, which was used by Jenyns in his account of Henslow's last days. Amongst his bedside visitors was Sedgwick, who arrived on 13 April, and found him 'calm, resigned, and quite happy; and though under bodily sufferings, full of peace and love'.

Darwin agonised over the news that Henslow was dying, and even wrote to Joseph Hooker two letters on 23 April, one an immediate offer to come to see Henslow 'if he would really like to see me'. The passage continues:

> The thought had once occurred to me to offer, & the sole reason why I did not was that the journey with the agitation would cause me probably to arrive utterly prostrated.
> I shd. be certain to have severe vomiting afterwards, but that would not

much signify, but I doubt whether I could stand the agitation at the time. I never felt my weakness a greater evil. I have just had specimen for I spoke a few minutes at Linn. Soc on Thursday & though extra well, it brought on 24 hours vomiting. I suppose there is some Inn at which I could stay, for I shd not like to be in the House (even if you could hold me) as my retching is apt to be extremely loud. –

I shd. never forgive myself, if I did not instantly come, if Henslow's wish to see me was more than a passing thought.

My dear old friend | Your affect | C. Darwin

P.S. Judge for me: I have stated exact truth: but remember that I shd. *never* forgive myself, if I disappointed the most fleeting wish of my master & friend to whom I owe so much.[42]

We do not know how Hooker replied to this extraordinary letter, but Darwin never came.

We close with two of Henslow's recorded sayings as he lay ill and increasingly facing the prospect of death with Christian equanimity. The first is regretful of his short life: 'I thought to have a few more years allowed me to carry out my trumpery designs: man purposes, God disposes.' The second was called forth by the busy Spring noises in the adjacent rookery: 'How pleasing to hear the rooks cawing. God has made the world full of pleasing sights and pleasant sounds. He might have made everything disagreeable: it is a very good world if people will only use it aright.'[43]

Epilogue

Henslow's funeral took place on 22 May 1861 at his church at Hitcham, and his curate, the Revd Richard Graves, conducted the service.[1] Jenyns tells us that Henslow had 'desired his funeral to be of the simplest description', and accordingly 'few besides his immediate relatives met at the Rectory to accompany his remains to their last resting-place'. They were, however, joined by 'a large number of the village labourers who, by special request, were permitted to pay this last mark of respect to their departed pastor and friend . . .' Near the churchyard the cortège was 'preceded by a deputation from the town of Ipswich, consisting of the Mayor and several other members of the Museum Committee'. During the funeral service 'the church and churchyard alike [were] filled with the parishioners, whose mournful silence and respectful behaviour testified to the impression made on them by the solemn scene'.[2]

The interment was in the same vault where Harriet was buried, and is marked by a coffin-shaped grave memorial. Next to it, with a similar memorial, is the grave of Henslow's mother, Frances, and his aunt Ann. These graves are in the western part of the churchyard not far from the base of the tower. The grass around the graves is not cut until after the wild flowers have flowered and set seed, so that Henslow's memorial is set in turf with ox-eye daisy (*Leucanthemum vulgare*) and other wild-flowers he would have known and used in his teaching. (Plate 11) Inside the church on the north wall of the chancel is the Henslow memorial plaque, and there is a smaller plaque on the wall of the south porch above the entrance to the church, recording the restoration of the porch in 1882–3 'by parishioners and friends in loving memory of the Rev. J.S. Henslow . . .' There is also a copy of the Maguire portrait of Henslow, at present hung on the north wall of the nave.[3]

It is one of the ironies of the Darwin–Henslow relationship that Darwin's enthusiasm for growing plants and botanical experiment in general, which led him to recall early instruction and influence from his teacher, grew very obviously soon after Henslow's death. It was especially enlarged by the building in 1863 of a hothouse at Down House, which greatly increased the range of plants that Darwin could keep for scientific investigations.[4] To some extent we can gather from his published correspondence how far the influence of Henslow's botany was still felt by Darwin. It is, for example, significant that in a letter to Joseph Hooker dated 3 August 1863 asking for advice on general botanical text-books, Henslow's text-book is first in a short list of what he already possesses: 'I have "Henslow's Botany", – "Asa Gray's Lessons", "Aug. St. Hilaire's Leçons de Botanique": can you name any good book with miscellaneous information for me to get.'[5] Darwin's copies of all these books, with copious annotations, are in the Cambridge University Library.

One comment on Henslow's text-book, which shows that Darwin is not averse to negative criticism of his mentor, comes in an earlier letter to Hooker (27 October 1862): 'By the way what a fault it is in Henslow's Botany that he gives hardly any references: he alludes to great series of experiments on absorption of poisons by roots, but where to find them I cannot guess. Possibly the all knowing Oliver may know.'[6]

Looking at the passage in Henslow's text-book (pp. 68–9, where the 'sensibility' of plants is under discussion) one feels much sympathy with Darwin. Perhaps Henslow's defence, if this sort of criticism had surfaced in his life-time, would have been that a student text-book should not be overloaded with references!

An example of how Darwin goes to Henslow's text-book first for his descriptive morphology is afforded by another letter to Hooker, 26 July 1863. Darwin is studying climbing plants, and having difficulties with ter-minology: 'I see Henslow says tendrils of Cucurbitaceae are stipules; Gray branches, Thomas leaves – : what is a poor devil to believe? Have you by any chance seed of *Lathyrus aphaca*: it would be good for my purpose.'[7] (Figure 48)

It was not only Henslow's botany that Darwin recalled. Several years later, there is a charming exchange in the Darwin–Hooker correspondence about the prehensile tails of mice. To a letter to Hooker on 23 July 1871, Darwin adds the following P.S.: 'Henslow used to keep tame field-mice, & I distinctly remember him telling me that they used their tails as prehensile organs, when climbing up a branched stick in their place of confinement. Will you ask Mrs Hooker whether she remembers them? I want to know what species of mouse they were & especially whether Henslow published in any of the popular journals an account of their habits?' Hooker's reply, apparently the same day, reads as follows: 'The mouse which I remember

DESCRIPTIVE BOTANY.

In the *Lathyrus aphaca* (*fig.* 69.) all the leaves become tendrils, except the first pair in the young plants, which are compound, and have two or three pairs of leaflets. Occasionally an odd leaflet (*b*) is developed on the tendrils, in a later stage of growth, which further indicates the origin of the organ on which it is seated. A provision is made for supplying the want of leaves in this plant, by an unusual development of the stipules (*a*), which are so large that they might readily be mistaken for real leaves.

FIGURE 48 *Lathyrus aphaca* woodcut from Henslow's text-book.

well was the Harvest mouse, M. Messorius & you are right about its caudal attributes . . . Henslow never published about it.'[8]

Henslow's geological observations feature in Darwin's correspondence in the following year. Here is Darwin writing to Amy Ruck on 24 February 1872:

> My dear Amy, I want you to observe another point for me, so you see that I treat you as my geologist in chief for North Wales.
>
> The late Prof Henslow, who was a very accurate man, said that he had often observed on very steep slopes, covered with fine turf, (such as may be found in mountainous countries & no where else) that the surface was marked by little, almost horizontal, sometimes sinuous & bifurcating ledges; or as he called them, wrinkles. These are commonly attributed to sheep walking in nearly horizontal lines along the sloping surface; & they are undoubtedly commonly used by sheep, but Henslow convinced himself that they did not thus originate. Dr Hooker, to whom Henslow made these remarks, has since observed such little ledges on the Himalayal & Atlas ranges, in parts where there were no sheep & few wild animals . . . Henslow speculated that the earth, beneath the turf, was in some manner gradually washed away; & he compared the wrinkles on the turf to those on the face of an old man whose face is shrunk. I cannot possibly believe in this notion.[9]

We do not usually think of Henslow as a student of mountain vegetation, nor is it obvious where he observed, and correctly interpreted, the phenomenon now generally known as solifluxion or 'soil creep'. He seems to have been well ahead of his time in interpreting this mountain phenomenon:

the term 'solifluxion' was apparently first used by T.G. Taylor in 1916, writing on Scott's Antarctic Expedition![10]

The 'all-knowing Oliver' referred to in Darwin's letter to Hooker is obviously Daniel Oliver, whom Hooker appointed as Assistant Keeper at Kew in 1858 at the age of 28, and who became Keeper in 1864. Oliver was persuaded by Hooker and George Henslow to prepare for publication Henslow's manuscript of which Jenyns tells us: '[Henslow] was engaged, at the time of his death, on yet another work on School Botany, the title of which was to have been "Practical Lessons in Systematic and Economic Botany, Educational and Instructional, for the use of Beginners in Village Schools and upwards." This little work had been long promised, and was left in a forward state, but not ready for publication.'[11]

Oliver did not have an easy task. He was working on the book in 1863, as is revealed in a letter from Hooker to him dated 12 August. Oliver has decided on his plan and has obviously sent it to Hooker to get his approval: 'Your formula is clear and sufficient ... All you can do is to preserve the spirit of Henslow's plan ...'[12]

As Oliver tells us in his Preface to the 'Lessons in Elementary Botany' which was published, under his name only, in 1864, he has used Henslow's material for 'the systematic portion (Part II) [based] upon the Type Lessons, which formed the largest and most valuable part of Professor Henslow's manuscripts'. Part I of the book, 'embracing the elements of Structural and Physiological Botany', is Oliver's own work. Since these six chapters are more than one-third of the book, Oliver's decision to call the book his own, and only mention the Henslow manuscript in the Preface, seems quite reasonable. The value of the book is greatly enhanced by the numerous small illustrations, which resemble those in Henslow's 1835 text-book. The reason is clear when we read the following in Oliver's Preface:

> Most of the excellent woodcuts employed in this work were drawn by Professor Henslow's daughter, Mrs. Barnard, of Cheltenham, from the admirable Sheet Diagrams designed by Professor Henslow, and executed by Mr. Fitch, for the Committee of Council on Education. They have been liberally placed at the disposal of the publishers by the Rev. George Henslow.[13]

Henslow would surely have been very pleased with the success of 'his book' transmuted by Daniel Oliver and copiously illustrated by his daughter

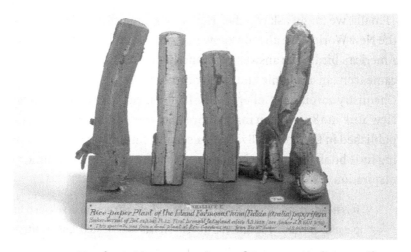

FIGURE 49 Henslow's Museum specimen of *Tetrapanax* (*Aralia*) *papyrifera*.

Anne. The book was reprinted six times between 1869 and 1876, when
Oliver prepared a second edition in which he made 'some little alterations,
chiefly in the chapter on minute structure and vital processes'. The second
edition itself was reprinted several times.

In spite of Henslow's great and abiding interest in museums, his own
'botanical museum' in Cambridge was inadequately housed and seriously
neglected both during and after his life-time. Babington seems to have had
no interest in rescuing Henslow's museum specimens, and Cambridge had
to wait until 1904, for the opening of the new Botany School, for both
Herbarium and Museum to be effectively housed in the new building.
Walter Gardiner's paper (1904) dedicated to Henslow as the 'Founder of the
University Museum of Botany', describes how he, in or around 1885, and his
colleague M.C. Potter, ignorant of any 'Henslow Botanical Museum', consulted Joseph Hooker at Kew on 'the desirability, or rather the necessity, for
the establishment of a Museum of Botany at Cambridge'; they were told by
Hooker that 'such a Museum . . . was actually in existence at Cambridge and
had already been founded by Henslow, during the tenure of his office as
Professor of Botany'. Armed with this information 'we two youngsters
invaded the recesses of the Professorial sanctum, and there discovered what
I can only describe as the wrecked remains of Henslow's Collection'.[14] This
revived and enlarged Botanical Museum lasted less than 50 years. History
repeats itself, and we have again now only 'the wrecked remains of
Henslow's Collection'. Ironically, more Henslow specimens seem to
survive in the Museum of Zoology than in his own Department[15] (Figure 49).

Finally, we might ask whether Henslow's fame *as a teacher* ever reached the New World. (We already know that he is commemorated in a small American bird!) The answer is definitely 'yes'. Eliza Ann Youmans, who came from an academic family (her brother Edward was Professor of Chemistry at Antioch College, Ohio) published *The First Book of Botany* in New York in 1870, which ran to more than one edition, including one published in London in 1872 with a second edition in 1880. This charming little book (Figure 50) contains a Preface explaining the author's inspiration:

> It is needful here to state that the method of instruction developed in these pages is no mere educational novelty; it has been tested, and its fitness for the end proposed has been shown in practice. The *schedule* feature which is here fully brought out, and which is its leading peculiarity as a mode of study, was devised and successfully used by Prof. J. S. Henslow, of Cambridge, England. My attention was first drawn to it as I was looking about in the educational department of the South Kensington Museum, in London. In a show-case of botanical specimens, I noticed some slates covered with childish handwriting, which proved to be illustrations of a method of teaching Botany to the young. They were furnished by Prof. Henslow for the International Exhibition of 1851. He died without publishing his method, but not without having subjected it to thorough practical trial. He had gathered together a class of poor country children, in the parish where he officiated as clergyman, and taught them Botany by a plan similar to the present, though less simplified.[16]

It seems that Youmans was unaware, when writing this, of Oliver's book. In her second book, published in London in 1872, she presents 'an essay' which is 'edited' by one Joseph Payne, who is interested in elementary education in general. His footnote on p. 28, questioning Youmans' choice of the *leaf* to begin her botany lessons, explicitly commends Oliver's book for choosing to start with 'an entire plant'.[17]

What of Henslow's influence remains into the twenty-first century? To any Cambridge botanist, his main memorial must surely be the Botanic Garden (Plate 14). Throughout the twentieth century, and particularly through the inter-war years, a succession of enthusiastic teachers ensured that students were at least presented with whole live plants as legitimate objects of study. No-one in this respect was more important than Humphrey Gilbert-Carter, first Scientific Director of the Botanic Garden who

FIGURE 50 Title-page of Eliza Youmans' book.

CHAPTER I.—THE LEAF.

THE pupil will see from the picture what is to be done first and how we are to proceed in commencing the study of plants Having collected some specimens, let us begin with the leaf. Or these printed leaves there is a language which children have already learned ; there is also a language written by Nature or the *leaves that grow :* we will now learn to read *that.*

was a plantsman who enjoyed the game of getting to know new species and, even more, of visiting again 'old friends' on field excursions. The mantle of Ray fitted him comfortably, and he made the local excursion in the Cambridge area once again a delightful, educative experience for generations of students after the First World War. Trees were his great love, and here again he fitted so well into a strong Cambridge tradition, which Henslow had founded and Marshall Ward and others had continued. We recall with pleasure the words with which he addressed his first meeting with the Botanic Garden Syndicate in 1921, recommending the planting of good specimen trees: 'Let us look ahead and think of those who will take our places. Neither shall we ourselves die altogether if the coming generations of students remember us when they study the great trees on the lawn, or the elders of the University and town bless our memories as they rest beneath their shade.'[18]

This is, of course, a parochial view. We must return to our title, and conclude with Darwin's own words. In 1873, in his *Autobiography*, Darwin again recalls his debt to his old teacher and friend, and assesses his ability:

> 'His strongest taste was to draw conclusions from long-continued minute observations. His judgment was excellent and his whole mind well-balanced, but I do not suppose that anyone would say that he possessed much original genius.'

This seems, a century later, a remarkably true assessment, to which we need only add that, for the progress of science as for the welfare of mankind, a flair for teaching which releases genius in others may be as important as the genius itself. Without Henslows there are no Darwins.[19]

Appendix 1
Genealogical tables

(These are set out in the pages following (pp 262–3).)

I: John Stevens Henslow's Family

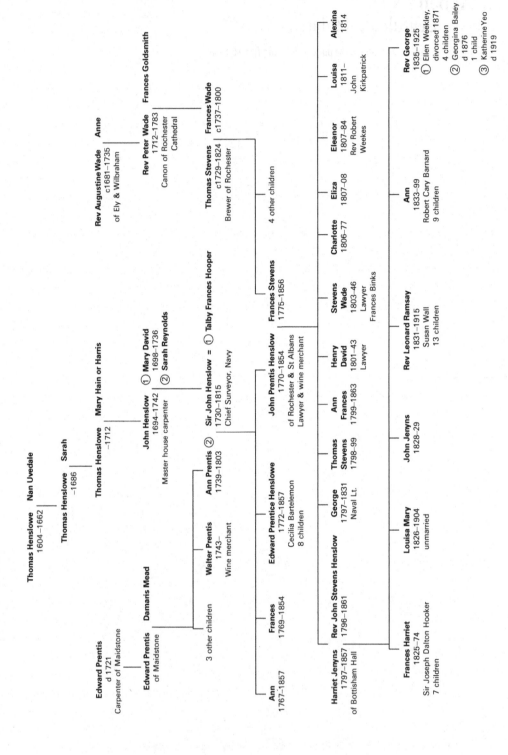

II: Harriet Jenyns' Family

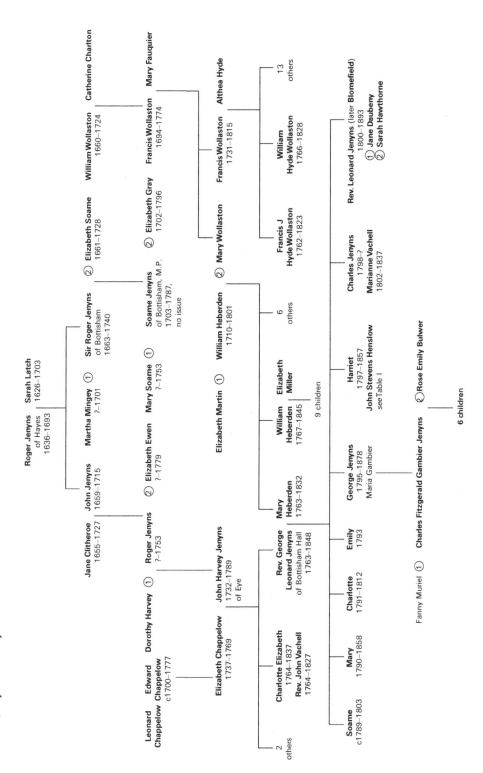

Appendix 2
Chronology

1796 6 February   Henslow born at Rochester, Kent

1797 8 February   Brother George born

1798 Brother Thomas Stevens born

1799 March   Brother Thomas Stevens died

3 December   Sister Ann Frances born

1800 September   Grandmother Frances Stevens died

1801 3 December   Brother Henry David born

1803 8 April   Grandmother Lady Ann Henslow died

October   Brother Stevens Wade born

1805 March   Enters Revd W. Jephson's School at Camberwell as a
boarder (formerly day scholar at Mr & Mrs Dillon's private School in
Rochester, then at Free School, Rochester kept by Mr Hawkins & Mr
Benjamin Hawkins)

Through George Samuel drawing master and amateur entomologist
introduced to Leach (BM) and Stephens (entomologist)

1806 7 May   Sister Charlotte born

1807 Sister Eliza born

1808 July   Sister Eliza died

1809 25 March   Sister Eleanor born

1810 Awarded prize of Levaillant's Travels in Africa

1811 13 July   Sister Louisa born

1814 7 June   Sister Alexina Frederica born and died in Sept

11 October   Entered St John's College Cambridge as a pensioner

1815 22 September   Grandfather Sir John Henslow died

His skill in museum work recognised by Adam Sedgwick

Freshman's prize at John's

1816 5 November   Admitted Gilbert Scholar

1817 Prizeman as a Senior soph

1818 January   BA 16th Wrangler

Fellow of Linnean Society

1819 March/April   Tour of Isle of Wight with Sedgwick

July   Tour of Isle of Man with students

November Helped found Cambridge Philosophical Society
Member of Geological Society

1820 Visited Isle of Anglesey
Returned to Isle of Man for Irish elk (without success!)
First meets Jenyns

1821 MA
May One of secretaries to CPS
July Visits Isle of Anglesey with students
November Read 1st paper to CPS on Anglesey
1 article published

1822 15 May Elected to Chair of Mineralogy succeeding E.D. Clarke
1 article published

1823 15 Feb Delivered 1st course of mineralogical lectures
16 December Married Harriet Jenyns
A syllabus of a course of lectures on mineralogy
1 article published

1824 11 April Ordained deacon by Bishop of Ely in St George's Church,
Hanover Square
October Grandfather Thomas Stevens died
7 November Ordained priest (by same Bishop) in Ely
Perpetual curate of Little St Mary's
Field excursions with Jenyns, esp. Gamlingay, 24 August
1 article published

1825 30 April Daughter Frances Harriet born
26 July Letters Patent Kings Reader [*sic*] in Botany [i.e. Professor]
10 October VC admits Henslow in the Senate House
25 November Appointed Walker's Reader by . . . the Governors . . .
of the said Readership

1826 6 October Daughter Louisa Mary born
1 political pamphlet printed with Lamb: *Remarks on the payment of*
expenses to out-voters

1827 March Gives up Chair of Mineralogy
Still lecturing in mineralogy in Lent Term (University Calendar)
30 April 1st lecture course in botany in Easter term advertised and
delivered
Sir John Richardson's Determination on the method of appointing
Professors

1828 17 August   Son John Jenyns born
      February   Commencement of Friday evening soirées
      7 March   Audubon present at a Henslow soirée
      Publication of *Syllabus of . . . Botanical lectures*
      2 articles published
1829 March   Son John Jenyns died
      Darwin attends lectures for the first time
      Michaelmas term: Darwin attends soirées
      Audubon names Henslow's sparrow
      Catalogue of British plants
      1 article published
      1 sermon printed
1830 2 articles published
1831 30 March   Act of Parliament for transfer of 38 acres from Trinity
      Hall to University for Botanic Garden received Royal Assent
      19 Jun   Son Leonard Ramsay born
      1 August   Marmaduke Ramsay dies
      Brother George (Naval Officer) dies
      Suggested Darwin for 'Beagle' voyage
      Elected Governor of Botanic Garden
      Joined British Association
      Visited Weymouth for long stay
      5 articles published
1832 Poll-book gives address as Regent St
      9 articles published
1833 23 June   Daughter Anne born
      Vicar of Cholsey-cum-Moulsford
      Received first box of Darwin specimens
      Sketch of a course of lectures on botany
      2 articles published
1834 1 political pamphlet printed
      1 article published
1835 23 March   Son George born
      Informed against Conservative agents guilty of bribery in election
      "Beagle letters" published by C.P.S.
      1 article published
1836 *Principles of descriptive and physiological botany*
      Catalogue of British plants, 2nd ed.

16 June   Brougham suggests exchange of living at Cholsey with
that of Michell at Histon
7 articles published

1837   17 May   Michell withdraws from Histon/Cholsey exchange
Advertises private classes in Botany
Poll-book gives address as St Andrew's Place (:Park Terrace)
Rector of Hitcham
1 political pamphlet printed
1 article published

1838   Visit to Bartlow
1 article published

1839   Gave up house in Cambridge and moved to Hitcham

1840   Village library
1 article published

1841   Village school
Report on the diseases of wheat
4 articles published

1842   Hadleigh Farmers Club
1 sermon printed
4 articles published

1843   Brother Henry David (lawyer) died in St Albans
Visited Felixstowe and discovered coprolites
Letters to the farmers of Suffolk
Opened Eastlow Hill tumulus at Rougham
"An account of the Roman antiquities found at Rougham"
3 sermons printed
4 articles published

1844   Investigated clover crop failure
2 political, 1 archaeological and 2 sermon pamphlets printed
4 articles published

1845   Village allotment scheme opposed by farmers
Failure of potato crop
2 political pamphlets printed
1 article published

1846   Brother Stevens Wade (lawyer – Clements Inn) died in London
Address . . . [on the] . . . Botanic Garden
Official opening of New Botanic Garden
Interested in move to establish a Natural Science Tripos

2 sermons printed
2 articles published

1847   Ipswich Museum founded
Daughter Frances meets Joseph Hooker again at BA meeting at Oxford
Subsidised excursions from annual tithe dinner money
1 article published

1848   Hitcham excursion to Ipswich by rail
Address . . . in the Ipswich Museum
Syllabus of a course . . . on botany . . . for a pass examination
2 articles published

1849   Secured 16 acres of Hitcham Charity land for allotments
2 Hitcham pamphlets printed
3 articles published

1850   First Hitcham Horticultural Show
Secured 20 more acres for allotments
Adult literacy classes started with help of daughter
President of Ipswich Museum
Controversy over dating Saxon pottery etc.
2 Hitcham pamphlets printed
3 articles published

1851   15 July   Daughter Frances married Joseph Hooker at Hitcham
Natural Science Tripos instituted
British Association meeting at Ipswich (Prince Albert attends)
3 day visit to Great Exhibition, Zoo and Kew
Questions . . . in botany . . . for a pass-examination
4 articles published

1852   Teaching botany in the village school
1 Hitcham pamphlet printed
1 article published

1853   24 January   Grandson William Henslow Hooker born at Hitcham
Syllabus . . . another edition
December   Period of illness

1854   17 April   Father John Prentis Henslow died in St Albans
Aunt Frances died
23 June   Grand-daughter Harriet Ann born at Hitcham
Hitcham party to Cambridge incl. pamphlet guide

Qualified teacher appointed to village school

3 articles published

1855 6 September   Sister Eleanor married Robert Weekes in
Woodbridge

Sent teaching models to Paris Exhibition

Visited Paris Exhibition with son and daughter

2 articles published

1856 January   Mother Frances died in Hitcham

3 articles published & 14 notes to the *Gardeners' Chronicle*

1857 15 May   Uncle Edward Prentice Henslowe died

9 Oct   Aunt Ann died in Hitcham

22 October   Sister Louisa married John Kirkpatrick in Woodbridge

13 November   Wife Harriet died in Hitcham

School qualifies for grant

End of excursions due to rail prices and regulations

"Botanical diagrams"

1 sermon printed

1 article published

1858 *A Dictionary of botanical terms*

Lectures to Royal children

Illustrations . . . in practical lessons on botany

2 Hitcham pamphlets printed

1 article published

1859 2 August   Daughter Anne marries Robert Cary Barnard

8 October   Son George marries Ellen Weekley

22 November   *Origin of species* published

Visited Hoxne

1 article published

1860 12 July   Grandson Charles Paget Hooker born at Richmond

12 July   Granddaughter Anna Mary Barnard born at Hitcham

Flora of Suffolk

Chairman of B.A. Natural History meeting in Oxford – famous
debate

Visited Amiens and Abbeville gravel pits

Senate agreed to credit Natural Science Tripos

1 sermon printed

3 articles published

1861 February   Lecture at Ipswich Museum
     March   In Cambridge
     Prepared first Honours paper for Natural Science Tripos
     Lectures cancelled
     1 article published
     16 May   Henslow died
     23 May   Buried at Hitcham

Appendix 3
Dramatis Personae

Biographical details have mainly been drawn from the *Dictionary of National Biography* (DNB) and the *Dictionary of Scientific Biography* (DSB). We have also consulted the Berkshire, Cambridgeshire, Kent and Suffolk Record Offices (BRO, CRO, KRO and SRO respectively), the Cambridge University Archives (UA), the National Maritime Museum, Greenwich (NMM), and the Family Records Centre in London (FRC) as well as Barnard and Henslow family records. Other sources are included in the Bibliography, Section 2.

Abinger, Lord (James Scarlett), 1769–1844, judge. Trinity College, Cambridge. BA 1789. Called to bar 1791; KC 1816 (DNB)

Airy, George Biddell, 1801–92, astronomer. Trinity College, Cambridge. BA 1823; Lucasian Professor of Mathematics 1826–8; Plumian Professor of Astronomy & Experimental Philosophy 1828–36; Astronomer Royal 1835–81. KCB 1872 (DNB)

Aiton, William, 1731–93, botanist. Botanic Garden, Chelsea 1754–9; manager of Kew Gardens 1759–93. *Hortus Kewensis*, 1789 (DNB)

Albert, Prince Consort, 1819–61. Chancellor of Cambridge University 1847. Original proposer of Great Exhibition 1851. President of the British Association for the Advancement of Science 1859 (DNB)

Audubon, John James, 1785–1851, ornithologist and illustrator. *The Birds of America*, 1827–38; *The Quadrupeds of North America*, 1849–54 (Ford, A., 1988)

Babington, Charles Cardale, 1808–95, botanist. St John's College, Cambridge. BA 1830; MA 1833. One of the founders of the Entomological Society 1833; *Manual of British Botany*, 1843; founder 1836, and for 55 years secretary of the Ray Club; assisted in founding Cambridge Antiquarian Society 1840; Professor of Botany at Cambridge 1861; *Flora of Cambridgeshire*, 1860 (DNB Suppl.)

Balfour, John Hutton, 1808–84, botanist. MA Edinburgh; MD 1832; FRCS Edinburgh 1833; Professor of Botany, Glasgow University 1841–5; Professor of Botany, Edinburgh University and Keeper of the Royal Botanic Garden 1845–79 (DNB)

Barnard, Anne (née Henslow), 1833–99, botanical artist. Henslow's youngest daughter and Sedgwick's god-daughter. Second wife of Robert Cary Barnard. Contributed plates to Curtis's *Botanical Magazine* (Desmond, R., 1994)

Barnard, Christina (née Porter), *c.* 1800–60. Daughter of Thomas Porter of Rockbeare in Devon, sister of William Porter, a student contemporary of Henslow, wife of Thomas Barnard and mother of Robert Cary Barnard (Family records)

Barnard, Robert Cary, 1827–1906. Winchester. Army career retiring as Major 1855. Emmanuel College, Cambridge. Fellow Commoner 1857, BA 1861. Botany master, Cheltenham College 1876. Married (1) Mary Reade, (2) Anne Henslow 1859. (Venn, J., 1940–54; Family records)

Barnard, Thomas, 1792–1877, Captain. Of Bartlow, Cambs and Prestbury

Beadon, Richard, 1737–1824. Master of Jesus College, Cambridge 1781–89; Bishop of Gloucester 1789; Bishop of Bath and Wells 1802–24 (DNB)

Beech, John Hawkesley, c. 1796–1855. St John's College, Cambridge. BA 1818. Died at Little Shelford, Cambs. (Venn, J., 1940–54)

Bentham, George, 1800–84, botanist. Hon. Secretary Horticultural Society 1829–40; President Linnean Society 1861–74. Hon. LL D, Cambridge 1874. *Genera Plantarum*, 1862–83 with Hooker (DNB)

Berkeley, Miles Joseph 1803–89, botanist. Christ's College, Cambridge. MA 1828. Rural dean of Rothwell and vicar of Sibbertoft, Northants, 1868; *Introduction to Cryptogamic Botany*, 1857, etc. (DNB Suppl.)

Biggs, Arthur, 1765–1848. ALS 1815; FLS 1815. Gardener to Isaac Swainson at Twickenham. Curator, Botanic Garden, Cambridge 1813–45. Raised 'Biggs' Nonsuch' apple (Desmond, R., 1994)

Brocklebank, John, d. 1843. Pembroke College, Cambridge 1804; BD 1814; Rector of Teversham 1817–43; Vicar of Melbourn 1817–24; Rector of Willingham 1824–43 (Venn, J., 1940–54)

Brongniart, Adolphe-Theodore, 1801–76, French palaeobotanist and taxonomist. Founded *Annales des sciences naturelles* (DSB)

Brongniart, Alexandre, 1770–1847, geologist (DSB)

Brougham, Henry, Baron Brougham and Vaux, 1778–1868. Advocate Edinburgh 1800. A founder of *Edinburgh Review*. Lord Chancellor 1830–4. Hon LL D, Cambridge 1862 (DNB)

Brougham, William 1795–1886. Jesus College, Cambridge. BA 1819. MP for Southwark 1831–5; a master in Chancery 1835–52. Succeeded brother Henry (qv) as second baron in 1868 (DNB, note under Henry Brougham)

Brown, Robert, 1773–1858, botanist. Librarian to Sir Joseph Banks 1810–20; Keeper of Botanical collections, British Museum 1827–58 (DNB)

Buckland, William 1784–1856, geologist. Corpus Christi College, Oxford. BA 1805. Professor of Mineralogy, Oxford, 1813– ; Dean of Westminster 1845– (DNB)

Bullen, Catherine, 1760–1844. Widow of Revd John Bullen of St Andrew's the Less, Barnwell, Cambridge and tenant of site of New Botanic Garden owned by Trinity Hall and sold to University of Cambridge in 1831. Lived in London with her son-in-law John Watson after her son George's death. The family refused to give up land until lease expired (CRO and UA)

Bullen, George, 1790–1830, farmer. Son of John and Catherine Bullen and tenant of land in Newtown, Cambridge, belonging to the Unversity and transferred to Trinity Hall in 1831 (CRO and UA)

Bunbury, Sir Charles James Fox, 1808–86, botanist. Trinity College, Cambridge 1829. Succeeded as 8th Bart 1860. Married Frances Joanna, dau. of Leonard Horner (Venn, J., 1940–54)

Bunbury, Sir Henry Edward, 1778–1860, 7th Bart., Lieut-General ret. Of Barton Hall & Mildenhall, Suffolk (DNB)

Calvert, Frederick, *c*. 1792–1852. Jesus College, Cambridge. BA 1815. Ordained 1823. Rector of Whatfield, Suffolk 1823–52 (Venn, J., 1940–54)

Candolle, Augustin-Pyramus de, 1778–1841, botanist and agronomist. Academy of Geneva 1794–6. *Physiologie végétale*, 1832. *Prodromus systematis naturalis regni vegetabilis*, 1824–39 (DSB)

Chafy, William, 1779–1843. Master of Sidney Sussex College, Cambridge 1813–43. Vice-Chancellor 1813, 1829 (Venn, J., 1940–54)

Clark, William, 1787–1869. Trinity College, Cambridge. BA 1808. Professor of Anatomy, University of Cambridge 1817–66. Rector of Guiseley, Yorks. 1826–59. Father of John Willis Clark (Venn, J., 1940–54)

Clarke, Edward, 1730–86, traveller. Father of Edward Daniel Clarke. St John's College, Cambridge. Vicar of Willingdon and Rector of Buxted, Sussex (DNB)

Clarke, Edward Daniel, 1769–1822, traveller. Jesus College, Cambridge. MA 1794. Professor of Mineralogy, Cambridge University 1808–22; Rector of Yeldham 1809–22; University Librarian 1817–22. Travelled round Europe with John Marten Cripps (DNB)

Clarke, Dr William Barnard, *c*. 1807–94, naturalist. First Curator of Ipswich Museum 1846–50 (Markham, R.A.D., 1990)

Copley, John Singleton, the younger, 1772–1863. Trinity College, Cambridge. BA 1794. Called to the bar 1804. Lord Chancellor, 1827–30, 1834–5, 1841–6. Baron Lyndhurst, 1827 (DNB)

Cordier, Pierre-Louis-Antoine, 1777–1861, geologist and mineralogist. Professor of Geology, Museum d'Histoire naturelle, 1819 (DSB)

Cottle, Thomas, 1805–*c*. 1883, cleric. Third son of Wyatt Cottle. Pembroke College, Oxford. BA 1827. Curate of Cholsey. Curate of East Lockinge, Berks. Rector of Petton, Salop. 1869–83 (Foster, J., 1888; BRO)

Cottle, Wyatt, 1756–*c*. 1833, cleric. Pembroke College, Oxford. BCL 1782. Vicar of Cholsey before Henslow (Foster, J., 1888; BRO)

Cripps, John Marten, d. 1853, traveller and antiquary. Jesus College, Cambridge. F.S.A. 1805. Travelled in Europe with E.D. Clarke (DNB)

Cumming, James, 1777–1861. Trinity College, Cambridge. BA 1801; Professor of Chemistry, Cambridge University 1815–60. *A Manual of Electro-Dynamics*, 1827 (DNB)

Darwin, Amy Richenda (née Ruck), 1850–76. Married Charles Darwin's son Francis, 1874 (Raverat, G., 1952)

Darwin, Caroline *see* Wedgwood, Caroline

Darwin, Charles Robert, 1809–82. Christ's College, Cambridge. Naturalist on the 'Beagle' 1831–6. Author of *Origin of species*, 1859 etc. (DNB)

Darwin, Emma (née Wedgwood), 1808–96. Youngest daughter of Bessy and Josiah Wedgwood II. Married her cousin Charles Darwin in 1839 (Darwin, C.R., 1985–)

Darwin, Erasmus Alvey, 1804–81. Charles Darwin's elder brother. Christ's College, Cambridge 1822. MB 1828. Edinburgh University 1825–6. Qualified but never practised as a physician (Venn, J., 1940–54)

Darwin, Sir Francis, 1848–1925, botanist. Charles Darwin's third son. Published *Life and letters of Charles Darwin*, 1887 etc. Married (1) 1874 Amy Ruck (2) 1883 Ellen Crofts (3) 1913 Florence Maitland (DNB 1901–50)

Darwin, Susan Elizabeth, 1803–66. Charles Darwin's sister. (Darwin, C.R., 1985–)

Davies, Richard, c. 1710–62, physician. Queens' College, Cambridge. BA 1730; MD 1740. Practised at Shrewsbury and Bath. Author of *The General State of Education in the Universities . . . 1759.* (DNB)

Davy, Sir Humphrey, 1778–1829, chemist. First Professor of Chemistry at Royal Institution. Inventor of safety lamp. (DNB)

Dawes, Richard, 1793–1867. Trinity College, Cambridge; MA 1820; bursar of Downing 1818; Rector of King's Somborne 1836–50; founded a model school in his parish, 1842; Dean of Hereford, 1850–67 (DNB)

Denson, John, c. 1800–c. 1870. ALS 1832. Gardener to N.S. Hudson at Bury St Edmunds. Curator, Botanic Garden, Bury St Edmunds, 1821–9. *Catalogue of Hardy Trees in Botanic Garden, Bury St Edmunds*, 1822. Botanical Assistant to Loudon. Edited Loudon's *Magazine of Natural History* from 1831. (Desmond, R., 1994)

Donn, James, 1758–1813, botanist. Curator of Cambridge Botanic Garden, 1790–1813. *Hortus Cantabrigiensis*, 1796. Named *Claytonia perfoliata* (DNB)

Downes, John, 1810–1890. Christ's College, Cambridge. BA 1833; MA 1836. Ordained priest 1834. Vicar of Horton with Piddington, Northants 1834–63; Rector of Hannington 1866–90. (Venn, J., 1940–54)

Fallows, Fearon, 1789–1831, astronomer. St. John's College, Cambridge. Director of astronomical observatory proposed for Cape of Good Hope (DNB)

Farish, William, 1759–1837. Magdalene College, Cambridge. Professor of Chemistry, 1794–1813. First sought to apply that science to the arts and manufactures and to combine with its study the practical adjuncts of mechanics and engineering. Jacksonian Professor of Natural & Experimental Philosophy 1813–[1837] (succeeded Francis John Hyde Wollaston) (DNB)

Fitch, Walter Hood, 1817–92. Botanical artist. At Kew from 1841. Illustrator of Curtis's *Botanical Magazine*, 1834–78 etc. (Lewis, J., 1992)

Fitton, William Henry, 1780–1861, geologist. MD, Cambridge 1816. Secretary Geological Society. Laid down proper succession of strata between oolite and chalk, 1824–36 (DNB)

Fitzroy, Captain Robert, 1805–65, vice-admiral, hydrographer and meteorologist. Commanded the 'Beagle' 1828–30 and 1831–6 while surveying the coast of South America. During his second command in which he also circumnavigated the world and visited the Galapagos, Charles Darwin accompanied him as naturalist (DNB)

Fitzwilliam, Hon. W.C. Wentworth, 1811–35. Trinity College, Cambridge; MA 1833. Later Viscount Milton (Venn, J., 1940–54)

Forbes, Edward, 1815–54, naturalist. Professor of Botany, King's College, London. *History of British Mollusca*, 1848, etc. (DNB)

Foster, Thomas Campbell, 1813–82, legal writer. Barrister, 1846. *Letters on the Condition of the People of Ireland*, 1846 (DNB)

Fox, William Darwin, 1805–80. Christ's College, Cambridge, BA 1829. Charles Darwin's second cousin and close friend at Cambridge. Rector of Delamere, Cheshire, 1838–73 (Darwin, C.R., 1985–)

Garnons, William Lewes Pugh, 1791–1863, entomologist and botanist. Sidney College, Cambridge; BA 1814; MA 1817; BD 1824. Ordained priest 1816. Vicar of Ulting, Essex 1848–63 (Desmond, R., 1994; Venn, J., 1940–54)

Gorham, George Cornelius, 1787–1857, divine and antiquary. Queens' College, Cambridge; BA 1808; MA 1812. Vicar of St Just, Cornwall 1846, and of Brampford Speke, Devon 1847–57 (DNB)

Graham, John, 1794–1865. Christ's College, Cambridge; MA 1819. Tutor of Darwin, then Master of Christ's College, 1830–48. Bishop of Chester 1848–65. Universities Commission (DNB)

Graham, Robert, 1786–1845, MD and botanist. Professor of Botany, Glasgow, 1818–20; Regius Professor at Edinburgh, 1820–45 (DNB)

Graves, Richard Drought. St John's College, Cambridge; BA 1855. Ordained deacon (Ely) 1858. Curate of Hitcham 1858–61; Curate of Mitcham, Surrey 1863–5; Perpetual Curate of Hanford, Staffs 1865–? (Venn, J., 1940–54)

Greenough, George Bellas, 1778–1855, geographer and geologist. Peterhouse, Cambridge. *Critical Examination of the First Principles of Geology*, 1819; *Geological map of United Kingdom*, 1820, etc. (DNB)

Greenwood, Revd William. Corpus Christi College, Cambridge; BA 1818; MA 1821. Ordained priest 1832. Rector of Thrapston, Northants 1828–37. (Venn, J., 1940–54)

Gregory, Tighe *see* Tighe-Gregory, Allott

Greville, Robert Kaye, 1794–1866, botanist. LL.D. Glasgow, 1824. *Scottish Cryptogamic Flora, Flora Edinensis*, 1824; *Icones Filicum* (with Hooker), 1829–31; *Algae Britannicae*, 1830 (DNB)

Gunning, Henry, 1768–1854. Christ's College, Cambridge; BA 1788. *Reminiscences of the University, Town and County of Cambridge*, 1854 (DNB)

Harper, John, Farmer and Churchwarden, Hitcham. Lived at Hitcham Hall from *c.* 1843 (SRO)

Haviland, John, 1785–1851. Professor of Anatomy 1814–17; Professor of Physic, Cambridge University 1817–51; Physician to Addenbrooke's Hospital 1817–39 (DNB)

Heberden, Mary (née Wollaston), 1763–1832. Daughter of William Wollaston, 1660–1724, 2nd wife of William Heberden the elder and Harriet Henslow's maternal grandmother (Jenyns, L., 1862)

Heberden, William, the elder, 1710–1801, physician. Harriet Henslow's maternal grandfather. St John's College, Cambridge; BA 1728 (DNB)

Heberden, William, the younger, 1767–1845, physician. Harriet Henslow's uncle. Fellow, St John's College, Cambridge 1788–96. Physician to George III (DNB)

Henslow, Lady Ann (née Prentis), 1739–1803. Daughter of Presbyterian parents, Edward and Damaris Prentis of Maidstone. Second wife of Sir John Henslow (KRO)

Henslow, George, 1835–1925. Henslow's third son. Christ's College, Cambridge; BA 1858. Ordained deacon 1859, priest 1861. Curate, Steyning, Sussex 1859–61. Head-master, Hampton Lucy Grammar School 1861–4, Grammar School, Store St, London 1865–74. Lecturer in Botany at St Bartholomew's Hospital 1866–80, also at Birkbeck and Queen's College. Curate, St John's Wood Chapel, 1868–70, St James', Marylebone 1870–87. Married (1) 1859 Ellen Weekley, divorced 1871; (2) 1872 Georgina Bailey who died 1876; (3) 1881 Katherine Yeo (Venn, J., 1940–54, family records, *The Times*)

Henslow, Harriet (née Jenyns) 1797–1857. Daughter of George Leonard Jenyns of Bottisham Hall, Cambs. Married John Stevens Henslow 1823. *A Practical application of the five books of Moses*, 1848; *John Borton . . .*, 1853 (Hailstone, E., 1873; family records)

Henslow, John Jenyns, 1828–9. Henslow's first son. Died in infancy (CRO)

Henslow, Sir John [1730]–1815. Henslow's grandfather. Surveyor to the Navy 1784–1806 (NMM)

Henslow, John Prentis, 1770–1854. Henslow's father. Solicitor in Enfield, wine merchant, etc. in Rochester and solicitor again in St Albans (Jenyns, L., 1862)

Henslow, Leonard Ramsay, 1831–1915. Henslow's second son. St John's College, Cambridge; BA 1854; MA 1857. Ordained deacon (Ely) 1854, priest 1855. Curate of Hitcham 1854–6; of Bangor-Monachorum, Flints 1856–60; of Great Chat, Kent 1860–3; Rector of St Mary Magdalene, Pulham, Norfolk 1863–70; of Zeals, Wilts 1870–1915. Married Susan Wall (Venn, J., 1940–54)

Henslow, Louisa Mary, 1826–1904. Henslow's second daughter. Unmarried (family records)

Henslowe, Philip, d. 1616, theatrical manager. Rebuilt and managed the Rose playhouse, and also the Swan on Bankside (DNB)

Herbert, John Maurice, 1808–82. St John's College, Cambridge; BA 1830. Called to Bar 1835. County Court judge 1847–82 (Venn, J., 1940–54)

Herbert, William, 1778–1847, naturalist. Exeter College, Oxford; BA 1798; Merton College, MA 1802 and DCL 1808. Rector of Spofforth 1814–40. Dean of Manchester 1840–7. Assisted in editions of White's *Selborne*, 1833 and 1837. *Amaryllidaceae*, 1837 (DNB)

Hildyard, William. Trinity College, Cambridge; BA 1817. Trinity Hall, Fellow and tutor 1824. Rector of Market Deeping, Lincs 1830–75 (Venn, J., 1940–54)

Hind, John, 1796–1866, mathematician. BA 1818. Migrated to Sidney Sussex College, Cambridge, 1819, MA 1821. Fellow 1823–4 (DNB)

Hooker, Frances Harriet (née Henslow), 1825–74. Henslow's eldest daughter. Married Joseph D. Hooker 1851. Translated E. Le Maout and J. Decaisne's *Traite Général de Botanique*, 1873 (Desmond, R., 1994)

Hooker, Sir Joseph Dalton, 1817–1911, botanist. Expeditions to Antarctica 1839–43 and the Himalayas 1847–50. Married Frances Harriet Henslow. Succeeded his father, Sir William Hooker as Director of Kew, 1865 (DNB Suppl.)

Hooker, Sir William Jackson, 1785–1865. Professor of Botany, Glasgow University 1820–41; Director of Kew Gardens 1841–65 (DNB)

Hope, John, 1725–86. King's Botanist for Scotland, Superintendent of the Royal Garden in the City and Professor of Botany at Edinburgh, 1761–? Created new Botanic Garden at Leith Walk. Edited Linnaeus' *Genera Animalium*, 1781 (Morton, A.G., 1986)

Horner, Frances Joanna and Leonora. Daughters of Leonard Horner, geologist. Frances Joanna married Sir Charles James Fox Bunbury

Humboldt, Freidrich Heinrich Alexander, Baron von, 1769–1859, German scientist and traveller. *Kosmos*, 1845–58 (DSB)

Huntley, John Thomas, 1790–1881. Trinity College, Cambridge; BA 1814; MA 1817. Ordained priest 1816. Rector of Swineshead and Vicar of Kimbolton 1819–45. Rector of Binbrooke St Mary with Vicar of Binbrooke St Gabriel, Lincs 1845–81. (Venn, J., 1940–54)

Hutton, William 1797–1860, geologist. Authority upon the coal measures and a collector of coal fossils. Took leading part in the establishment of mechanics' institutes of the north of England. *The Fossil Flora of Great Britain*, 1831–7 (with John Lindley) (DNB)

Huxley, Thomas Henry, 1825–95, biologist and scientist. Exponent of doctrine of evolution. Hon. LL D Cantab. 1879 (DNB)

Jenyns, George Leonard, 1763–1848. Caius College, Cambridge; BA 1785; MA 1788. Ordained priest 1787. Inherited Bottisham Hall 1787. Vicar of Swaffham Prior 1787–1848; Prebendary of Ely 1802–48. Chairman of Board of Agriculture in London (Venn, J., 1940–54; Hailstone, E., 1873)

Jenyns, Leonard, 1800–93, naturalist. St John's College, Cambridge; BA 1822. Vicar of Swaffham Bulbeck, 1828–49. Founder and First President of Bath Natural History and Antiquarian Field Club. *Memoir of the Rev. John Stevens Henslow*, 1862. Changed name to Blomefield as condition of inheritance. (DNB Suppl. under Blomefield)

Jenyns, Mary (née Heberden) 1763–1832. Harriet Henslow's mother (Hailstone, E. 1873)

Jenyns, Soame, 1704–87, author. St John's College, Cambridge. MP for Cambridge, 1742–80. *Free Enquiry into the nature and origin of evil*, 1757; *View of the internal evidence of the Christian Religion*, 1776 (DNB)

Jephson, Thomas, 1784–1864. St John's College, Cambridge; BA 1806; MA 1809; Fellow 1808–64. Candidate for Chair of Mineralogy, 1822 (Venn, J., 1940–54)

Jephson, William, 1775–1848, schoolteacher. St John's College, Cambridge; BA 1796; MA 1799. Headmaster of Wilson's Grammar School, Camberwell 1803–43. Brother to Thomas. (Venn, J., 1940–54)

Kerrich, Dr Thomas, 1748–1828, librarian and antiquary. Magdalene College, Cambridge; MA 1775. University Librarian 1797. (DNB)

Kirby, Henry, 1794–1858. Clare College, Cambridge 1813; BA 1817; MA 1820; Fellow 1817; Senior Proctor 1829–30; Curate of Oakley, Suffolk 1841–2; Rector of Gt Waldringfield 1842–58. (Venn, J., 1940–54; *Gentleman's Magazine*, 1858, I. 332)

Kirby, William, 1759–1850, entomologist. Caius College, Cambridge; BA 1781. Rector of Barham, Suffolk 1782–1850. First President Ipswich Museum. *Introduction to Entomology* (with William Spence), 1815–56 in various editions (DNB)

Knight Bruce, James Lewis, 1791–1866, judge. Barrister, 1817. Lord Justice of Appeal, 1851 (DNB under Bruce)

Knights, George, d. 1872. Curator of Ipswich Museum 1853–72 (Markham, R.A.D., 1990)

Lamb, John, 1789–1850. Master of Corpus Christi College, Cambridge 1822–50; Dean of Bristol 1837–50 (DNB)

Lapidge, Edward, d. 1860, architect. Father was chief gardener at Hampton Court Palace. Built new bridge over Thames at Kingston, 1825–28; St Peters, Hammersmith, 1827–30; Chapel of St Andrew on Ham Common, 1832. Unsuccessful competitor for Houses of Parliament, 1836, and Fitzwilliam Museum, 1837. Surveyor of bridges and public works for Surrey (DNB)

Leach, William Elford, 1790–1836, naturalist. Assistant Librarian and Assistant Keeper, British Museum 1813–21. *Malacostraca Podophthalma Britanniae*, 1815–16, etc. (DNB)

Lefevre, Sir John George Shaw *see* Shaw-Lefevre, Sir John George

Leighton, William Allport 1805–89. St John's College, Cambridge; BA 1833. *Flora of Shropshire*, 1841; *Lichen Flora of Great Britain*, 1871. "Henslow found in Leighton one of his most zealous pupils" *Shrewsbury Chronicle*, 8 March 1889; *Journal of Botany*, 1889, p. 111 (DNB)

Lemann, Charles Morgan, 1803–52. Trinity College, Cambridge; MB 1828; MD 1833. Collected plants in Madeira and Gibraltar (Venn, J., 1940–54; Desmond, R., 1994)

Lindley, John, 1799–1865, botanist and horticulturist. Assistant Librarian to Sir Joseph Banks. Professor of Botany in the University of London, 1829–60. *The Fossil Flora of Great Britain*, 1831–7 (with William Hutton). *The Vegetable Kingdom*, 1846 (DNB)

Linnaeus, Carolus, 1707–78, Swedish botanist. Professor of Botany, Uppsala. His *Species Plantarum*, 1753, and his system of classification introduced binomial nomenclature. Collections bought by Sir James Edward Smith (q.v.) (DSB)

Loudon, John Claudius, 1783–1843, horticultural writer and editor. Founded the *Gardener's Magazine*, and *Magazine of Natural History* (Desmond, R., 1994)

Lowe, Richard Thomas, 1802–74, naturalist. Christ's College, Cambridge; BA 1825. English chaplain at Madeira, 1832–54. *A Manual Flora of Madeira*, 1857–72. (DNB)

Lubbock, Sir John William, 1803–65, astronomer and mathematician. Trinity College, Cambridge (DNB)

Lyell, Sir Charles, 1797–1875, geologist. Exeter College Oxford. *Principles of Geology*, 1830–3; *Elements of Geology*, 1838 (DNB)

Macleay, Alexander, 1767–1848, entomologist and colonial statesman (DNB)

Mapletoft, Robert. Christ's College, Cambridge; BA 1804. Owner of Hitcham Hall. Magistrate on Bury St Edmunds bench (Venn, J., 1940–54)

Marsh, William, 1775–1864, divine. Vicar of St Peters, Colchester 1814–29; Rector of St Thomas, Birmingham 1829–39; St Mary, Leamington 1839–60; Rector of Beddington, Surrey 1860–4. Known as 'Millennial Marsh' he was an impressive evangelical preacher; friend and correspondent of Charles Simeon (DNB)

Martin, Thomas, 1779–1867 FRCS. Doctor in Reigate 1810–67. Formed a local mechanics institute, 1830. Encouraged local societies, including Cottage Gardeners' Society and a savings bank. Established National Schools in Reigate and Redhill (Plarr, V.G., 1930)

Martyn, John, 1699–1768. Professor of Botany, University of Cambridge 1732–62 (DNB)

Martyn, Thomas, 1735–1825. Son of John Martyn. Professor of Botany, University of Cambridge 1762–1825 (DNB)

Mathews, John Staverton, 1776–1837. Trinity College, Oxford; BA 1799. Rector of Hitcham 1801–37 (SRO)

Maund, Benjamin, 1790–1863, botanical writer, druggist and bookseller, Bromsgrove, Worcs. *The Botanic Garden*, 1825–50; *The Botanist*, 1836–42 (DNB)

Michell, Thomas Pennuddock, c. 1800–66. Merton College, Oxford. Ordained priest 1821. Vicar of Histon 1821–66. Admitted Downing College, Cambridge 1832 (VCH Cambridge, v.9)

Miller, Charles, 1739–1817. Son of Philip Miller (q.v.). First Curator of Cambridge Botanic Garden, 1762–70 (Desmond, R., 1994)

Miller, Philip, 1691–1771. Gardener at Physic Garden, Chelsea, 1722–70. *The Gardeners' Dictionary*, 1731 (Desmond, R., 1994)

Murray, Andrew, c. 1810–50. Curator of Cambridge Botanic Garden 1845–50 (Walters, S.M., 1981)

Murray, George, 1784–1860. 2nd son of Lord George Murray. Christ Church College, Oxford; BA 1806; MA 1810; DD 1814. Archdeacon of Man 1808; Bishop of Sodor and Man 1814; Bishop of Rochester 1827–60 (also Dean of Worcester 1828–45). Involved there in the 'Whiston affair' (DNB)

Oliver, Daniel 1830–1916. Aberdeen LLD 1891. Assistant Keeper, Kew Herbarium 1858; Keeper 1864–90. Professor of Botany, University College, London 1861–88. *Lessons in Elementary Botany*, 1864 (Desmond, R., 1994)

Owen, Sir Richard, 1804–92, anatomist and palaeontologist. Edinburgh University MRCS. Museum, Royal College of Surgeons 1826–56. Hunterian Professor of

Anatomy 1836–56. Superintendent, Natural History Department, British Museum 1856 (DNB)

Palmerston, Viscount, Henry John Temple, 1784–1865. St John's College, Cambridge. MP for Cambridge University 1811–31; Foreign Office and Foreign Secretary 1830–41; Home Secretary 1853–5; Prime Minister 1855–8, 1859–65 (DNB under Temple)

Peacock, George, 1791–1858, mathematician. Lowndean Professor of Astronomy 1836–58; Dean of Ely 1839–58; instigated restoration of Ely cathedral by Scott (DNB)

Pinniger, Richard Broome, 1803–87. Pembroke College, Oxford; BA 1825. Curate of Cholsey 1836–7. Curate of Hitcham 1837–9. Rector of Whichford, Warks. 1839–85 (BRO, SRO; Foster, J., 1888)

Playfair, Sir Lyon, 1818–98, chemist. St Andrews. Professor of Chemistry, Royal Institution of Manchester 1842; Chemist, Geological Survey and Professor, School of Mines 1845. Organiser, Great Exhibition, 1851. Professor of Chemistry, Edinburgh, 1858 (DNB)

Porter, Christina *see* Barnard, Christina

Porter, William, c. 1796–1820. Son of Thomas Porter, Esq of Rockbeare House, near Exeter. Born in W. Indies. Schools: Blundells, Tiverton & Edinburgh. St John's College, Cambridge; BA 1818 (Venn, J., 1940–54; family records)

Prentis, Ann *see* Henslow, Lady Ann

Pryme, George, 1781–1868, political economist. Trinity College, Cambridge; BA 1803. Called to Bar 1806. Professor of Political Economy 1828–63. MP for Cambridge 1832–41 (DNB)

Pulling, William, 1786–1860. Emmanuel College, Cambridge 1808. Migrated to Sidney Sussex College 1809; BA 1813; MA 1817. Ordained priest 1813. Curate of St Andrew the Less, Cambridge 1821–6; Rector of Dymchurch and Rector of Blackmanstone, Kent, 1835–60. Married Mary Elizabeth, eldest daughter of Revd R. Relhan (Venn, J., 1940–54)

Pulteney, Richard, 1730–1801, botanist. MD Edinburgh (DNB)

Quetelet, Lambert Adolphe Jacques, 1796–1874, Belgian statistician and astronomer. Founder and first director of Royal Observatory, Ghent. *Sur l'homme et le Développement de ses Facultés*, 1835 (DSB)

Ramsay, Marmaduke, d. 1831. Jesus College, Cambridge; BA 1818; MA 1821; Fellow 1819–31 Tutor. FLS Died at Perth (Venn, J., 1940–54)

Ransome, George, 1811–76. Grandson of founder of the Ipswich agricultural implement manufacturers. Hon. Secretary of Ipswich Museum to 1852 (Markham, R.A.D., 1990)

Ray, John, 1627–1705, botanist. Trinity College, Cambridge. *Catalogus Plantarum circa cantabrigiam nascentium*, 1660. *Synopsis Methodica Stirpium Britannicarum*, 1670. *Historia Plantarum Generalis*, 1686–1704 (DNB)

Richardson, Sir John, 1771–1841, judge. University College, Oxford. Puisne Judge of the Court of Common Pleas 1818–24 (DNB)

Rodwell, John Medows, 1808–1900, orientalist. Caius College, Cambridge; MA 1833; Rector of St Peter's, Saffron Hill, London 1836–43; Rector of St Ethelburga's, Bishopsgate, London 1843–1900 (DNB)

Romilly, Joseph, 1791–1864. Trinity College, Cambridge, MA 1816. Registrary of the University 1832–61 (DNB)

Rose, Hugh James, 1795–1838, theologian. Trinity College, Cambridge; BA 1817. (DNB)

Ruck, Amy *see* Darwin, Amy

Samuel, George, drawing master (possibly the landscape painter, d. 1823). Teacher at Wilsons Grammar School, Camberwell (Jenyns, L., 1862)

Schultes, Josef Auguste, 1773–1831, Austrian botanist. Dr. Med Wien, 1796. Professor of General Natural History and Botany, University of Landshut, Bavaria 1809–31 (Stafleu, F.A. and Cowan, R.S. 1976–   )

Sedgwick, Adam 1785–1873. Trinity College, Cambridge. Woodwardian Professor of Geology, University of Cambridge 1818–73. Prebendary of Norwich 1834 (DNB)

Shaw-Lefevre, Sir John George, 1797–1879, civil servant. Trinity College, Cambridge; BA 1818. Barrister 1825 (DNB)

Simeon, Charles, 1759–1836. King's College, Cambridge; BA 1783. Incumbent of Holy Trinity, Cambridge 1783–1836. Influential evangelical leader (DNB)

Skepper, Edmund, 1825–67, druggist. Harwich and Bury St Edmunds. Co-author with Henslow of *Flora of Suffolk* (Desmond, R., 1994)

Smith, Sir James Edward, 1759–1828, botanist. Purchased Linnaean collections 1784. Founded Linnean Society 1788. *English Botany*, illustrated by Sowerby, 1790–1814 (DNB)

Sowerby, James, 1757–1822, botanical artist and engraver, botanist, zoologist. *English Botany*, 1790–1814 (text by Sir James Edward Smith); *Coloured Figures of English Fungi*, 1795–1815 (DNB)

Sowerby, James de Carle, 1787–1871, naturalist and artist. Eldest son of James. Assisted father with the plates for *English Botany* and illustrated the four volume Supplement by Sir William Hooker, 1831–1840 (DNB)

Spineto, Elizabeth (née Campbell), Marchesa di Spineto. Second wife of the Marquis de Spineto, teacher of Italian in the University (Darwin, C.R., 1985–   )

Spring Rice, Thomas, 1790–1866. Trinity College, Cambridge; BA 1811. Chancellor of the Exchequer, 1835–9. First Baron Monteagle of Brandon in Kerry, 1839 (DNB)

Stanley, Edward, 1779–1849. St John's College, Cambridge; BA 1802. Incumbent of Alderley, 1805–37. Bishop of Norwich 1837–49. *Familiar history of Birds*, 1836 (DNB)

Starkie, Thomas, 1782–1849. St John's College, Cambridge; BA 1803. Barrister 1810. Q.C. Downing Professor of Law, Cambridge, 1823–49 (DNB)

Stephens, James Francis, 1792–1852, entomologist. In Admiralty 1807–45. Assisted in arranging insects at British Museum from 1818. *Catalogue of British Lepidoptera*, 1850–2 (DNB)

Stevens, Thomas, *c.* 1729–1824. Henslow's maternal grandfather. Ostler at Star Hotel, Rochester. Married (1) widow of landlord (2) Frances Wade. Brewer, councillor and mayor of Rochester. Portrait in Guildhall, Rochester. Built Gad's Hill. (KRO)

Stow, Jane. Schoolmistress at Hitcham. Sister to Bildeston blacksmith (Suffolk Directory, 1840; FRC)

Stratton, James. Curator of Cambridge Botanic Garden 1851–64 (Walters, S.M., 1981)

Swainson, William, 1789–1855, naturalist. Clerk H.M. Customs; ret'd 1815. Wrote and illustrated books on zoology, including 11 volumes for Lardner's *Cabinet Cyclopaedia*. (Natusch, S., 1988)

Tatham, Ralph, 1778–1857. St John's College, Cambridge; BA 1800; DD 1839. Vice-Chancellor 1839–40, 1845–6. Rector of St Mary Colkirk, Norfolk 1816–57 (Venn, J., 1940–54)

Thackeray, George, 1777–1850. King's College, Cambridge; BA 1802; MA 1805; DD 1814; Fellow 1800–03; Provost 1814–50 (Venn, J., 1940–54)

Thompson, John Elijah, 1831–1920, cleric. St John's College, Cambridge; BA 1854. Ordained Deacon 1856, priest 1857. Curate of Hitcham 1856–8. Vicar of Offton with Little Bricet, Suffolk 1858–1903 (Venn, J., 1940–54)

Tighe-Gregory, Allott (formerly Gregory, Richard Allott Tighe), *c.* 1820–? Trinity College, Dublin; BA 1840; LL B 1843. Vicar of Bawdsey 1847–1910. (Crockford's Clerical Directory, 1870)

Turner, Henry, *c.* 1810–76. Curator Botanic Gardens, Bury St Edmunds until 1857 (Desmond, R., 1994)

Underwood, Thomas Richard, 1772–1836, water colour painter, geologist. Visited France and Italy in 1802–03 and was imprisoned on his return journey. Much of latter part of life spent in France (Mallalieu, H.L., 1986)

Walker, Richard, 1679–1764. Trinity College, Cambridge. Professor of Moral Philosophy 1744. Founded University Botanic Garden (DNB)

Wallich, Nathaniel, 1786–1854, botanist and plant collector. MD, Copenhagen. Superintendent Calcutta Botanic Garden 1815–50. *Plantae Asiaticae Rariores*, 1830–2. Settled in England 1847 (DNB)

Watson, John, lawyer. Son-in-law to Catherine Bullen who lived with him in Paddington, London after her son George died (CRO, FRC)

Wedgwood, Caroline Sarah (née Darwin), 1800–88. Darwin's sister. Married Josiah Wedgwood III, her cousin, 1837 (Darwin, C.R., 1985–)

Whewell, William, 1794–1866, philosopher of science. Trinity College, Cambridge; BA 1816. Master of Trinity College, Cambridge. Succeeded Henslow as Professor of Mineralogy 1828–32. Knightsbridge Professor of Moral Philosophy, 1838–55. Vice-Chancellor 1842–3 and 1855–6 (DNB)

Wilberforce, Samuel, 1805–73. Rector of Brighstone, Isle of Wight 1830–40; of Alverstoke, Hants 1840–3; Dean of Westminster 1843–5; Bishop of Oxford 1845–69 (DNB)

Wollaston, Francis, 1694–1774. Sidney Sussex College, Cambridge; LL B 1717. FRS 1723

Wollaston, Revd Francis, 1731–1815, author. Sidney Sussex College, Cambridge; LL B 1754. Ordained priest 1755. Rector of Chislehurst 1769–1815 (DNB)

Wollaston, Francis John Hyde, 1762–1823, natural philosopher. Sidney College, Cambridge; BA 1783, MA 1786. Jacksonian Professor 1792–1813. Rector of South Weald, Essex 1794–1823. Prebendary of St Paul's 1802–23 (DNB)

Wollaston, Mary *see* Heberden, Mary

Wollaston, William, 1660–1724, moral philosopher. Sidney Sussex College, Cambridge; BA 1677–8, MA 1681 (DNB)

Wollaston, William Hyde, 1766–1828, physiologist, chemist, physicist. Caius College, Cambridge; MB 1788, MD 1793 (DNB)

Wood, James, 1760–1839. St. John's College, Cambridge; BA 1782, MA 1785, Fellow 1782–1815, Master 1815–39. Vice-Chancellor 1816–17. Ordained priest 1787. Dean of Ely 1820–39. Rector of Freshwater, Isle of Wright 1823–39. (Venn, J.A., 1940–54)

Wooster, David. Curator of Ipswich Museum 1851–3 (Markham, R.A.D., 1990)

Worsley, Thomas, 1797–1885. Master of Downing College, Cambridge 1836–85. Vice-Chancellor 1837–8 (Venn, J.A., 1940–54)

Appendix 4
Eponymous taxa

Like many nineteenth-century European biologists, Henslow had genera and species of plants and animals named after him. This practice had been widely used by Linnaeus in the previous century – *Magnolia*, after Pierre Magnol (1638–1715), is a very familiar example – and the Linnaean model has continued to be followed to the present day. It may come as a surprise how many *animals* were named after Henslow, until we remember that, as a boy and a young man, Henslow's collecting activities were devoted to zoological rather than botanical material. He supplied many specimens of animals to Leach at the British Museum (see Chapter 1), and several of these were considered by Leach to be hitherto undescribed species.

All the eponymous zoological taxa are (or were) at the specific level: we have to wait for 'Henslowias' until the Professor of Botany had established his reputation as a botanist, as we shall see later. Taking the 'Henslow animals' in chronological order of date of publication, his Swimming Crab, *Polybius henslowii*, was described by Leach (1820) and beautifully illustrated by James Sowerby (see Figure 2, p. 20). The text accompanying 'tab. ix' informs us that 'this species was first communicated to [Leach] by John Henslow, Esq. who found it in the net of a Herring fisher, on the northern coast of Devon, in 1817'. It is reasonable to suppose that the young Henslow, visiting his college friend William Porter (see Chapter 2), took the opportunity of pursuing his collecting passion in Devon and sending the results to Leach, as was his practice.

Other zoological specimens, particularly fresh-water (from the Cambridge area) and marine (from summer visits) found their way to Leach from the young Henslow; some of these are mentioned in Leach 1852 (pp. 321–6), where Henslow is one of the people thanked for providing material of the fresh-water mussel, and styled as 'J.S. Henslow Esq., Professor of Natural History, Cambridge' [*sic*]. It seems that Leach, knowing Henslow as a zoologist, found the real title as Professor of Botany incredible! In chronological order, the next eponymous animal is the small fresh-water mollusc first published in 1823 as *Tellina henslowana* by Sheppard, and transferred by Jenyns to the genus *Pisidium* in 1832. The name *Pisidium henslowanum* (Sheppard) survives for this common river shell to the present day: as Killeen, 1992, p. 78 states – 'The original description of this species . . . was based on shells sent to Sheppard by Dr. Leach, who obtained them from Professor Henslow.' The suggestion by Morley (1938) that Henslow's specimens came from Hitcham seems quite erroneous: according to Jenyns, in the MS *Fauna Cantabrigiensis* (1869, p. 148) the animal was 'first discovered by Prof. Henslow in a creek of the river at Baits-bite near Cambridge'. Leach, receiving Cambridge material from Henslow around 1820, called it *Pera henslowiana* in his unpublished MS, a name which Jenyns assigned to synonymy in his paper in 1832. The inclusion or omission of the 'i' in *henslow(i)ana* can be treated as an orthographic variant.

Before leaving the Henslow–Leach names, we should mention the sea-cucumber (Holothurian) called by Leach in MS in 1819 *Jemania henslowana* based upon a specimen sent by Henslow from Aberystwyth in 1819. This species name was eventually published by J.E. Gray in 1848 as *Synapta henslowana*, but has sunk into synonymy under the modern name *Leptosynapta inhaerens* (O.S. Mueller 1776). See Woodward & Barrett 1858.

Henslow's Sparrow, first published by Audubon in 1829 as *Emberiza Henslowii*, is the best-known of the eponymous taxa. We have discussed the association of Audubon with Henslow in Chapter 6, and need only add here that the modern name, as used by Mearns 1992, is *Ammodramus henslowii* (Audubon).

This account of the eponymous zoological taxa would not be complete without a mention of fossils commemorating Henslow's interest in palaeontology and geology. In Chapter 3 we describe the young Henslow's Isle of Man field trip. Fossils collected on that excursion are preserved in the Sedgwick Museum, and include *Nautilus complanatus*, which we have chosen to illustrate as Figure 4. Henslow's specimen is the holotype of that species, described by James Sowerby in 1820; it is now called *Discites complanatus* (J.Sow.). Another fossil collected by Henslow on the same trip is *Ammonites henslowi*, described by Sowerby in the same volume (vol. 3) of his seven-volume work entitled *The Mineral Conchology of Great Britain*. This eponymous Henslow fossil is described by Sowerby on p. 111 and illustrated in tab. 242: the text includes the following:

'This is one of the many curious petrifactions found by J.S. Henslow, Esq. during a visit to the Isle of Man in 1819.' The eponymous basionym survived in *Agonides h.*, *Ceratites h.* and *Goniotites h.* in different later publications: it is now called *Prolecanites henslowi* (J.Sow.). The holotype is preserved (as E.13406) in the Sedgwick Museum.

Eponymous botanical taxa all date from after Henslow's election as Professor of Botany in 1825. He was first honoured in this way by Nathaniel Wallich, a Danish botanist who specialised in Indian and Malayan plants, and held the post of Superintendent of the Calcutta Botanic Garden, with interruptions, from 1817 to 1846. Henslow was largely responsible for Wallich's election to the Cambridge Philosophical Society in 1832, and a letter from Wallich to Henslow dated 20 March 1832, which is in the Kew Archives (Henslow Letters, 217–19), thanks Henslow for the 'high honour' and tells him of a present that is on its way: 'a coloured [engraving] of *Henslowia* and a sheet of my 10th number (published yesterday) with a description of that genus'. This beautiful coloured illustration and the accompanying description are in Wallich's *Plantae Asiaticae Rariores* 3: 13–14, Plate 221 (1832). Later botanists have judged Wallich's genus to be the same as *Crypteronia* Blume, published five years earlier in 1827; under the priority rule that name must be used for the genus, and *Henslowia* Wallich goes into synonymy.

The second *Henslowia* was due to Richard Thomas Lowe, author of the *Manual Flora of Madeira*, whom we have mentioned in Chapter 11 as a Christ's man who was helped by Henslow. Lowe gave the name to an endemic tree of Madeira and the Canaries which we now call *Notelaea excelsa* Webb et Berthelot. This new *Henslowia*, published in 1844 in de Candolle's *Prodromus* viii, 288, is unfortunately a synonym of *Picconia* DC, an earlier name for the same Atlantic Island endemic, but has in any case been judged for over a century to be part of the genus *Notelaea* Vent. which dates from 1803.

A third *Henslowia*, published by Carl Ludwig Blume in 1850 in Leiden, was for an entirely different kind of Tropical plant: it was a semi-parasitic shrub belonging to the family Santalaceae. Henslow, we might guess, would have been quite proud to be commemorated with such an interesting plant, because, even as early as 1835, he discusses in his text-book the semi-parasitic mistletoe *Viscum* (also see below). In fact, this *Henslowia* persisted in the taxonomic literature for more than a century after Henslow's death, but eventually the rules of the International Code caught up, and in 1984 A.S. George correctly transferred all *Henslowia* species to Miguel's genus *Dendrotrophe*, and the third, and last, *Henslowia* disappears into synonymy.

The spelling *Henslovia*, which occurs in several important nineteenth-century works such as Hooker's *Flora of British India*, has to be treated under the rules as an 'orthographic variant' and does not affect the issue.

Henslow has been luckier with eponymy at the level of species. His son-in-law, Joseph Hooker, was responsible for two new species based on the Galapagos collections of Charles Darwin. One is the fern *Adiantum henslovianum* published in 1845; the other, *Viscum henslovii*, is a green parasite related to our mistletoe *Viscum album*, based upon Darwin's herbarium specimen bearing his field number 3244, and collected on Charles Island with his accompanying field note: 'Parasite – growing on various kinds of trees.' This plant was transferred to the genus *Phoradendron* in 1902, but the eponymous specific epithet survives in the name *Phoradendron henslovii* (Hook.fil.) Robins. For more detail on both these species, see Porter, D.M. 1980 (pp. 95 & 120), and Wiggins, I.L. & Porter, D.M. 1971 (pp. 87 & 252).

Appendix 5
Local botanical records

Henslow's botanical discoveries in the British flora are numerous, but consist almost entirely of what botanists now call 'new county records' (NCR). The exception, which we discuss in Chapter 7 (p. 122), is his recognition of *Fumaria vaillantii* as a native British plant, first in Kent, then later on the chalk hills in Cambridgeshire, where it still occurs as an arable weed. See Hanbury & Marshall 1899, and Henslow, J.S. 1832, for details.

In addition to *Althaea hirsuta*, mentioned in Chapter 7 as an NCR for Kent, Henslow was also responsible for the first record of *Vicia bithynica* for that County (see Hanbury & Marshall 1899, p. 102). A specimen in CGE of *Aster tripolium* labelled 'Frindsbury, Kent 7 Sept. 1825 H.D. Henslow' is, incidentally, the only evidence we have that Henslow's brother Henry David, who died in 1843 aged 44, ever collected plants.

Henslow's earliest botanical discoveries now recognised as NCRs were two plants collected in the Isle of Man in 1819 – the pondweed *Potamogeton polygonifolius* and the eyebright *Euphrasia micrantha*: we discuss these records in Chapter 3 (p. 27).

First records for Cambridgeshire (the 'old' County, which is vice-county 29 in the system used by botanists) were made by Henslow in the 1820s. In the list that follows we have included only those species that are fully documented and backed by a herbarium specimen in CGE. The data are supplied or checked by Mrs G. Crompton from her records for the forthcoming *Historical Flora of Cambridgeshire*.

Bunias orientalis   'Roadside near Cambridge, 16 June 1825'
This large alien Crucifer has been known to Cambridge botanists at 'roadside by railway off Brooklands Avenue', the locality where Humphrey Gilbert-Carter, first Director of the Botanic Garden, recorded it in 1932. It seems likely that this is Henslow's locality, but curiously there is *no* record between Henslow's of 1825 and Gilbert-Carter's more than a century later. Babington, 1860, ignores the plant completely.

Centaurea solstitialis   'Swaffham Prior, 4.8.1828'
An introduced annual or biennial, in Henslow's time naturalised in arable fields, and nowadays occurring as a casual from wood shoddy or bird seed.

Luzula sylvatica   'Wood Ditton, 5.6.1829'

The Henslow herbarium sheet in CGE contains also a specimen of *Luzula pilosa*. Both Babington 1860, and Perring *et al.* 1964, attribute the first record to J. Downes in 1832. John Downes (see Appendix 3 and Chapter 5) was one of Henslow's keenest pupils, and it seems very likely that he was introduced to the flora of Ditton Park Wood by Henslow himself. *Luzula sylvatica*, common in N. and W. Britain, is rare throughout East Anglia. Recent records for Cambridgeshire are confined to the boulder-clay woods on the S.E. edge of the County, including Ditton Park Wood.

Myosotis laxa subsp. *caespitosa* (M. caespitosa)   'Gamlingay, 4.8.1828'

This plant is rare in Cambridgeshire and several nineteenth-century records which lack voucher specimens must be considered doubtful. Henslow obviously knew the plant at Gamlingay, because his herbarium specimen in CGE is certainly correct. Welch 1961, who studied the Cambridgeshire material of the two Water Forget-me-nots, *M. scorpioides*, the common plant, and *M. laxa* subsp. *caespitosa*, suggests that the latter may be restricted to the more acid soils.

Ornithogalum nutans   'abundant but accidental, between Trumpington & Cambridge, 2.5.1821'

This record is discussed in Chapter 5 (p. 71). The plant is an 'infrequent garden escape which sometimes persists' (Perring *et al.* 1964).

Veronica persica (V. buxbaumii)   'Cambridge, 10.10.1826'

Native of W. Asia, this annual Speedwell, now abundant throughout lowland Britain, was first recorded as a British plant in 1825. Henslow's record, in the following year, is a tribute to his 'eye' for a new plant or animal. By 1860, Babington had recorded it (as *V. buxbaumii*) in more than a dozen localities in the County and, a century later, Perring *et al.* 1964, state that it is 'an abundant weed of arable land and waste places throughout the County.'

Henslow's first records for Suffolk (vice-counties 25 and 26) begin in 1840, the first season after he took up residence in the Rectory at Hitcham. As for Cambridgeshire, we are only listing here species for which there is a Henslow voucher specimen: all of these are in Herb. Henslow in the Ipswich Museum, with one exception, which is in CGE. The records have been compiled with the help of Martin Sanford, BSBI Recorder for Suffolk, to whom we are very grateful. We have not included some species attributed to Henslow in Hind 1889, because, even where Hind cites 'Herb. Henslow', we have been unable to trace a specimen. It could be that Hind had access to Henslow material of which we know nothing.

There are nine totally substantiated records.

Calamagrostis canescens (C. lanceolata)   'Hitcham Wood, 17.7.1852'
A relatively rare grass, perhaps also in Henslow's time.

Cannabis sativa   'Hitcham, April 1858'
This plant had none of its present-day notoriety in Henslow's time!

Carex caryophyllea (C. praecox)   'Hitcham, 1.6.1853'

C. sylvatica   'Hitcham, 23.6.1840'

C. vesicaria   'Hitcham Wood, 10.6.1853'
Three first county records for *Carex* species emphasise how Henslow took a special interest in the sedges.

Elodea canadensis (Anacharis alsinastrum)   'Hitcham Rectory Pond. August 1853'
This notorious aquatic, introduced into Britain from North America in the 1840s, is associated with the names of Babington and Murray (see Chapter 8, p. 147). It is tempting to suppose that Henslow himself brought material from Cambridge and liberated it into his own pond at Hitcham.

Elymus caninus (Triticum caninum)   There are two specimens, 'Hitcham, 30.7.1838' and 'Hitcham, 8.7.1853'.
Henslow was already familiar with this grass in Cambridge, as we can see from two specimens in CGE labelled by him 'H.C. 23.6.1826' and 'H.C. 25.6.1828'. Presumably H.C. means Hort. Cantab., the old Walkerian Botanic Garden. Both these specimens are called by Henslow *Elymus caninus* – a fine example of a Linnaean name now re-used exactly as Henslow had it, after nearly two centuries!

Hieracium maculatum (H. sylvaticum var. maculatum)   'Plantations, Hitcham, 10 June 1840'
This specimen is in CGE, not in Ipswich. Originally a garden plant, occasionally recorded as naturalised on old walls or outside gardens.

Juncus compressus   'Hitcham, July 1853'
A rather rare plant, almost certainly known to Henslow on the local
'moors' around Cambridge and at Gamlingay. Very rare now in Suffolk.

The first confirmed records of the following bryophytes in Cambridgeshire
(v.c. 29) are based on specimens collected by Henslow and checked by the
late Dr H.L.K. Whitehouse. With one exception the specimens are in CGE
and if not dated were collected between 1821 and 1835. The specimen of
Sphagnum auriculatum is in LTR.

   Nomenclature follows Hill *et al.* 1991–94 with the name used in Perring *et
al.* 1964 in parentheses if it differs.

   Bryum rubens   1834
   Dicranoweisia cirrata
   Drepanocladus revolvens   1827
   Fissidens incurvus   1821
   Isothecium myurum   1821
   Orthotrichum affine   1821
   Orthotrichum diaphanum   1821
   Pogonatum aloides   (*Polytrichum aloides*)
   Pottia davalliana   1821
   Rhynchostegiella tenella   1821
   Riccia fluitans
   Schistidium apocarpum (*Grimmia apocarpa*)   1821
   Sphagnum auriculatum (*S. subsecundum* var. *auriculatum*)
   Weissia longifolia (*W. crispa*)   1827

We are grateful to Chris Preston for providing this up-to-date information
on Henslow's bryophyte records.

CGE and LTR are the international *Index Herbariorum* abbreviations for the
Cambridge University Herbarium and the Leicester University Herbarium
respectively.

Endnotes

Abbreviations

C.U.L. Add: Cambridge University Library. Additional manuscripts

C.U.L. DAR: Cambridge University Library. Darwin Papers

C.U.L. EDR: Cambridge University Library. Ely Diocesan Records

C.U.L. UA: Cambridge University Library. Cambridge University Archives

DAR. Corr.: *The Correspondence of Charles Darwin*. Ed. F. Burkhardt, S. Smith et al. Vols. 1– . Cambridge University Press, 1985–

Jenyns Corr.: Jenyns Correspondence, Vols. 1–4 in the Jenyns Library, Bath Royal Literary & Scientific Institution

Kew Dir. Corr.: Royal Botanic Gardens, Kew, Archives. Directors' Correspondence

Kew, Letters to Henslow: Kew Archives. Letters to Henslow.

Kew Lindley Letters: Kew Archives. Lindley Letters.

S.R.O.: Suffolk Record Office, Bury St Edmunds

CHAPTER 1

1.1 Browne, J., 1995, chapter 1, is especially recommended.

1.2 The use of the final 'e' happens arbitrarily until the two sons of Sir John Henslow(e). Thereafter the John Prentis Henslows and the Edward Prentice Henslowes are consistent. Oddly, Charles Darwin seems to have been unsure at times whether his teacher should or should not have a final 'e'. In one letter to Fox in August 1830 he spells 'Henslow(e)' both ways, and again uses the final 'e' in another Fox letter in November 1830 (DAR. Corr. 1: 105–6 & 111). Of course there was no such doubt in any official use in the University, or in Henslow's own use.

1.3 Sir John Henslow used the shield and crest from this coat of arms for his bookplate, surmounting it on a ship of the line in dry dock with all flags flying.

1.4 The family Bible, which passed to John Stevens Henslow and then to his eldest son Leonard, has not been traced.

1.5 This information is taken from a typescript life of J.S. Henslow by A.R. Milne, dated 1961, S.R.O., Bury St Edmunds. FL/586/13/3. Milne says Sir John was 'a widely-travelled navy surveyor', misunderstanding his title.

1.6 *Naval Chronicle* **34**: 440 (1815).

1.7 Jenyns, L. 1862 p. 4.

1.8 *Ibid.* p. 4.

1.9 *Ibid.* pp. 4–5. In this passage we learn also that Thomas Stevens built himself a 'country seat', Gad's Hill House, near Rochester, to which he retired. After his death in 1844 this house was sold and eventually bought by Charles Dickens in 1857. Dickens wrote his later novels there, among them *A Tale of Two Cities* and *Great Expectations*.

1.10 *Ibid.* p. 6.

1.11 *Ibid.* p. 7.

1.12 Ibid. p. 6.

1.13 Ibid. p. 8.

1.14 Ibid. pp. 8–9.

1.15 Ibid. p. 11; for the date 1810, see Henslow, G. 1900–21: 301.

CHAPTER 2

2.1 Leedham-Green, E. 1996 p. 131.

2.2 Garland, M.M. 1980 p. 18.

2.3 Leedham-Green, E. 1996 p. 122.

2.4 Henslow, J.S. 1854 p. 9. In this passage Henslow is acting as guide to his parishioners: 'We then turn till we come to All Saints' Church on the left, and to the principal entrance to St John's College on the right. Entering the first court we see the Chapel on the right, opposite which are the rooms in which I *kept* (as we say in Cambridge) during the time I was a student.'

2.5 Details can be found in the *Cambridge University Calendar* from 1816 onwards.

2.6 Jenyns, L. 1862 pp. 12–13.

2.7 Ibid. p. 20.

2.8 Otter, W. 1824 is the source of much of our information on Clarke. See also Dolan, B.P. 1995.

2.9 Clarke, E.D. 1807; ed. 2, 1818.

2.10 Otter, W. 1824 2, p. 234.

2.11 William Farish was Professor of Chemistry 1794–1813, then Jacksonian Professor of Natural and Experimental Philosophy from 1813 to his death in 1837. He sought to apply chemistry to manufacturing and to study practical aspects of mechanics and engineering.

2.12 Wright, J.M.F. 1827 2: 30–31.

2.13 Dolan, B.P. 1995 p. 53. There is some doubt about the 'working model of Vesuvius' in Clarke's lectures, *fide* K. Edgar (pers. comm.).

2.14 McKie, D. & Mills, J. 1998 p. 49.

2.15 Evidence that Henslow as an undergraduate did not totally neglect botanical study exists in the form of a herbarium sheet dated 'Oct. 1817' and inscribed in Henslow's writing 'Plantago major var. pyramidalis (monstrosa): "spicate thyrse" a panicle'. This sheet, together with another made by Henslow in
1830 of an abnormal 'Rose Plantain' from 'garden, Cambridge', are in the Herbarium of the Royal Horticultural Society at Wisley, and came there via George Henslow's Herbarium which he left to the Society. On the 1817 sheet are mounted five inflorescences which show the 'monstrosity' of multiple branching. Such abnormalities in the common species of *Plantago* were familiar to Gerard as early as the sixteenth century, and Henslow was to develop his interest in their significance in interpreting the floral and the vegetative axis in his text-book of 1835 (pp. 79–81).

2.16 Leedham-Green, E. 1996 pp. 125–31.

2.17 *Cambridge University Calendar* (St John's) for 1818.

2.18 Barnard, Robert Cary MS, in possession of Barnard family.

2.19 Winstanley, D.A. 1940, p. 32 erroneously states that Henslow was a Fellow of St John's, and Russell-Gebbett, J. 1977 p. 15 repeats the error.

2.20 Jenyns, L. 1862 pp. 19–20

CHAPTER 3

3.1 Clark, J.W. & Hughes, T.M. 1890, the standard biography, makes good reading, and has been used by us as a primary source. Speakman, C. 1982 is written primarily from the angle of Sedgwick's devotion to his native Dent, but is particularly good also at assessing the significance of Sedgwick's teaching and his difficulties with Darwin's writings.

3.2 John George Shaw-Lefevre (his full name) graduated as Senior Wrangler in 1818, gained a Fellowship in Trinity the following year, and FRS in 1820. He had an illustrious career as a civil servant and was knighted in 1857.

3.3 Sedgwick visited Henslow in April 1861 during his terminal illness. See Jenyns, L. 1862 pp. 261–2.

3.4 Clark, J.W. & Hughes, T.M. 1890 1: 204.

3.5 Ibid. p. 281.

3.6 Ibid. p. 283.

3.7 Ibid. 2: 140.

3.8 Ibid. 1: 204.

3.9 The holder of the Woodwardian Chair was

expressly forbidden to marry, and Sedgwick's predecessor, Hailstone, had in fact resigned the Chair on marrying, thus creating the vacancy. See Clark, J.W. & Hughes, T.M. 1890 1: 152, 182.

3.10 Letter from Sedgwick to Mrs Lyell, 16 October 1837. See Clark, J.W. & Hughes, T.M. 1890 2: 488–9.

3.11 Interestingly, Sedgwick's obsession with his own health makes him much more like Darwin than Henslow. We are not aware of any re-assessment of this side of Sedgwick's character – unlike the case of Darwin, for whom attempts to explain his increasing hypochondria in later life became part of the 'Darwin industry'. See Clark, J.W. & Hughes, T.M. 1890 2: 498 and, for Darwin, see e.g. Desmond, A. & Moore, J.R. 1992 pp. 286–90.

3.12 Jenyns, L. 1862 p. 249

3.13 Clark, J.W. & Hughes, T.M. 1890 1: 160–2.

3.14 Entitled 'Supplementary Observations to Dr Berger's account of the Isle of Man', and published in *Transactions of the Geological Society of London* 1821 5(2): 482–505, this paper clearly established Henslow's claim to be a geologist. See Challinor, J. 1970 p. 225.

3.15 There were no Fellows of the Geological Society until 1825, when the Society received its Charter. See Woodward, H.B. 1907 pp. 68–70.

3.16 See Allen, D.E. 1984 p. 31; also Jenyns 1862 pp. 14–15.

3.17 Letter from Swainson to Henslow, 15 September 1826. C.U.L. Add. 8176.

3.18 Clark, J.W. 1891 p. ii.

3.19 The house still exists, substantially altered, and has been occupied by various tenants over the years.

3.20 The Giant Irish Elk was well known in the 18th century as an extinct animal whose bones and antlers were preserved in bogs. See Mitchell, G.F. & Ryan, M. 1997 pp. 87–9.

3.21 Jenyns, L. 1862 p. 17.

3.22 Curiously, Henslow makes a mistake in the initials of T.R. Underwood: the second paragraph of the Anglesey paper reads: 'I have to acknowledge my obligations to L.P. Underwood Esq., whose previ-

ous visits to Anglesea had enabled him to collect many interesting facts connected with its Geology . . .' There seems no doubt that this should read 'T.R. Underwood'.

3.23 Our attention was drawn to the Henslow references in the Sedgwick correspondence by Rosemary and David Gardiner, who were preparing an account of Underwood for the New D.N.B. The relevant letters are in the Sedgwick Ms in C.U.L. Add. 7652.

3.24 Clark, J.W. & Hughes, T.M. 1890 1: 234.

3.25 See the introduction by Secord to his edition of Lyell's *Principles of Geology* (Lyell, C. 1997 ed. Secord).

3.26 Lyell, K.M. (ed.) 1906 2: 108–9.

3.27 Whewell apparently first coined the word 'scientist' as late as 1834. Before that time, if a special designation were required, 'men (or more probably 'gentlemen') of science' would be used. Indeed, the word 'scientist' was liable to be stigmatised as an 'ignoble Americanism' or 'a cheap and vulgar product of transatlantic slang'! See Jesperson, O. 1938, and Ross, S. 1962.

3.28 Becher, H.W. 1986. This paper explores how Henslow, Sedgwick, Whewell and others laid the foundations of science in Cambridge in the second half of the 19th century. See also Garland, M.M. 1980, especially Chapter 2.

3.29 *Ibid.* pp. 61–2.

CHAPTER 4

4.1 R.C.H.M. N.E. *Cambs* pp. 6–8.

4.2 Blomefield (late Jenyns), L. 1889 p. 9.

4.3 See Walters, S.M. 1981 p. 41.

4.4 A copy of this portrait is at Bottisham Hall: the original was sold in 1917 to a buyer in the United States.

4.5 Jenyns, L. 1862 pp. 263–4.

4.6 Blomefield (late Jenyns), L. 1889 p. 11.

4.7 The M. Alexandré referred to was a ventriloquist whose recent act in 'the Theatre' in Cambridge is described in some detail in the *Cambridge Chronicle* for 21 March 1823. The sarcophagus lid, in which Harriet's mother had shown interest, was a recent acquisition by the Fitzwilliam Museum, described in the *Cambridge Chronicle* for 4 April 1823. The volumes of Schiller's works, 'in 12 volumes' of which Henslow sends three, must have been in the original German, since no multi-volume edition in English or French was published at the time. *Memoirs of the Life of Mary Queen of Scots*, by Elizabeth Ogilvy Benger had just (1823) been published in 2 volumes.

4.8 Blomefield (late Jenyns), L. 1889 p. 21.

4.9 Kew Dir. Corr. 9: 211.

4.10 Kew Lindley Letters vol. A–K: 387–9.

4.11 Henslow, Mrs J.S. 1848.

4.12 Romilly's Diaries cover the years 1818 to 1864, and the extracts given come from the originals in the Cambridge University Library, C.U.L. Add. 6804–49.

4.13 Romilly's phrase 'g.s. of the Botc Garden' to explain Charles Darwin, is unexpected, but emphasises how the young Darwin, back from the 'Beagle' voyage, was not yet widely known. The phrase obviously means 'grandson of Erasmus Darwin, author of *The Botanic Garden*'.

4.14 See also references to the Marchese di Spineto and Henslow's parties in Trevelyan, R. 1978, esp. pp. 10–15.

4.15 DAR. Corr. 1: 105, 109, 110. The originals of these letters are in Christ's College.

4.16 Litchfield, H.E., ed. 1904 pp. 460–1.

4.17 DAR. Corr. 2: 124.

4.18 DAR. Corr. 2: 140–1.

4.19 Browne, J. 1995 p. 400.

4.20 DAR. Corr. 2: 173. Henslow's playful address to 'Brother Benedict' (the character in Shakespeare's *Much Ado About Nothing*) implies a newly-married man.

4.21 Frances, Henslow's eldest daughter.

4.22 DAR. Corr. 6: 495.

CHAPTER 5

5.1 There is a very detailed account, explaining in particular Sedgwick's rôle in the controversy, in Clark, J.W. & Hughes, T.M. 1890 1: 235–45. See also Winstanley, D.A. 1940 pp. 32–41.

5.2 Henslow, J.S. 1823 pp. v–vi.

5.3 DAR. Corr. 1: 7.

5.4 Jenyns, L. 1862 pp. 27–8.

5.5 *Ibid.* p. 29.

5.6 *Ibid.* pp. 55–6.

5.7 Looking back on his appointment as Professor of Botany, Henslow writes in 1846: 'When appointed to the Botanical Chair I knew very little indeed about Botany, my attention having before been devoted chiefly to other departments; though I probably knew as much of the subject as any other resident in Cambridge.' (Henslow, J.S. 1846 p. 16)

5.8 Jenyns, L. 1862 p. 29.

5.9 *Cambridge Independent Press* 11 June 1825.

5.10 C.U.L. Add. 8176, no. 30.

5.11 Speech by Henslow in Ipswich Museum archive (George Ransome scrapbook no. 2), by courtesy of Ipswich Borough Council Museums and Galleries. See also Freeman, J. 1852 pp. 344–5.

5.12 C.U.L. UA O.XIV.109. See also Clark, J.W. 1904 pp. 194–5.

5.13 The Senate Grace requesting Sir John Richardson 'to determine . . . the manner in which the Professors of Mineralogy, Botany and Anatomy are in future to be determined' is dated 22 December 1825. 'The determination' took nearly two years, being finally published on 1 December 1827. See Clark, J.W. 1904 pp. 219–20.

5.14 C.U.L. UA U.P.7 f29.

5.15 *Ibid.* U.P.4 f173.

5.16 *Ibid.* U.P.4 f303.

5.17 Jenyns, L. 1862 p. 29.

5.18 Letter from Whewell to 'his aunt and sisters' dated 4 April 1828, reproduced in Stair-Douglas 1881 pp. 121–2.

5.19 For more detail, see Becher, H.W. 1986.

5.20 See Ransford, C. 1846, quoted in Fletcher, H.R. & Brown, W.H. 1970 p. 110.

5.21 C.U.L. Add. 8176 no. 47.

5.22 Kew Dir. Corr. 1: 166–7.

5.23 Henslow, J.S. 1846 p. 19.

5.24 C.U.L. UA O.XIV. 261.

5.25 Analysis of the attendance lists for undergraduates and others cannot be very precise, because for several of the years covered the total figures given in the original lists do not quite add up to the analysis total. Thus, for 1833, the analysis gives 71 undergraduates and 10 'others', making a total of 81, but the original is 78. The discovered discrepancies are not large, and do not affect the general picture.

5.26 Babington, C.C. 1897 pp. 296–7.

5.27 For more detail on Babington, see Walters, S.M. 1981 pp. 65–70.

5.28 See Walker, M. 1988 pp. 45–8.

5.29 For an account of the life, work and international influence of John Hope, bringing out his abilities both as a botanist and an administrator, see Morton, A.G. 1986.

5.30 Smith, J.E. 1795, from the Preface.

5.31 Jenyns, L. 1862 pp. 38–9. Many of Henslow's original 'wall paintings' are in the teaching collection of wall diagrams in the Whipple Museum of the History of Science, on permanent loan from the Department of Plant Sciences, Cambridge (Plate 3).

5.32 Gascoigne, J. 1988 pp. 182–3.

5.33 De Candolle invented the word 'taxinomie' in 1813 for the study of the classification of plants and animals. The translated term 'taxinomy' does not appear in Henslow's time, but the etymologically incorrect spelling 'taxonomy' seems to be first used in English in de Candolle, A.P. & Sprengel, K. 1821, a work translated anonymously 'from the German' and published in Edinburgh. Presumably Henslow knew this book though he does not cite it anywhere. It may be that the Edinburgh translator (probably Jameson) is therefore responsible for the consistent use of the spelling 'taxonomy' in English. Modern French dictionaries give both spellings.

5.34 An interleaved copy of the second edition, inscribed 'Frances Harriet Henslow, from her affte Papa: February 22nd 1840. Hitcham Rectory', is in the possession of the Barnard family, to whom we are most grateful for showing it to us. The many neat annotations confirm records for plants seen at Hitcham in 1840: whether they are in Frances' own hand seems doubtful, as the writing is rather mature for a girl of 15.

5.35 See, for example, an end-note in Morton, A.G. 1981 p. 442, an excellent book on the history of botany.

5.36 Proctor, M.C.F. et al. 1996 pp. 18–19. Note that the author is Christian Konrad Sprengel, not Kurt Sprengel (see note 5.33). The fourth edition of Kirby, W. & Spence, W. 1822 1: pp. 292–6, surely familiar to Henslow, contains an acceptance of C.K. Sprengel's ideas.

5.37 See Lloyd, F.E. 1942.

5.38 Jenyns, L. 1862 p. 43.

5.39 Henslow's text-book was first published in 1835 and appeared in identical form, but with an additional title-page, in 1836 as part of Lardner's *Cabinet Cyclopaedia*, a series of books totalling 133 separate volumes covering 'History, Biography, Literature, The Sciences, Arts, and Manufactures, comprising contributions from "the most eminent writers of the age in the various departments"' (to quote from the 15-page Catalogue).

5.40 Copies of these Annual Reports are in the University of Cambridge Department of Plant Sciences, Miscellaneous Botany +AD6.

5.41 Copies are also in Miscellaneous Botany +AD6.

5.42 Henslow, J.S. 1832. It is probable, but not certain, that the 1500 specimens (some preserved in Cambridge University Herbarium) were collected at Coton. See Porter, T.M. 1986 for a detailed treatment of the history of statistics. Quetelet, who contributed much to the subject, was an invited foreign guest at the Cambridge meeting of the newly-formed British Association in 1833; it is interesting to speculate whether he and Henslow discussed meristic variation, but, alas, we have no information!

5.43 Text Ms designed to accompany the *Fasciculus* dated 'Cambridge 30 Apr. 1834' is in Miscellaneous Botany +AD6 in the Department of Plant Sciences Library. No printed copy of this text has been found, but most if not all of the actual specimens are represented in Cambridge University Herbarium.

CHAPTER 6

6.1 Browne, J. 1995 p. 97.

6.2 Gruber, H.E. 1974 pp. 73–93, can be consulted for

an assessment of the rôles of Henslow and other teachers in influencing the young Darwin.

6.3 Browne, J. 1995 p. 118.

6.4 de Beer, G., ed. 1974 p. 33.

6.5 See Garland, M. 1980; also, for detail of the formal changes in the examination system in Cambridge, see Leedham-Green, E. 1996, especially Chapter 4.

6.6 Leedham-Green, E. 1996 p. 141.

6.7 de Beer, G., ed. 1974 pp. 32–3.

6.8 See letter from Darwin to Fox, dated 3 November 1829: 'Professor Henslow's parties go on very well. The last was the pleasantist I was ever at.' DAR Corr. 1: 95.

6.9 Browne, J. 1995 p. 123.

6.10 Ibid.

6.11 The Jenyns papers are preserved in the Bath Royal Scientific & Literary Institution as 'Correspondence relating to Memoir of Professor Henslow'. We are grateful to Roger Vaughan, who has kindly made available much unpublished material relevant to Henslow.

6.12 Berkeley has not quite remembered the name correctly. See note 4.14.

6.13 Jenyns Corr.; see note 6.11.

6.14 Henslow presented an account of this study as a paper read to the Cambridge Philosophical Society on 14 November 1831, and published it in the *Transactions* of the Society in Vol. 4(2) p. 257–78 in 1832. A facsimile edition, with a Preface by S.M.W. and an Introduction by V.H. Heywood, was published by the University Botanic Garden in 1981, on the occasion of the Sesquicentenary of the Act of Parliament by which the University acquired land to make the present Garden.

6.15 Browne, J. 1995 p. 109.

6.16 de Beer, G., ed. 1974 p. 37.

6.17 Quoted by Jenyns, L. 1862 p. 52.

6.18 Quoted by Browne, J. 1995 p. 131: the original is C.U.L. DAR 112 (ser. 2): 120.

6.19 Babington, C.C. 1897 p. 8.

6.20 Quoted by Browne, J. 1995 p. 131. C.U.L. DAR 112 (ser. 2): 118.

6.21 Ibid. pp. 131–2. C.U.L. DAR 112 (ser. 2): 95–6.

6.22 Ibid. p. 132. Darwin's dislike of Babington persisted; in letters to Joseph Hooker after Henslow's death he says:
[18 May 1861] 'I suppose Babington will be Professor at Cambridge. What a contrast!' followed by [24 May 1861] 'Pray give my kindest remembrances to L. Jenyns, who is often associated with my recollection of those old happy days. It is really fearful to think of Babington in that Chair.' Both extracts are from DAR Corr. 9: 133 & 137. If Darwin and Babington were mutually incompatible and remained so, the same could not be said for Darwin's relations with the other enthusiastic local insect collector, Leonard Jenyns himself. Darwin recalls in later years that:

At first I disliked him from his somewhat grim and sarcastic expression; and it is not often that a first impression is lost; but I was completely mistaken and found him very kind-hearted, pleasant and with a good stock of humour. I visited him at his parsonage on the borders of the Fens, and had many a good walk and talk with him about Natural History. Darwin, C.R. 1887

6.23 Jenyns Corr. (3) 6.

6.24 Audubon, M.R. & Coues, E. 1898 1: 285–90.

6.25 The initial is wrong: the Fellow of Corpus who acted as host to Audubon must have been the Revd William Greenwood, Fellow from 1821 to 1828.

6.26 See Chalmers, J. 1993.

6.27 These deliberations took nearly two years; the Minutes of the Council Meeting of the Philosophical Society on 14 July 1924 record that: 'It was agreed to accept the offer made by Mr Gabriel Wells of £400 for the purchase of the Society's copy of "Audubon's Birds of America", consisting of five volumes of plates and five volumes of text.'

A fourth copy of the set of *Birds of America* published by 1828, which clearly dates back to the Audubon visit, was in the Library of King's College until 1923, when the College sold it. Perhaps it was the Philosophical Society's discussions (which included the fact that there were duplicate copies in

the University Library and in King's College
Library) that alerted the College to the possible sale
value. Be that as it may, the set, which included the
Ornithological Biography, realised £397 net at
Sotheby's – a very similar yield!

6.28 Mearns, B. & R. 1992 pp. 234–6.

6.29 Audubon, J.J. 1831 1: 360–1.

6.30 Jenyns' 'Aves Notebook' (*CUL Newton Papers*, 48)
contains many records of birds 'caught' by
Henslow in 1832–33.

6.31 Jenyns, L. 1862 pp. 20–1. *Macroplea equiseti* is a
Chrysomelid beetle, now called *M. appendiculata*.
See also Appendix 4 for *Pisidium henslowanum*.

6.32 Jenyns, L. 1832.

6.33 Jenyns Corr. 1(1).

6.34 Experts in the Museum of Zoology inform us that
Henslow's Snail was probably a juvenile form of a
common species, either *Trichia hispida* or
Acanthinula aculeata.

6.35 Jenyns, L. 1869 Ms *Collections towards a Fauna
Cantabrigiensis: Vertebrata & Mollusca* in Museum of
Zoology, University of Cambridge, Notice [i.e.
Introduction].

6.36 *Ibid.* p. 129.

6.37 See Marr, J.E. & Shipley, A.E., eds. 1904 pp. 130–1;
also Pollard, E. 1974 pp. 239–42 and 1975 pp.
305–29.

6.38 R. Preece, Museum of Zoology, Cambridge, *verb.
comm.*

6.39 Jenyns, L. 1862 p. 23.

6.40 This and other extracts (mainly but not exclusively
botanical) have been recently published by G.
Crompton in a paper entitled *Botanizing in
Cambridgeshire in the 1820s. Nature in Cambridgeshire*
39: 59–73. The original Jenyns Natural History
Notebooks are no. 24 in the Newton Papers in the
Cambridge University Library.

6.41 DAR Corr. 1: 125–6.

6.42 Barlow, N. 1967 pp. 28–9.

6.43 DAR Corr. 1: 139.

6.44 Browne, J. 1995 p. 152.

6.45 Jenyns, L. 1862 p. 50.

6.46 Browne, J. 1995 p. 136.

6.47 DAR Corr. **3**: 55.

6.48 For an assessment of the importance of this field
trip with Sedgwick to the young Darwin, see
Browne, J. 1995 pp. 140–3.

6.49 DAR Corr. 1: 238.

6.50 Browne, J. 1995 pp. 144–61 gives a detailed account
of Darwin's acceptance of the 'Beagle' offer.

6.51 This correspondence is most conveniently studied
in Barlow, N. 1967, which also contains an excellent introduction summarising the relationship of
Darwin and Henslow.

6.52 Henslow's first letter to Darwin, dated 6 February
1832, is given in full in DAR Corr. 1: 199–200, but
unaccountably omitted from Barlow, N. 1967;
Darwin's reply, of 16 June 1832, is given in both
sources.

6.53 Paley's *Evidences of Christianity*, a set book for BA
examination.

6.54 Richard Thomas Lowe, 1802–74, a keen pupil of
Henslow, who wrote a *Flora of Madeira*.

6.55 C.U.L. DAR 204.6.2.

6.56 DAR Corr. 1: 344–5. The 'footnote²' to this letter is
incorrect in saying that 'there is no record of when
the earlier letter [i.e. Henslow's letter of 6 February
1832] was received'. Darwin received this letter
when the 'Beagle' arrived in Rio de Janeiro on 4
April 1832, and reports this in his letter to Henslow
posted on 16 June 1832 (DAR Corr. 1: 237).

6.57 No letter from Henslow dated 12 December 1833
survives. Darwin has noted the error in the date
written by Henslow on the important letter –
though he does not exactly clarify it. See Barlow, N.
1967 pp. 65 & 91.

6.58 See Browne, J. 1995 pp. 279–80.

6.59 *Ibid.* pp. 283–8.

6.60 Barlow, N. 1967 pp. 89–91.

6.61 DAR Corr. 1: 461–3. The footnote explains that the
date on the letter should be 12 <u>August</u>. In fact, the
'Beagle' did not sail from South America until 6
September.

6.62 Browne, J. 1995 pp. 296–7.

6.63 See Browne, J. 1995 pp. 316–20.

6.64 DAR Corr. 1: 512–13.

6.65 There is a plaque marking the house in Fitzwilliam Street where Darwin stayed.

6.66 Henslow, J.S. 1837b (Galapagos Opuntia) and 1838 (Flora Keelingensis).

CHAPTER 7

7.1 All traces of Gothic Cottage, Regent Street have disappeared. The Henslows' second house in Park Terrace, which they occupied only from 1836 to 1839, was one of the two smaller houses, now nos. 7 and 8, which had been built in 1831 and was leased from Jesus College. The name 'St Andrew's Place' was used for these houses in the 1841 census.

7.2 Jenyns, L. 1862 p. 57.

7.3 A 'perpetual curate' officiated in a parish to which he was nominated by the Lay Rector and licensed by the Bishop. Once licensed, the curacy became permanent and could only be revoked by the licensing Bishop. Payment was by agreement. We are grateful to Mrs P.M. Hall for this information on Henslow's Little St Mary's curacy from the files of Peterhouse, the Lay Rector.

7.4 Henslow, J.S. 1846a pp. 16–17.

7.5 Barlow, N. 1967 p. 61.

7.6 Ibid. p. 90.

7.7 Knight, F. 1995 pp. 131–2.

7.8 Bennett, R.M. 1974 pp. 632–4.

7.9 According to V.C.H. Cambridge 9: 90–107, the Patronage of Histon Church had been in the hands of the Michell family since 1732 and T.P. Michell himself was Patron and Vicar. This may be one of the reasons why he decided finally not to move.

7.10 Kew Dir. Corr. 8: 7 & 9: 218.

7.11 Jenyns, L. 1862 p. 20.

7.12 See Leedham-Green, E. 1996, Chapters 4 & 5, for an excellent, concise account of reform of the University.

7.13 Pettit-Stevens, H.W. 1899 p. 168.

7.14 See Winstanley, D.A. 1955, especially Chapter 6, and Pettit-Stevens, H.W. 1899 p. 169.

7.15 DAR Corr. 1: 200.

7.16 Babington, C.C. 1897 pp. 80, 99. Thomas Starkie was Downing Professor of the Laws of England 1823–49.

7.17 C.U.L. Add. 8339.

7.18 Clark, J.W. & Hughes, T.M. 1890 1: 277–9: the original pamphlet is Lamb, J. & Henslow, J.S. 1826.

7.19 For the political background to the Reform Bill and the new Whig government, Adams, G.B. ed. 2 1963, Chapter 18 can be recommended. See also Smyth, C. 1939.

7.20 DAR Corr. 1: 175.

7.21 DAR Corr. 1: 199.

7.22 Jenyns, L. 1862 p. 62.

7.23 The Cambridge Independent Press, from which the extract is taken, and the rival Cambridge Chronicle both report the cases heard on 19 March 1835. It is interesting to read both. The Chronicle had Tory sympathies, and reported that Mr Sergeant Storks, defending, 'severly commented' on the conduct of the Rev. Prof Henslow in mixing himself up in such affairs and observed that in addition to his other honours the learned and distinguished Professor had now acquired the honourable title of Common Informer.

7.24 It is tempting to suppose that the wall of Corpus Christi College became the target for the 'ruffians' to inscribe the prominent graffito because of the joint action of Lamb, Master of the College, and Henslow; but it may be simply that the brand-new building provided a perfect wall for defacement!

7.25 Leedham-Green, E. 1994 pp. 182–3.

7.26 There are three 'Committal Books' covering the years 1823 to 1894: C.U.L. UA Tvii nos 1–3.

7.27 Holbrook, M. 1999 p. 6.

7.28 Allen, D.E. 1996 p. 108.

7.29 Ibid. pp. 112–13.

7.30 C.U.L. Add. 8176: 77.

7.31 Henslow, J.S. 1828c. See also Hanbury, F.J. & Marshall, E.S. 1899 p. 66 for detail of the Althaea hirsuta record: surprisingly, they completely ignore Henslow's 1828 record, which antedates that of Woods (1830).

7.32 Henslow, J.S. 1832c. See also Hanbury, F.J. & Marshall, E.S. 1899 p. 66.

7.33 Published in Magazine of Natural History 4: 383.

7.34 Ibid. 5: 301–2.

7.35 Ibid. 294.

7.36 *Ibid.* 515–16.

7.37 Kew Dir. Corr. **9**: 218. See also Humphreys, J. 1926. The *Cactus speciosus* note is in *The Botanist* **1** no. 12. This familiar group of day-flowering cacti is nowadays generally referred to as *Heliocereus speciosus* and hybrids, but the nomenclature is complex (David Hunt, *pers. comm.*).

7.38 Henslow, J.S. 1849c.

7.39 Jenyns devotes a whole chapter to the subject (Chapter 8, pp. 220–31). See also Henslow, G. 1900, who describes how in 1843 he 'used to explore with his father the cliffs near [Felixstowe] where were the remains of a Roman or Romano-British "kitchen-midden"'.

7.40 Edward Martin, pers. comm. For an earlier assessment of Henslow's papers, together with a complete reprinting of the two papers, see Babington, Churchill, 1874.

7.41 Gage, J. 1838.

7.42 See Coates, R. 1999, which gives many literature references.

CHAPTER 8

8.1 See Walters, S.M. 1981 pp. 45–6.

8.2 See Le Rougetel, H. 1990 pp. 146–8.

8.3 Gorham, G.C. 1830 p. 108.

8.4 See Walters, S.M. 1981 pp. 36–7.

8.5 Martyn, T. 1807.

8.6 W.J. Hooker published this paper in Hooker's *Botanical Miscellany* **1**: 48–78 (1830). The original article, in German, by Schultes is in *Flora* **8**(1) Beiheft. **1**: 1–48 (1825).

8.7 Ackermann, R. 1815. The artist was W. Westall.

8.8 Illustration by Le Keux, J. in Cooper, C.H. 1860–66.

8.9 Illustration from Harraden, R. 1809–11.

8.10 Illustration from Henslow's article on the Botanic Garden in Smith, J.J. 1840 pp. 81–5.

8.11 For details of Thomas Martyn's career see Walters, S.M. 1981 pp. 36–46.

8.12 J.D., in *Magazine of Natural History* 1833 **6**: 397. J.D. was probably John Denson, Loudon's editorial assistant.

8.13 Henslow, J.S. 1846a p. 15.

8.14 All Botanic Garden Trustees and Syndicate Minutes quotations are from the originals in C.U.L. UA Char. 11.13 and C.U.L. UA CUR 25.1.

8.15 Illustration from Harraden, R.B. 1809.

8.16 This draft agreement, though dated 2 November 1839, carries the names of William Chafy (Vice-Chancellor) and *George* Bullen, though George had been dead for six months. The account in Walters, S.M. 1981 pp. 60–1 is incorrect in attributing to George Bullen the failure of negotiations to terminate the lease.

8.17 The five-page original manuscript report, dated 'September 1830', and obviously the one presented by Lapidge to the Syndicate between 4 and 6 October of that year, is preserved in the University Library archives (C.U.L. UA CUR 25.1.17). One of the two accompanying plans – the design for the original 'ground plan' for the Garden – does not seem to have survived. Very fortunately, however, this was published in Henslow's article on the Botanic Garden in Smith, J.J. 1840.

8.18 It is clear that from the first the University did not envisage, after accepting Lapidge's report and plans, that the whole site purchased from Trinity Hall the following year should necessarily be developed as a botanic garden. Indeed, the implication of Lapidge's report is that the new Botanic Garden (in the strict sense) should occupy only *half* (19 of the 38 acres) of the purchased land. To that extent, the statement made in Walters, S.M. 1981 (p. 62) concerning Murray's rôle in developing only the western half of the site is somewhat misleading. What Lapidge strongly recommended – that the whole site should be bought to prevent inappropriate building development and protect the Garden itself – could be said to be the policy consistently held to throughout the new Garden's history.

8.19 The 'island' and the lake eventually constructed, and still essentially unaltered today, incorporated most of these suggestions – but 'vaults underneath for Fungi and Mosses' presumably proved too difficult!

8.20 Kew Dir. Corr. **1**: 166.

8.21 C.U.L. Add. 8176. 51.

8.22 The Syndicate Minute is undated, and even the year has to be deduced from the reference to the lapse of three years since the lease was valued in 1830.

8.23 No. 247 in Maund's journal *The Botanist*, and bound in *The Floral Keepsake* (see Bibliography). The modern name for Henslow's plant is *Sinningia douglasii*.

8.24 Henslow, J.S. 1840b. This paper is cited in Walters, S.M. 1981 as 'Smith, J.J. 1840'. Smith was the *editor* of this short-lived Journal *The Cambridge Portfolio*. It is clear from the 'List of Signatures' in vol. 1. that Henslow is the author of the article entitled 'The Botanical Garden', though the entry 'J.H. J.S. Henslow, M.A. Professor of Botany 87' is misleading, since the article is on pp. 81–5 (not p. 87), and the initials 'J.H.' are not printed at the end of the article. Jenyns, L. 1862 seems to be unaware of the paper.

8.25 C.U.L. UA CUR 25.1. 25(3).

8.26 Trinity College Wren Library Add. Ms. a65/30 and 65/31.

8.27 Walters, S.M. 1981 p. 61. The precise reference to the form and conduct of the examination to select Murray has not been traced.

8.28 Kew, Letters to Henslow 153.

8.29 Kew Dir. Corr. **23**: 309.

8.30 Babington, C.C. 1897 pp. 124–40. It is particularly surprising that there is no mention in the Journal of the official opening of the New Botanic Garden by the Vice-Chancellor on 2 November 1846, although Babington does tell us on the 11th that he has been appointed to the Syndicate 'to superintend the new Botanic Garden'. It is, of course, possible that Babington's widow, who made the selection of material to publish, omitted the entry for the 2nd – for obscure reasons!

8.31 The 'collection of plants in the garden' refers to the Old Botanic Garden, which continued to function until 1846.

8.32 C.U.L. UA CUR 111 (1845:6a). The original letter is damaged, and [] in the extract indicate that the exact words are uncertain.

8.33 In Walters, S.M. 1981 pp. 61–2 the then recently rediscovered plan of Murray was assumed to be the

original one which he prepared for the appointment examination. This cannot be the case. The plan is obviously the one made and used by Murray during his Curatorship, and incorporates agreed features arising from his discussions with Henslow during 1846.

8.34 Murray, A. 1848 and 1850.

8.35 A letter dated 6 November 1846 from Henslow to Hooker (Kew Dir. Corr. **24**: 265) says '20 men are deeply trenching 7 acres for the reception of trees for the Arboretum.' The metal label at the foot of the Common Lime tree ('*Tilia vulgaris*', now T. *europaea*) which the Vice-Chancellor planted was unfortunately stolen in the 1980s. There is an illustration of this label in Walters, S.M. 1981 p. 57. The tree, now more than 150 years old, still grows immediately to the south of the steps from the formal entrance on Trumpington Road. Curiously, the stolen label clearly bore the date 9 November 1846, whereas the ceremony was performed on the 2nd. The reason may be that the Vice-Chancellor, Ralph Tatham, was reaching the end of his term of office, and therefore moved the ceremony forward by one week. The metal commemorative label bore the original date and could not be changed!

8.36 *Gardeners Chronicle* 1850, p. 438.

8.37 Babington complains twice to Balfour about Henslow, first in a letter dated 12 April 1846 (Edinburgh Botanic Garden Balfour Corr. **2**: 48) and again in a letter dated 2 June 1846 and published in Babington, C.C. 1897 p. 297. The quotation in Chapter 5 is from the June letter.

8.38 *Report of Her Majesty's Commissioners appointed to inquire into the state . . . of the University and Colleges of Cambridge . . .* (The Graham Commission) Parliamentary Papers 1852–53 House of Commons xliv. See also Leedham-Green, E. 1996 pp. 153–5.

8.39 The Plate (Figure 31) accompanying the booklet shows a rather different set of tender plants from those on the list in the booklet. Perhaps Henslow used a plate he had made for some other purpose, and hoped his parishioners would not notice.

8.40 For the history of the Garden after Henslow, see Walters, S.M. 1981.

CHAPTER 9

9.1 Darwin, F., ed. 1885 p. 52. As an example of the misuse of this quotation, see the footnote on DAR Corr. 1: 110 about Henslow's 'curious religious opinions' of which Darwin has heard rumours. The footnote fails to connect this Darwin remark with the notorious Henslow sermon in Great St Mary's Church, and says simply: 'Apparently the rumour was groundless', quoting as evidence the passage from the Darwin Autobiography only!

9.2 Jenyns, L. 1862 p. 46.

9.3 Ann Prentis was baptised in Maidstone Earl St Presbyterian Church on 9 March 1739.

9.4 See Gow, H. 1928, especially pp. 50–2.

9.5 For the 'Cambridge network' centred on Whewell and Trinity College, see especially Cannon, S.F. 1978 pp. 48–51.

9.6 See end-note 4.12.

9.7 See Hitchin, N. in Taithe, B. & Thornton, T., ed. 1997 pp. 133–4.

9.8 Jenyns, in the Preface to the *Memoir*, says that Henslow 'left behind him no journals or other papers' [that could be used to write a biography] and no diary has been traced. The statement in Bridson, G. et al. 1980 that Henslow's diaries are at Kew seems quite erroneous.

9.9 Jenyns, L. 1862 p. 132.

9.10 It is possible that Henslow's earlier sermon in Great St Mary's Church was the occasion mentioned by Audubon in his European Journal, p. 290. On Sunday 9 March 1828 Audubon tells us that 'he went to Church with Mr. Whewell at Great St Mary's, and heard an impressive sermon on Hope from Mr. Henslow'. (Audubon, M.R. & Coues, E. 1898 p. 290).

9.11 Wolffe, J., ed. 1995, especially the Preface pp. 12–13. Recent studies suggest that the extent and variety of millenarian and other potentially 'dangerous' or heretical ideas within the Anglican Church in the eighteenth century have been remarkably neglected in the history of science. In particular, there is a tendency to ignore the alchemical writings of Isaac Newton and the heretical views of his pupil and successor as Lucasian Professor of Mathematics, William Whiston. Dobbs, B.J.T. 1975 on Newton, and Force, J.E. 1985 on Whiston are strongly recommended. Henslow sent a copy of his printed sermon to 'Millennial Marsh' (the Revd William Marsh) at Colchester, and received an appreciative acknowledgment from him (C.U.L. Add. 8176:106).

9.12 See Vidler, A. 1961 p. 36.

9.13 Kew Dir. Corr. 1: 170.

9.14 See Vidler, A. 1961 p. 36.

9.15 See Nockles, P.B. 1994, which contains much information about the fore-runners of what became known in late Victorian England as the 'Oxford Movement'.

9.16 Darwin to Fox 10 May 1830: 'I have seen a good deal of Henslow lately & the more I see of him the more I like him. I have some thoughts of reading divinity with him the summer after next.' (DAR Corr. 1: 104).

9.17 DAR Corr. 1: 110.

9.18 See Roberts, M. 1998; even this author, whose paper is a richly-documented refutation of the simplistic view that most Churchmen in the early nineteenth century were still interpreting Genesis literally, seems to accept at its face value Darwin's '39 Articles' assessment of Henslow (p. 241).

9.19 Herbert, S. 1992 p. 69. This paper provides a detailed assessment of the early religious controversies over geological quesions at the time of Darwin.

9.20 *Ibid.* p. 75.

9.21 Henslow, J.S. 1823b.

9.22 Although Henslow attributes his explanation of Noah's Flood to Greenhough only, he could hardly have been ignorant of the fact that Halley in particular had put forward essentially the same idea more than a century earlier. See Force, J.E. 1985 p. 178, end-note 99.

9.23 Quoted by Herbert, S. 1992 p. 79. The introduction by Secord to the Penguin edition of Lyell's *Principles of Geology*, 1997, should also be consulted.

9.24 Henslow, J.S. 1830a.

9.25 Quoted in Raven, C.E., ed. 2 1950, repr. 1986 p. 189.

9.26 A letter from his friend J.S. Huntley, dated 18

February 1824 (C.U.L. Add. 8176) alerts Henslow to the fact that Biggs is to receive 'a plant of the *Primula vulgaris & P. veris* on the same root'. This is clear evidence that Henslow was already interested, and probably already growing *Primula* plants for investigation, before 1826.

9.27 The locality is Westhoe Park, near Bartlow. See Preston, C.D. 1993. The specimen in the Cambridge University Herbarium cited by Preston on p. 42 is *Primula elatior*, and fixes the date of Henslow's visit as 8 April 1826. Henslow's description suggests that there was a large population of the cowslip–oxlip hybrid 'in fields and copses' there, with few true cowslips. Preston discusses the problem of interpreting these early records, in view of the confusion, which persisted until Babington's *Flora of Cambridgeshire* (1860), between the 'true oxlip', *Primula elatior*, and the 'false oxlip', which is the hybrid between cowslip, *P. veris*, and primrose, *P. vulgaris*.

9.28 The *Potentilla* paper appeared in Maund's *The Botanic Garden* **5**: no. 385 (1833). For an account of the history of biosystematics, see Briggs, D. & Walters, S.M., ed. 3 1997, Chapters 1–5.

9.29 See DAR Corr. **2**: 179–86.

9.30 The full text of Darwin's letter to Henslow dated 3 November 1838 is now available in the Darwin Manuscript Collection, C.U.L. DAR **249**: 83–5, but is not yet published. We are grateful to Duncan Porter and David Kohn for information about this letter and Henslow's reply.

9.31 The *Lophospermum* paper is in Maund's *The Floral Keepsake* No. 242.

9.32 Trinity College, Wren Library Add. Ms a206/68.

9.33 See Desmond, A. & Moore, J. 1991 pp. 476–7.

9.34 DAR Corr. **7**: 370.

9.35 DAR Corr. **8**: 78 (footnote).

9.36 *Macmillan's Magazine* (1861) **3**: 336. See also Jenyns, L. 1862 pp. 212–13.

9.37 C.U.L. Add 8177: 198–9.

9.38 DAR Corr. **8**: 87–8.

9.39 *Ibid.* 97.

9.40 *Ibid.* 200–2.

9.41 *Ibid.* 208–9.

9.42 For a recent account of the Oxford Meeting, see Desmond, A. & Moore, J. 1991 pp. 492–8. See also DAR Corr. **8**: xx–xxi and pp. 270–2.

9.43 DAR Corr. **8**: 270. 'Burked' means 'prevented from speaking'.

9.44 From a wealth of recent literature, the book by the Regius Professor of Divinity at Oxford University (Ward, K. 1996) can be strongly recommended.

CHAPTER 10

10.1 Knight, F. 1995 p. 1.

10.2 Austen, J. 1814 (1949 ed.) p. 85.

10.3 See Cole, G.D.H., ed. 3 1947 p. 227. Cobbett died in 1835, before the passing of the Tithe Redemption Act of the following year, but the influence of his writings on the complex movement of working-class agitation was enormous and lasted long after his death. For a more recent, and less obviously left-wing, assessment of Cobbett's life, see Spater, G. 1982.

10.4 See Russell, A. 1986 p. 221. Anthony Russell's book provides a very readable account of the life of the English country parish over the centuries. Unfortunately, the only reference to Henslow (p. 228) confuses Henslow and Jenyns. Two other books by the same author (Russell, 1980 & 1993) contain a great deal of material that is relevant to Henslow's problems at Hitcham: Chapter 5 in his most recent book (1993) is particularly recommended.

10.5 David Turner, *pers. comm.*

10.6 Jenyns, L. 1862 pp. 69–70.

10.7 *Ibid.* p. 70.

10.8 Several books survey the relationship between squire and parson in English village life in Georgian and Victorian England. One in particular has become a deserved success: Owen Chadwick's *Victorian Miniature* (1960). Based upon the unusually fortunate survival of both the squire's and the parson's diaries for 31 years in the same Norfolk parish of Ketteringham, the book describes easily and vividly the relationship between these two local dignatories, representing the secular and the religious worlds. The period covered, 1838–69, is roughly contemporary with ours, and the story unfolded by Chadwick has many rich resonances with the Hitcham scene. (A good, if trivial, example, is the evangelical parson's doubt whether a lecture sponsored and hosted by the squire on *The Structure of Plants* by a Revd Mr Sidney was suitable, and religious enough, for a minister of Christ (p. 97).)

10.9 David Turner, *pers. comm.*

10.10 Morsley, C. 1961 p. 11. This article, published for the centenary of Henslow's death, is perhaps too optimistic and laudatory of Henslow's Hitcham life, but provides a helpful view of his main concerns for the parishioners.

10.11 Jenyns, L. 1862 pp. 70–1.

10.12 Extracts from Romilly's Diaries in Cambridge University Library, C.U.L. Add. 6804–49

10.13 Kew Dir. Corr. **13**.7. Henslow refers to William Hutton, who collaborated with John Lindley in producing *The Fossil Flora of Gt Britain*, 1831–37. Nothing has been found to support Henslow's claim.

10.14 Russell-Gebbett, J. 1977 p. 75.

10.15 See Mitch, D.F. 1992, especially p. 20 and the references given in end-notes.

10.16 Quoted by Russell-Gebbett, J. 1977 p. 35.

10.17 Dated 5 November 1850, this privately-printed Address covers a single sheet. Its full title is: 'Address to the subscribers to the Parish School; and to the Parents who send their children there.' A copy is in the Henslow papers in the Suffolk Record Office, Bury St Edmunds.

10.18 Quoted by Russell-Gebbett, J. 1977 p. 29.

10.19 Henslow papers in Suffolk Record Office, Bury St Edmunds FL586/13/1.

10.20 Quoted from a newspaper cutting in the Henslow Papers, *loc.cit.*

10.21 Henslow Papers, *loc.cit.*

10.22 Russell, A. 1986 p. 218. (See note 10.4)

10.23 Jenyns, L. 1862 pp. 133–4.

10.24 David Turner, *pers. comm.*

10.25 A complete list of all printed sermons available to us is given in the Bibliography.

10.26 There is a specimen of the Wood Forget-me-not, *Myosotis sylvatica*, in Henslow's Herbarium in Ipswich Museum, labelled as collected in Hitcham Wood on 3 June 1853.

10.27 Anon. (n.d.) All Saints' Church, Hitcham Church Guide. Available in the Church.

10.28 C.U.L. Add. 6826.

CHAPTER 11

11.1 The Deanery of Sudbury, of which Hitcham is a parish, was transferred from the Diocese of Norwich to the Diocese of Ely in 1837. Papers which record returns of questionnaires from Hitcham exist for the years 1794, 1806, 1813 and 1820. The last three were answered by Mathews, who became Rector in 1801. These earlier papers are in the Norfolk & Norwich Records Office under DN/VIS 33/8, 40/8, 46/1, and 52/4. After 1837, the papers are in Cambridge University Library under Ely Diocesan Records EDR.

11.2 This Minute Book, begun by Henslow in March 1840, was in continuous use for well over a century to record the Minutes of what became in later years the Parochial Church Council of Hitcham. It is now in the Suffolk Record Office.

11.3 Jenyns, L. 1862 p. 69.

11.4 See Jones, D. 1976 pp. 12–13, and references given there.

11.5 Glyde, J. 1851 p. 126.

11.6 Letter from Henslow to Whewell in Trinity College Wren Library: Whewell Papers Add. Ms a65/31.

11.7 From the account of the trial in *The Bury and Norwich Post*, Wednesday 23 March 1832, p. 2.

11.8 Henslow, J.S. 1842a.

11.9 Quoted by Knight, F. 1995 p. 21: original in C.U.L. EDR C1/6 1825.

11.10 See *Suffolk Directory* 1840 p. 293.

11.11 Glyde, J. 1851 pp. 283–4; see also Virgin, P. 1989 pp. 120–5.

11.12 Henslow was not the magistrate involved in the case, but he clearly supported his fellow-magistrates in the performance of their duties.

11.13 Mitch, D.F. 1992 contains abundant material on the rapid spread of literacy in England in Henslow's time. Much of the paragraph is based on it: the particular reference is on p. 48.

11.14 Quoted and discussed by Jenyns, L. 1862 pp. 76–81; the original pamphlet is Henslow, J.S. 1843.

11.15 Jones, D. 1976 p. 7.

11.16 Jenyns, L. 1862 p. 87.

11.17 Henslow, J.S. 1844a.

11.18 Sir Henry Bunbury's son, Sir Charles, gives an interesting comment on Henslow's press correspondence in a letter to Leonard Horner on 7 December 1844:

> Henslow's articles in the Bury paper are excellent in feeling and much to be commended for the courage with which he attacks some of the worst evils of our social state; but they are lengthily, heavily, and confusedly written, and not likely, I am afraid, to attract so much attention as they really deserve. (Bunbury, F.J. (ed.) 1894 p. 283)

11.19 Jones, D. 1976 gives a detailed account of the involvement of *The Times* in the problem of the violent agricultural unrest in East Anglia.

11.20 Glyde, J. 1851 pp. 359–60. See also Knight, F. 1995 pp. 68–9.

11.21 Glyde, J. 1851 p. 183.

11.22 *Ibid.* p. 186.

11.23 Henslow, J.S. 1844a p. 13.

11.24 *Ibid.* p. 16.

11.25 *Ibid.* p. 26.

11.26 Blythe, R. 1969 p. 113. This 'portrait of an English village' is based on a small village not far from Hitcham, and the real recorded interviews the author made there in the 1960s. The name 'Akenfield' is imaginary.

11.27 Henslow, J.S. 1845b p. 3.

11.28 Henslow, J.S. 1845a.

11.29 See Salaman, R. 1949 pp. 522–9.

11.30 Henslow, J.S. 1845b p. 6.

11.31 Henslow, J.S. 1849a pp. 11–15.

11.32 Henslow, J.S. 1850a pp. 11–13.

11.33 See Ingram, D. & Robertson, N. 1999 pp. 11–13.

11.34 A copy of this leaflet, with a signed note by Henslow to 'Sir W.', is in the Director's Correspondence at Kew. The leaflet itself, dated 'Hadleigh October 10th 1845', is anonymous, but obviously Henslow's work. Kew Dir. Corr. **23**: 307.

11.35 Henslow, J.S. 1841a, 1841d, 1842b are on the diseases of cereals, and 1844e is on the failure of the clover crop.

11.36 Henslow, G. 1900 gives a detailed account of Henslow's work on coprolites.

11.37 Henslow, J.S. 1843d, 1845c, 1847b, 1848c & 1851d are all devoted to coprolites.

11.38 Jenyns, L. 1862 p. 203. For a recent scientific account of coprolites in the Cambridge Greensand, see Fraser, N.C. 1990.

11.39 Green, V.H.H. 1969 p. 101.

11.40 Jenyns, L. 1862 pp. 161–3. Hooker's letter to Jenyns, preserved in the Jenyns Corr. at Bath, is a 15-page reply to eight questions about Henslow's life and works. It obviously supplied Jenyns with much material (not always attributed) for the *Memoir*.

11.41 Henslow, J.S. 1846a.

11.42 See Nash, R. 1990 for a biography of Lowe.

11.43 Kew Bentham Corr. **5**: letter from Bentham to Henslow, 6 October 1852.

11.44 *Ibid.* letter from Henslow to Bentham, 11 October 1852.

11.45 Gilmour, J.S.L. & Tutin, T.G. 1933 p. 8.

11.46 Letter from Henslow to Whewell in Trinity College Wren Library: Whewell Papers Add. Ms a65/46.

CHAPTER 12

12.1 Markham, R.A.D. 1990 p. 11

12.2 Clark, J.W. & Hughes, T.M. 1890 **2**: 150.

12.3 Cooper, P.H.M. 1999.

12.4 Report from a local newspaper (?*Ipswich Chronicle*) in George Ransome scrapbook no. 2 in Ipswich Museum.

12.5 Markham, R.A.D. 1990 p. 30.

12.6 Jenyns, L. 1862 p. 159.

12.7 Henslow, G. 1902. The seventh article, on *Village Excursions*, contains much material used here. Since these *Reminiscences* are written by a man in his late sixties recalling events in his childhood, there may be inaccuracies. In particular he seems to have conflated the excursions of 1849 and 1850, but the passage quoted here is obviously describing the 1850 excursion.

12.8 Russell-Gebbett, J. 1977 pp. 84–7.

12.9 *Ibid*. pp. 86–7.

12.10 The Vice-Chancellor of the time was Thomas Geldart, Master of Trinity Hall, not of Downing College.

12.11 In Henslow papers, Suffolk Record Office, Bury St Edmunds: FL/586/13/1.

12.12 Russell-Gebbett, J. 1977 p. 90.

12.13 *Ibid*. pp. 42–59, and references given.

12.14 *Ibid*. pp. 44–5.

12.15 Huxley, L. 1918, 1: 393.

12.16 DAR Corr. **5**: 74–5.

12.17 Quoted by Russell-Gebbett, J. 1977 pp. 46–7.

12.18 DAR Corr. **5**: 389.

12.19 Russell-Gebbett, J. 1977 p. 53.

12.20 The named schedule is illustrated in Henslow, G. 1862 p. 677.

12.21 Information on Walter Barton from David Turner, pers. comm. It is based on 'Grant's Log', a log book of his parishioners kept by Henslow's successor as Rector, Alexander Ronald Grant (1861–1903).

12.22 Henslow, G. 1902. *Reminiscences . . . VIII The Minister* p. 354.

12.23 C.U.L. EDR C1/10.

12.24 Bull, A. 1977. The comparison is between Henslow's *first* printed list of Hitcham plants, produced in 1851, which contains just over 400 species, and Bull's own records for the parish made between 1944 and 1960. Neither we nor Alec Bull were aware of the fact that there are *three* editions of the Henslow list, dated 1851, 1856 and 1859, until they were found by accident in the Library of the Department of Plant Sciences in Cambridge at a late stage in the preparation of the text. The third edition contains about 500 species, and reflects the efficiency with which Henslow and his pupils searched the parish. In one respect Bull's paper may be too pessimistic. A species list for Church Meadow, in the grounds of Hitcham Hall just behind the Church, made by Rachel and Robin Hamilton in June 2000, records nearly 50 species, including c. 20 grasses. Two of the grasses, *Briza media* and *Hordeum secalinum*, are on Bull's list of grassland 'casualties' and the list also contains Pyramidal Orchid, *Anacamptis pyramidalis*, of which

Bull wrote, gloomily, 'may persist'. It is good to be able to report that this 'unimproved' meadow is protected by the owners, and a traditional hay crop is taken from it.

12.25 See Rackham, O. 1986. On a special visit to the site on 28 June 1999, when the party included Dr Oliver Rackham, we were able to verify that surviving hedgerows still contained some woody species that were relics of the original wood. Examples were field maple, *Acer campestre*, hornbeam, *Carpinus betulus*, dogwood, *Cornus sanguinea* and aspen, *Populus tremula*. Dr Rackham was especially interested to find that two straight hedgerows within the site, which he was expecting to be 'new' hedges planted after the wood had been grubbed out, were actually species-rich and could only be interpreted as directly descended from original woodland rides.

12.26 See the introduction to McAlpine, D. 1989 ed.

12.27 Lewis, J. 1992 p. 17. Long after Henslow's death, Darwin recalled, in correspondence with Arnold Dodel-Port, who sought his advice on publication of his botanical wall-diagrams, how impressed he had been by Henslow's own teaching diagrams. In a letter date 6 July 1877, Darwin writes:

I know from experience how useful large diagrams are to a lecturer, for when {I was} at Cambridge above 40 years ago the Professor of Botany there [Henslow] made a large number of diagrams at the cost of much time; & he used to say that the great success & popularity of his lectures was largely due to these diagrams, which were nearly on the same scale as yours, but not nearly so well executed.

12.28 Henslow, J.S. 1860d; see also Jenyns, L. 1862 pp. 207–10.

12.29 Henslow, J.S. & Skepper, E. 1860. The six-page Preface is written by Henslow.

12.30 Dawes' letters to Henslow are in C.U.L. Add. 8177, nos. 107–12.

12.31 Thomas Martin evidently kept all Henslow's letters, pamphlets, etc. and on Martin's death these MS were given to the Ipswich Museum. They are now in the Suffolk Record Office at Bury St Edmunds, FL/586/13/1. There is a useful obituary in

Plarr's *Lives of the Fellows of the Royal College of Surgeons*, pp. 35–6.

12.32 Jenyns, L. 1862 p. 250.

12.33 Russell-Gebbett, J. 1977 p. 116.

12.34 Ibid. p. 112.

12.35 Kew Dir. Corr. **35**: 220.

12.36 Allan, M. 1967 p. 181.

12.37 The rush broom supplied by Henslow is still in the Kew Museum of Economic Botany (catalogue no. 34619).

12.38 Kew Dir. Corr. **36**: 234.

12.39 Jenyns, L. 1862 pp. 214–19.

12.40 Wymer, J.J. & West, R.G. 2000. We are very grateful to Richard West, 12th Professor of Botany in Cambridge, now retired, for this reference and much other helpful information about the Hoxne site and the 'celts'.

12.41 The term 'celt', defined as 'a prehistoric edged implement of bronze or stone', is no longer used by archaeologists.

12.42 DAR. Corr. **9**: 98–9. From DAR Corr. **9**: 196 we learn that Henslow wrote to Darwin from his deathbed, but the letter has not been found.

12.43 Jenyns, L. 1862 pp. 256–7. An MS diary kept by Leonard Henslow during his father's terminal illness records many of his utterances. It is in the possession of the Henslow family, and has been kindly made available to us.

EPILOGUE

E.1 From 1854 onwards Henslow had employed curates again: his son Leonard from 1854 to 1856, John Elijah Thompson 1856–58 and finally Richard Drought Graves, 1858–61.

E.2 Jenyns, L. 1862 p. 264.

E.3 David Turner, pers. comm.

E.4 DAR Corr. **11**: 740.

E.5 Ibid. 579–80.

E.6 Ibid. **10**: 486–8.

E.7 Ibid. **11**: 564.

E.8 DAR 94: 199–200. To appear in DAR Corr. 12 (in press).

E.9 DAR 185. Amy Ruck married Darwin's son Francis: Darwin is here writing to her affectionately as a future daughter-in-law. She was born and bred near Machynlleth in North Wales, and must have known Snowdonia well. Unfortunately she died aged 26 at the birth of their only child, Bernard. See Raverat, G. (1952), 1960 edn, pp. 13 & 191.

E.10 O.E.D. definition and first use. T.G. Taylor, *With Scott*, 1916 p. 115. For a description of the phenomenon, see Pearsall, W.H. (1950) 1971 edn, pp. 26–9.

E.11 Jenyns, L. 1862 p. 108.

E.12 Kew. Letters Hooker, J. to Oliver, D. 12 Aug. 1863.

E.13 Oliver, D. (1864) ed. 2 1878, p. viii (Preface). The reprinted Preface to the first edition is here dated September 1868: this seems to be an error for 1864.

E.14 Gardiner, W. 1904. This pamphlet in the Dept. of Plant Sciences Library, of which only 200 numbered copies were 'printed for private circulation only', is a mine of information about the rescue of 'Henslow's Botanical Museum'. Much of the 'spirit material' of the original collection put together by Henslow had dried out and was thrown away, but the lists of 'additions' to the Museum from 1886 to 1904 contain a good many original Henslow specimens.

E.15 Henslow and Jenyns were jointly responsible for the formation of the Philosophical Society's Museum. Henslow began his donations in 1820 with insects and shells, and in 1833 Henslow and Jenyns added a collection of fish, largely made by Henslow on his visits to Weymouth. The Museum was offered to the Univerity in 1865, and became the nucleus of the present-day Museum of Zoology. See Clark, J.W. 1891, pp. ix–xvi.

E.16 Youmans, E.A. (1870) 2nd London edn 1880, pp. vi–vii.

E.17 Ibid. 1872, pp. 28–9. The footnote is by Payne.

E.18 Walters, S.M. 1981 pp. 96–7.

E.19 Ibid. p. 64.

Bibliography

I. Works by Henslow
(Unless otherwise stated copies are available in the University
Library, Cambridge)

Books

1829 *A Catalogue of British plants, arranged according to the natural system* . . .
40pp. Cambridge. 2nd edn 1835 [62pp.]

1835 *The Principles of descriptive and physiological botany.* (The Cabinet
Cyclopaedia: Natural history, 1836) viii, 322pp. London: Longmans.

1840 *The botanical portion of "Le Bouquet des Souvenirs: a wreath of friendship".*
216pp. London. [NB Only Henslow's contribution of botanical
description is acknowledged in this compilation of flower illustration, general comment and poems. The Preface states that the book
was compiled by three friends and Henslow addresses the introduction to his contribution to 'My dear sister'.]

1843 *Letters to the farmers of Suffolk, with a glossary of terms used, and the address
delivered at the last anniversary meeting of the Hadleigh Farmers' Club.*
114pp. London and Hadleigh: Groombridge & Hardacre.

1849 *A Dictionary of botanical terms.* 218pp., woodcuts. London. [NB Earlier
versions were published in parts with Maund's *The Botanist*, 1837–46
and *Botanic Garden*, 1825–50 but never completed.]

1860 *Flora of Suffolk: a catalogue of the plants (indigenous or naturalized) found in
a wild state in the county of Suffolk.* By J.S. Henslow and Edmund
Skepper. 140pp. London and Bury St Edmunds.

1864 Oliver, D. *Lessons in elementary botany. The part on systematic botany based
upon material left in manuscript by . . . Professor Henslow, etc.* viii, 317pp.
London. [NB Woodcuts taken from drawings by Anne Barnard,
Henslow's daughter.]

1837– Co-editor with B. Maund of *The Botanist* 1837–46, 5 vols. [NB
Henslow's contributions were mainly to Vol. 1. Later contributors
included Bentham. *The Botanist* was issued in parts. A collection of 60
of these may be found in the Cory Library, Cambridge University
Botanic Garden under the title 'The Floral Keepsake'.]

Pamphlets and Articles

1821　Supplementary observations to Dr Berger's account of the Isle of Man. *Transactions, Geological Society of London*, 5, 482–505, 1 plate.

1822　Geological description of Anglesea. *Transactions, Cambridge Philosophical Society*, 1, 359–452, 7 plates.

1823a *A syllabus of a course of lectures on mineralogy*. xxii, 119pp. Cambridge.

—b　On the Deluge. *Annals of Philosophy* ns, 6, 344–8.

1824　Remarks upon Dr Berger's reply (to Prof. Henslow's "Observations on Dr Berger's account of the Isle of Man"). *Annals of Philosophy* ns, 7, 407–9.

1826　*Remarks upon the payment of the expenses of out-voters at an University election in a letter to the Vice-Chancellor of Cambridge, by the Rev. John Lamb . . . and the Rev. J.S. Henslow*. 19pp. Cambridge.

1828a *Syllabus of a course of botanical lectures*. 16pp. Cambridge (There were later and enlarged editions).

—b　On the crystallization of gold. *Magazine of Natural History and Journal of Zoology*, 1, 146–7; *Journal, Royal Institution of Great Britain*, I, 1831, 176–7.

—c　Botanical Museum of Cambridge, Statement respecting. *Magazine of Natural History and Journal of Zoology*, 1, 82–3 (Appeal for specimens).

1829a *A sermon on [John vii, 17] the first and second resurrection, preached at Great St Mary's Church, Cambridge, Feb. 15, 1829*. xiii, 14pp.

—b　On the leaves of Malaxis paludosa. *Magazine of Natural History and Journal of Zoology*, 1, 441–2; Sur les feuilles du Malaxis paludosa. *Annales des sciences naturelles*, 19, 1830, 103–4.

1830a On the specific identity of the primrose, oxlip, cowslip, and polyanthus. *Magazine of Natural History and Journal of Zoology*, 3, 406–9.

—b　On the specific identity of Anagallis arvensis and A. coerulea. *Magazine of Natural History and Journal of Zoology*, 3, 537–8.

1831a On the examination of a hybrid digitalis. *Transactions, Cambridge Philosophical Society*, 4, 1833, 257–78, 4 plates.

—b　The Breathing-tube of the boa; snakes in the Fens of Cambridgeshire. *Magazine of Natural History and Journal of Zoology*, 4, 279.

—c　Rooks detecting grubs. *Magazine of Natural History and Journal of Zoology*, 4, 280–1.

—d　The Portuguese man-of-war. *Magazine of Natural History and Journal of Zoology*, 4, 282.

—e   Specific relations of Anagallis arvensis and coerulea. *Magazine of Natural History and Journal of Zoology*, **4**, 466.

1832a *Sketch of a course of lectures on botany for 1832. 4pp. Cambridge {Bot}*

—b   A Variety of the common groundsel (Senecio vulgaris). *Magazine of Natural History and Journal of Zoology*, **5**, 87–8.

—c   Fumaria Vaillantii, a British plant. *Magazine of Natural History and Journal of Zoology*, **5**, 88.

—d   On variations in the cotyledons and primordial leaves of the sycamore (Acer pseudoplatanus). *Magazine of Natural History and Journal of Zoology*, **5**, 346–7.

—e   On the fructification of the genus Chara. *Magazine of Natural History and Journal of Zoology*, **5**, 348–50.

—f   On the varieties of Paris quadrifolia, considered with respect to the ordinary characteristics of monocotyledonous plants. *Magazine of Natural History and Journal of Zoology*, **5**, 429–34.

—g   On the specific identity of Anagallis arvensis and coerulea. *Magazine of Natural History and Journal of Zoology*, **5**, 493–4.

—h   Collectors and collections in Nat. Hist., in the University, Town, and County of Cambridge. *Magazine of Natural History and Journal of Zoology*, **5**, 545–6.

—i   Review of De Candolle's "Physiologie Vegetale". *Foreign Quarterly Review*, **11**, no. 22, 334–82. (The review is not signed.)

—j   On the geographical distribution of the plants of Cambridgeshire. *Report of the British Association for the Advancement of Science*, 1832, sects. p. 596 (14 lines).

1833a *Sketch of a course of lectures on botany for 1833. 8pp. Cambridge.*

—b   Facts in relation to the reproductive economy of the mistletoe (Viscum album). *Magazine of Natural History and Journal of Zoology*, **6**, 500.

—c   On a monstrosity of the common mignonette. *Transactions, Cambridge Philosophical Society*, **5**, 1835, 95–100, 2 plates.

1834a *Address to the reformers of the town of Cambridge. 8pp.* (Originally published as a letter dated 17 June 1834 in the *Cambridge Independent Press* for 21 June 1834).

—b   Some of the habits and anatomical conditions of a pair of hybrid birds, obtained from the union of a male pheasant with hens of the bantam fowl; and an incidental notice of a hybrid dove. *Magazine of Natural History and Journal of Zoology*, **7**, 153–5. (Birds were presented to the Cambridge Philosophical Society by Henslow's father J.P. Henslow, and article includes comments by him and Leadbetter.)

1835 Observations concerning the indigenousness and distinctness of certain species of plants included in the British flora. *Magazine of Natural History and Journal of Zoology*, **8**, 84–8.

1836a An Enumeration of species and varieties of plants which have been deemed British, but whose indigenousness to Britain is considered to be questionable. *Magazine of Natural History and Journal of Zoology*, **9**, 88–91.

—b A Notice of the fact, and of particulars on the mode, of sugar-candy being produced in the flowers of Rhododendron ponticum; and a notice of the effect on the germination of the seeds of an Acacia . . . by boiling them variously. *Magazine of Natural History and Journal of Zoology*, **9**, 476–9.

—c Notice of crystals of sugar found in Rhododendron ponticum. *Report of the British Association for the Advancement of Science*, 1836, sects. p. 106. (7 lines on an exhibit)

—d Method of preventing the decomposition of the Sheppey fossils. *Magazine of Natural History and Journal of Zoology*, **9**, 551.

—e On the disunion of contiguous layers in the wood of exogenous trees. *Magazine of Zoology and Botany*, **1**, 32–5.

—f On the requisites necessary for the advance of botany. *Magazine of Zoology and Botany*, **1**, 113–25.

—g On the structure of the flowers of Adoxa moschatellina. *Magazine of Zoology and Botany*, **1**, 259–61.

1837a *A Reformer's duty: an address to the reformers of the town of Cambridge* . . . 27pp. Cambridge.

—b Description of two new species of Opuntia (O. Darwinii and O. Galapageia); with remarks on the structure of the fruit of Rhipsalis. *Magazine of Zoology and Botany*, **1**, 466–9, 1 plate.

1838 Florula Keelingensis. An account of the native plants of the Keeling Islands. *Annals of Natural History*, **1**, 337–47; *Notizen aus dem Gebiete der Natur- und Heilkunde*, **10**, 1839, col. 257–9. (Cocos Keeling Islands; plant specimens presented to J.S.H. by Darwin.)

1840a Report of the Committee appointed to try experiments on the preservation of animal and vegetable substances. *Report of the British Association for the Advancement of Science*, 1840, 421–2.

—b   The Botanical Garden. In J.J. Smith (ed.) *The Cambridge Portfolio*, **1**
     81–5.

1841a Report on the diseases of wheat. *Journal, Royal Agricultural Society*, **2**,
       pt 1, 1–25.

—b   On the specific identity of the fungi producing rust and mildew.
     Journal, Royal Agricultural Society, **2**, pt 2, 220–4.

—c   On Cecidomyia Tritici. *Report of the British Association for the
     Advancement of Science*, 1841, sects. (pt 2), 72–3.

—d   On the occurrence of the animalcule of Vibrio tritici in blighted
     grains of the ears of wheat, constituting what is termed earcockle,
     purples and peppercorns. *Microscopical Journal and Structural Record*, **1**,
     36–9.

1842a *Sermon preached 20th March 1842. On the occasion of three persons of the
       parish being condemned to fifteen years' transportation for sheep stealing …*
       [For private distribution only] 24pp. Hadleigh: Hardacre.

—b   Observations on the Wheat-midge. *Journal, Royal Agricultural Society*,
     3, pt 1, 36–40.

—c   To destroy wasps' nests [: turpentine]. *Gardeners' Chronicle*, 1842,
     637–8.

—d   On Primula veris and allied species. *Annals and Magazine of Natural
     History*, **9**, 153–4. (Comment on paper by J.E. Leefe to Botanical
     Society of Edinburgh meeting, 10.2.1842)

—e   Second Report of a Committee consisting of Mr H.E. Strickland,
     Prof. Daubeny, Prof. Henslow and Prof. Lindley appointed to make
     experiments on the growth and vitality of seeds. *Report of the British
     Association for the Advancement of Science*, 1842, 34–8.

1843a *An account of the Roman antiquities found at Rougham, near Bury St.
       Edmund's, Sept. 15, 1843.* 12pp. Hitcham.

—b   *A sermon [on Col ii 6] on improved attention to the sacraments, etc.* 23pp.
     Hadleigh.

—c   *Two sermons preached First & Eighth of January 1843 – In aid of a parish
     school, and certain clubs established for the assistance and relief of the poor.*
     [For private distribution only] 54pp. Hadleigh: Hardacre. {Bury
     R.O.}

—d   On concretions in the Red Crag at Felixstow, Suffolk. *Proceedings of the
     Geological Society of London*, 1843, 281–5. (Henslow presented the
     paper in Dec. 1843. Proceedings appeared separately and also in the

Journal of the Geological Society of London in 1845 with H's article on pp. 35–6. There are supplementary papers by Richard Owen and Edward Charlesworth).

—e   On fixing ammonia. *Gardeners' Chronicle*, 1843, 135–6. (To ascertain whether the addition of gypsum to a common dunghill improves the manure: experiments with 43 farmers).

—f   [Three notes on dodder and broomrape, the trefoil dodder, and the clover dodder] *Gardeners' Chronicle*, 1843, 607, 644, 710.

—g   'To the public' and 'To the farmers of Suffolk': letters. *Bury and Norwich Post.* 11 Jan 1843ff.

1844a *Suggestions towards an inquiry into the present condition of the labouring population of Suffolk, 3rd July 1844.* 29pp. Hadleigh: Hardacre. {Bury R.O.}

—b   *The Roman tumulus, Eastlow Hill, Rougham. Opened on Thursday, the 4th of July, 1844.* 8pp. Bury St Edmunds. (Reprinted from the Bury Post.)

—c   *Two sermons in aid of Parish charities, preached January the 7th and 14th, 1844.* [For private distribution] 21pp. Hadleigh: Hardacre. {Bury R.O.}

—d   On experimental co-operation among agriculturists. *Gardeners' Chronicle*, 1844, 9–10. (Continuing the ammonia experiments)

—e   On the failure of the red clover crop. *Gardeners' Chronicle*, 1844, 529–30.

—f   On the registration of facts tending to illustrate questions of scientific interest. *Gardeners' Chronicle*, 1844, 659. (The coincidence of outbreaks of rust and blight in wheat in various parts of the world including samples from South America provided by Darwin)

—g   On the action of gypsum, with remarks on misstatements of the writer's views. *Gardeners' Chronicle*, 1844, 691–2. (see earlier articles)

1845a *An address to landlords, on the advantages to be expected from the general establishment of a spade tenantry from among the labouring classes.* 36pp. Hadleigh and London.

—b   *Letter to his parishioners[: on condition of labourers]*, 8 July. 8pp. Hadleigh: Hardacre. {Bury R.O.}

—c   On nodules, apparently coprolitic, from the Red Crag, London Clay, and Green Sand. *Report of the British Association for the Advancement of Science*, 1845, sects. pp. 51–2.

1846a *Address to the members of the University of Cambridge, on the expediency of improving, and on the funds required for remodelling and supporting the Botanic Garden.* 20pp. Cambridge.

—b *A Sermon in aid of Parish Charities and the Suffolk General Hospital, preached January 4th, 1846.* [For private distribution] 16pp. Hadleigh: Hardacre. {Bury R.O.}

—c Roman British remains. On the materials of the sepulchral vessels found at Warden, Co. Beds. *Publications, Cambridge Antiquarian Society.* [4to ser.] no. 10, 1–2, 2 plates. (Questioning earlier assumptions that they were of oak)

—d Mummy wheat. *Gardeners' Chronicle,* 1846, 757–8. (Doubts cast on age of specimens and their uniqueness)

1847a *A Sermon on the occasion of the Irish famine, preached January 24th, 1847.* 18pp. Hadleigh: Hardacre. {Bury R.O.}

—b On detritus derived from the London Clay, and deposited in the Red Crag. *Report of the British Association for the Advancement of Science,* 1847, sects. p. 64.

1848a *Address delivered in the Ipswich Museum, on the 9th March, 1848.* 19pp. Ipswich.

—b *Syllabus of a course of lectures on botany, suggesting matter for a pass-examination at Cambridge in this subject.* 29pp. Cambridge.

—c On native phosphate of lime. *Gardeners' Chronicle,* 1848, 180–1.

—d On the parasitical habits of Scrophularineae. *Annals and Magazine of Natural History* 2s., **2**, 294–5.

1849a *Address to the inhabitants of the parish of Hitcham, Easter week, 1849.* 16pp. Hadleigh: Hardacre. {Bury R.O.}

—b *To the parish of Hitcham, 10th Sep. 1849.* 4pp. Hitcham Rectory {Bury R.O.}

—c Parasitic larvae observed in the nests of hornets, wasps, and humble bees. *Zoologist,* **7**, 2584–6.

—d On the awns of Nepaul barley (Hordeum coeleste, vars. trifurcatum and aegiceras). *Hooker's Journal of Botany,* **1**, 33–40; *Notizen aus dem Gebiete der Natur- und Heilkunde,* 3s 10, col. 102–4.

—e On the structure of the pistil in Eschscholtzia Californica. *Hooker's Journal of Botany,* **1**, 289–91.

1850a *Address to the inhabitants of the parish of Hitcham, Easter week, 1850.* 15pp. Hadleigh: Hardacre. {Bury R.O.}

—b An address to the subscribers to the parish school; and to the parents who send their children there. 2pp. Hitcham, 5th Nov. 1850. {Bury R.O.}

—c Gnawing power of caterpillar of the goat moth (Xyleutes cossus). Zoologist, **8**, 2897.

—d The Way in which toads shed their skins. Gardeners' Chronicle, 1850, 373 and 422; Zoologist, **8**, 2881–.

—e Village excursions. Gardeners' Chronicle, 1850, 596, 629, 661 and 691. (Organising parties of villagers to Harwich, the railway and the sea)

1851a Questions of the subject-matter of sixteen lectures in botany, required for a pass-examination. 22pp. Cambridge.

—b Way to kill fleas on dogs. Gardeners' Chronicle, 1851, 749. (Using sweet oil and afterwards common soda)

—c Correspondence [with J.B. Lawes] on the clover failure. Gardeners' Chronicle, 1851, 764.

—d On the phosphate nodules of Felixstow, in Suffolk. Gardeners' Chronicle, 1851, 764.

—e On alternate culture. Gardeners' Chronicle, 1851, 779. (Alternate strips of wheat and fallow with references to De Candolle)

1852a Address to the parish of Hitcham, chiefly in reference to an attack upon the allottees by some of the farmers. 38pp. Hadleigh. {Bury R.O.}

—b Food of micro-lepidoptera. Zoologist, **10**, 3358.

1853a Syllabus of a course of lectures on botany: with an appendix, containing copious demonstrations of fourteen common plants for the illustration of terms. 85pp. Cambridge.

—b Address [to the parish of Hitcham]. 2pp. Hitcham Rectory, April 9th 1853. {Bury R.O.}

1854a Programme of Village excursion from Hitcham to Cambridge. 11pp. illus.

—b Note on two brown eagles. Zoologist, **12**, 4251.

—c Suggestions for the consideration of collectors of British shells. Zoologist, **12**, 4264–5.

—d Hitcham Horticultural Shows reported. Gardeners' Chronicle, 1854, 1856, 1857, 1859.

1855a On typical series of objects in natural history adapted to local museums. Report of the British Association for the Advancement of Science, 1855, 108–26.

—b On the anomalous oyster-shell. Annals and magazine of natural history, 2s, **15**, 314 and 385.

1856a On typical forms of minerals, plants, and animals for museums. *Report of the British Association for the Advancement of Science*, 1856, 461–2 (9 lines. Suppl. to 1855 article)

—b On the triticoidal forms of Oegilops, and on the specific identity of Centaurea nigra and C. nigrescens. *Report of the British Association for the Advancement of Science*, 1856, sects. pp. 87–8. (10 lines)

—c Example of botany in village education. *Gardeners' Chronicle*, 1856, 453–4.

—d Practical lessons in botany for beginners of all classes, nos. 1–14. *Gardeners' Chronicle*, 1856, pp. 468, 484, 500, 516, 532, 565, 596, 613, 629, 676, 724, 740, 772, 837 and 853.

1857a *A Sermon preached in Hitcham church . . . October 18, 1857, the day on which a collection was made for the benefit of the survivors of the Indian mutiny.* 16pp. Hadleigh.

—b The locust in Suffolk. *Zoologist*, **15**, 5787. (1 line)

1858a *Illustrations to be employed in practical lessons on botany; adapted to beginners of all classes.* Prepared for the South Kensington Museum. 31pp., woodcuts. London: Chapman & Hall. {Linnean Society}

—b *Appendix to Hitcham allotment report for 1857.* 16pp. Hadleigh, Suffolk. {Bury R.O.}

—c *To the parish of Hitcham [: Moral obligation of rate payers to pay Church rate].* March 31st, 1858. 8pp. {Bury R.O.}

—d On a monstrous development in Habenaria chlorantha [1857]. *Journal and Proceedings of the Linnean Society*, **2**, (Botany) 104–5.

1859 Celts in the drift. *Athenaeum*, 1859, no. 1673, p. 668; no. 1678, p. 853.

1860a *Sudden Death: a sermon preached at Hitcham Church, on March 11th.* 16pp. Hadleigh: Hardacre {Linnean Society}

—b *A sermon preached at Hitcham, Suffolk, on September 30th, 1860 in aid of the "British Syria Relief fund" [For private distribution in the parish]* 12pp. Hadleigh: Hardacre {Bury R.O.}

—c Flint weapons in the drift. *Athenaeum*, 1860, no. 1685, p. 206; no. 1721, p. 516, and no. 1723, p. 592.

—d On the supposed germination of mummy wheat. *Report of the British Association for the Advancement of Science*, 1860, sects. pp. 110–11.

—e Botanical lectures to the Royal family [: syllabus of lectures lately delivered . . . by . . . Henslow.] *Gardeners' Chronicle*, 1860, 647–8.

1861 Letter to the Editor of "Macmillan's Magazine" on Darwin's Origin of the species. *Macmillan's Magazine*, **3**, 336.

Diagrams and other single sheet publications

[1834] [Two illustrations to accompany paper to Cambridge Philosophical Society on cone-scales of Spruce. Paper not published] 2sh. Cambridge: W. Metcalfe. {Bot}

[1851] List of native plants, (with a few trees introduced into plantations,) growing in the parish of Hitcham, Suffolk s.sh.Ipswich. {Bot}

[1855] [2nd edn] British plants, growing wild (and a few common trees in plantations) in the parish of Hitcham, Suffolk.] s.sh. Ipswich. [1859] [3rd edn] s.sh. {Bot}

1857 Professor Henslow's Botanical diagrams. Drawn by W. Fitch for the Committee of Council of Education, Dept of Science & Art. Day & Sons, London. 9 sheets, obl. 102 × 76 cm. and Linnean classes and orders . . . 1 sheet, fol.

{Bot}: Department of Plant Sciences Library, Cambridge
{Bury R.O.}: Suffolk Record Office, Bury St Edmunds
{Linnean Society}: Linnean Society, London

II: Reference

Ackermann, R. (1815) A History of the University of Cambridge, its Colleges, Halls and Public Buildings. 2 vols. London: R. Ackermann.

Adams, G.B. (1963) Constitutional History of England rev. edn. London: Jonathan Cape.

Allan, M. (1967). The Hookers of Kew. London: Michael Joseph.

— (1977) Darwin and his Flowers. London: Faber & Faber.

Allen, D.E. (1978) The Naturalist in Britain: a Social History. London: Pelican.

— (1984) Flora of the Isle of Man. Douglas: Manx Museum.

— (1986) The Botanists: a History of the Botanical Society of the British Isles through a Hundred and Fifty Years. London: St Paul's Bibliographies.

— (1996) The struggle for specialist journals. Archives of Natural History, **23**, 107–23.

Audubon, J.J. (1827–38) The Birds of America. 4 vols. London: printed by the author.

— (1931–39) The Ornithological Biography. 5 vols. Edinburgh: Adam & Charles Black.

Audubon, M.R. & Coues, E. (1898) Audubon and his Journals, vol. 1. London: John C. Nimmo.

Austen, J. (1814; 1949 ed.) Mansfield Park. London: Zodiac Press.

Babington, C.C. (1860) *Flora of Cambridgeshire.* London: J. Van Voorst.

— (1897) *Memoirs, Journal and Botanical Correspondence . . .* Cambridge: Macmillan & Bowes.

Babington, Churchill (1874) Roman antiquities found at Rougham in 1843 & 1844. *Proceedings of the Suffolk Institute of Archaeology,* **4**, 257–81.

Barber, L. (1980) *The Heyday of Natural History.* London: Cape.

Barlow, N. (1967) *Darwin and Henslow: the Growth of an Idea.* London: John Murray.

Barrett, P.H. et al. (eds.) (1987) *Charles Darwin's Notebooks, 1836–1844.* Cambridge: Cambridge University Press.

Bebbington, D.W. (1989 repr. 1995) *Evangelicalism in Modern Britain.* London: Hyman repr. Routledge.

Becher, H.W. (1986) Voluntary science in nineteenth century Cambridge University to the 1850's. *British Journal for the History of Science,* **19**, 57–87.

Bennett, R.M. (1974) John Stevens Henslow, East Anglian. *East Anglian Magazine,* 1974, 632–4.

Bettey, J.H. (1979) *Church & Community: the Parish Church in English Life.* Bradford-on-Avon: Moonraker Press.

Blomefield (formerly Jenyns), Leonard (1887, repr. with additions 1889) *Chapters in my Life.* Bath: privately printed.

Blythe, R. (1969) *Akenfield: Portrait of an English Village.* London: Allen Lane Penguin Press.

Bowler, P.J. (1990) *Charles Darwin: the Man and his Influence.* Oxford: Blackwell.

Bridson, G. et al. (1980) *Natural History Manuscript Sources in the British Isles.* London: Mansell.

Briggs, D. & Walters, S.M. (1997) *Plant Variation and Evolution,* 3rd edn. Cambridge: Cambridge University Press.

Brooke, J.H. (1991) Indications of a creator: Whewell as apologist & priest. In *William Whewell: a Composite Portrait,* ed. M. Fisch & S. Schaffer, pp. 149–73. Oxford: Clarendon Press.

Browne, J. (1995) *Charles Darwin, Voyaging: Volume 1 of a Biography.* London: Jonathan Cape.

Buckland, W. (1823) *Reliquiae Diluvianae.* London: John Murray.

Bull, A. (1977) A Century of change. *Suffolk Natural History,* **17**, 220–4.

Bunbury, F.J. (ed.) (1894) *Life, Letters & Journals of Sir Charles J.F. Bunbury.* 3 vols. London: private circulation.

Burnham, J. (1999) *Great Comets*. Cambridge: Cambridge University Press.

de Candolle, A.P. (1819) *Théorie élémentaire de la botanique*. 2nd edn. Paris: Deterville.

— (1827) *Organographie végétale*. 2 vols. Paris: Deterville.

— (1832) *Physiologie végétale*. 3 vols. Paris: Bechet jeune.

de Candolle, A.P. and Sprengel, K. (1821) *Elements of the Philosophy of plants ... Translated from the German*. Edinburgh: W. Blackwood.

Cannon, S.F. (1978) *Science in Culture: the Early Victorian Period*. New York: Dawson & Science History Publications.

Chadwick, O. (1960) *Victorian miniature*. Cambridge: Cambridge University Press.

Challinor, J. (1970) The Progress of British geology during the early part of the nineteenth century. *Annals of Science*, **26**, 177–234.

Chalmers, J. (1993) Audubon in Edinburgh. *Archives of Natural History*, **20**, 157–66.

Clapham, A.R. et al. (1960) *Flora of the British Isles illus. by S.J. Roles*. Cambridge: Cambridge University Press.

Clark, J.W. (1891) The Foundation and early years of the [Philosophical] Society. *Proceedings of the Cambridge Philosophical Society*, **7**, i–l.

— (1904a) *Endowments of the University of Cambridge*. Cambridge: Cambridge University Press.

— (1904b) *A Concise Guide to the Town and University of Cambridge*. Cambridge: Macmillan & Bowes.

Clark, J.W. & Hughes, T.M. (1890) *The Life and Letters of the Reverend Adam Sedgwick*. 2 vols. Cambridge: Cambridge University Press.

Clarke, E.D. (1807) *Syllabus of lectures on mineralogy*. Cambridge: Cambridge University Press. 2nd edn. 1818. London: Cadell & Davies.

Coates, R. (1999) Box in English place-names. *The Year's work in English Studies* **80**, 2–45.

Cobbett, W. (1830) *Rural rides*. London: W. Cobbett. (repr. in Penguin Classics, 1985 with introduction by George Woodcock)

Cole, G.D.H. (1947) *The Life of William Cobbett*, 3rd edn. London: Home & Van Thal.

Cooper, C.H. (1860–66) *Memorials of Cambridge*. New edn.: an enlargement of Le Keux. Cambridge: William Metcalfe.

Cooper, P.H.M. (1999) *Fossils, faith and farming: newspaper portraits of Little Cornard in the Darwinian Age together with Some Account of the Rev Edwin Sidney*. Great Cornard, Suffolk: Little Cornard Conservation Society.

Cranston, M. (1957) *John Locke: a Biography* (2nd edn. 1985) Oxford: University Press.

Crompton, G. (1997) Botanizing in Cambridgeshire in the 1820s. *Nature in Cambridgeshire*, **39**, 59–73.

Curtis, John (1824–39) *British Entomology*. London: Printed for the author.

Darwin, C.R. (1859) *On the origin of species by means of natural selection or the preservation of favoured races in the struggle for life*. London: John Murray.

— (1876a) *Insectivorous plants*. London: John Murray.

— (1876b) *The Effects of Cross- and Self fertilization in the Vegetable Kingdom*. London: John Murray.

— (1877) *The Different Forms of Flowers on Plants of the Same Species*. London: John Murray.

— (1887) Autobiography. In *Charles Darwin: Thomas Henry Huxley: Autobiographies*, ed. G. de Beer (1974). London: Oxford University Press.

— (1985–) *The Correspondence of Charles Darwin*. Vols. 1– . Ed. F. Burkhardt, S. Smith et al. Cambridge: Cambridge University Press.

Darwin, F. (ed.) (1888) *The Life and Letters of Charles Darwin including an Autobiographical Chapter*. 3 vols. London: John Murray.

—(ed.) (1958) *The Autobiography of Charles Darwin and Selected Letters*. New York: Dover.

De Beer, G. (ed.) (1974) *Charles Darwin: Thomas Henry Huxley: Autobiographies*. London: Oxford University Press, etc.

Dell, R. (1996) Charles Simeon of King's & Holy Trinity, Cambridge. *Cambridge: the Magazine of the Cambridge Society*, **39**, 75–9.

Desmond, A. & Moore, J.R. (1992) *Darwin*. London: Michael Joseph.

Desmond, R. (1994) *Dictionary of British and Irish Botanists and Horticulturalists* ... London: Taylor & Francis and Natural History Museum.

Dobbs, B.J.T. (1975) *The Foundations of Newton's Alchemy*. Cambridge: Cambridge University Press.

Dolan, B.P. (1995) *Governing Matters: the values of English Education in the earth sciences*. Ph.D. Thesis, University of Cambridge.

Douglas, Mrs Stair *see* Stair Douglas.

Fisch, M. & Schaffer, S. (eds.) (1991) *William Whewell: a Composite Portrait*. Oxford: Clarendon Press.

Fletcher, H.R. & Brown, W.H. (1970) *The Royal Botanic Garden, Edinburgh, 1670–1970*. Edinburgh: H.M. Stationery Office.

Force, J.E. (1985) *William Whiston, Honest Newtonian*. Cambridge: Cambridge University Press.

Foster, J. (1888) *Alumni Oxoniensis, 1715–1886.* Oxford: Parker & Co.

Fraser, N.C. (1990) The fossil fauna of the mid-Cretaceous of Cambridgeshire with particular attention to the reptiles. *Nature in Cambridgeshire,* **32,** 17–26.

Freeman, J. (1852) *Life of the Rev. William Kirby, M.A.* London: Longman, Brown, Green & Longman.

Frere, J. (1800) Account of flint weapons discovered at Hoxne in Suffolk. *Archaeologia,* **13,** 204–5.

Gage, J. (1838) A Letter . . . containing an account of further discoveries of Roman sepulchral relics at the Bartlow Hills. *Archaeologia,* **28,** 1–6, plates I–II.

Gardiner, W. (1904) *An Account of the Foundation and Re-establishment of the Botanical Museum of the University of Cambridge.* 27pp. Cambridge: Printed at the University Press for private circulation only.

Garland, M.M. (1980) *Cambridge before Darwin.* Cambridge: Cambridge University Press.

Gascoigne, J. (1988) *Cambridge in the Age of the Enlightenment.* Cambridge: Cambridge University Press.

George, A.S. (1984) Santalaceae, 7: Dendrotrophe. In George, A.S. et al. *Flora of Australia,* **22,** 59. Canberra: Australian Government Publishing Service.

Gilmour, J.S.L. & Tutin, T.G. (1933) *A List of the More Important Collections in the University Herbarium, Cambridge.* 40pp. Cambridge: Cambridge University Press.

Glen, W. (ed.) (1994) *The Mass-extinction Debates.* Stanford, Ca.: Stanford University Press.

Glyde, J. (1851) *Suffolk in the Nineteenth Century.* London: Simpkin, Marshall & Co.; Ipswich: J.M. Burton & Co.

Gorham, G.C. (1830) *Memoirs of John Martyn, F.R.S. and of Thomas Martyn, B.D., F.R.S., F.L.S.* London.

Gow, H. (1928) *The Unitarians.* London: Methuen.

Graham, J. (chairman) (1852) *Report of Her Majesty's Commissioners Appointed to Inquire into the State, Discipline, Studies and Revenues of the University and Colleges of Cambridge, with the Evidence and Correspondence.* Parliamentary Papers, 1852–53: House of Commons, xliv. London: H.M.S.O.

Green, V.H.H. (1969) *The Universities.* London: Pelican.

Gruber, H.E. (1974) *Darwin on Man: a Psychological Study of Scientific Creativity.* London: Wildwood House.

Gunton, C.E. (ed.) (1997) *The Cambridge Companion to Christian Doctrine.* Cambridge: Cambridge University Press.

Hailstone, E. (1873) *History and Antiquities of the Parish of Bottisham.* Cambridge: Cambridge Antiquarian Society.

Hall, A.R. (1969) *The Cambridge Philosophical Society: a History, 1819–1969.* Cambridge: C.P.S.

Hampson, N. (1968) *The Enlightenment.* London: Pelican.

Hanbury, F.J. & Marshall, E.S. (1899) *Flora of Kent.* London: F.J. Hanbury.

Harraden, R. (1809–11) *Cantabrigia Depicta: a series of engravings* . . . Cambridge: Harraden & Son; London: R. Cribb & Son.

Henslow, G. (1862) Practical lessons in systematic & economic botany. *The Leisure Hour*, 676–9.

— (1913) John Stevens Henslow, 1796–1861. In *Makers of British Botany*, ed. F.W. Oliver, pp. 151–63. Cambridge: Cambridge University Press.

— (1900–2) Reminiscences of a scientific Suffolk clergyman, I–VIII. *Eastern Counties Magazine*, **1**, 22–30, 106–14, 195–203, 300–7, **2**, 24–31, 147–54, 226–34, 353–60.

Henslow, Mrs J.S. (1843) *John Borton: or a Word in Season.* London: Joseph Masters.

— (1848) *A Practical Application of the Five Books of Moses, Adapted to Young Persons.* London: B. Wertheim.

Herbert, S. (1992) Between Genesis and Geology: Darwin and some contemporaries in the 1820s and 1830s. In R.W. Davis & R.J. Helmstadter (eds.) *Religion and Irreligion in Victorian Society.* pp. 68–84. London, New York: Routledge.

Herbert, W. (1837) *Amaryllidaceae* . . . London: J. Ridgway & Sons.

Hill, M.O., Preston, C.D. & Smith, A.J.E. (eds.) (1991–94) *Atlas of the Bryophytes of Britain and Ireland.* 3 vols. Colchester: Harley Books.

Hind, W.M. (1889) *The Flora of Suffolk* . . . London: Gurney & Jackson.

Hitchin, N. (1997) The Evidence of things seen: Georgian churchmen & biblical prophecy. In B. Taithe & T. Thornton (eds.) *Prophecy: the Power of Inspired Language, 1300–2000*, pp. 119–39. Stroud: Allen Sutton.

Holbrook, M. (1999) The Spinning House. *Review, Cambridgeshire Local History Society* (new ser.), **8**, 3–25.

Hooker, W.J. (ed.) (1830) Schultes's Botanical visit to England. *Botanical Miscellany*, **1**, 48–78.

Humphreys, J. (1926) A Great Bromsgrovian: Benjamin Maund, world-famed botanist. *The Bromsgrove Messenger*, Sat. 12 Jun.

Hunt, F. & Barker, C. (1998) *Women at Cambridge: a Brief History*. Cambridge: Cambridge University Press.

Huxley, L. (1918) *Life & Letters of Sir Joseph Dalton Hooker*, 2 vols. London: John Murray.

Ingram, D. & Robertson, N. (1999) *Plant Diseases* (New Naturalist, 85) London: Harper-Collins.

Irving, W. (1955) *Apes, Angels, & Victorians: a Joint Biography of Darwin & Huxley*. London: Weidenfeld & Nicolson.

Jenyns, L. (1833) A Monograph on the British species of Cyclas & Pisidium. *Transactions of the Cambridge Philosophical Society*, **4**, 289–312.

—— (1862) *Memoir of the Reverend John Stevens Henslow*. London: Van Voorst.

Jesperson, O. (1938) *Growth and Studies of the English Language*. Oxford: Blackwell.

Jones, D. (1976) Thomas Campbell Foster and the rural labourer; incendiarism in East Anglia in the 1840s. *Social History*, **1**, 5–37.

Killeen, I.J. (1992) *The Land and Freshwater Molluscs of Suffolk*. Ipswich: Suffolk Naturalists Society.

Kirby, W. & Spence, W. (1815–26) *An Introduction to Entomology*. 4 vols. London: Longman, Hurst, Rees, Orme, and Brown.

Knight, F. (1995) *The Nineteenth Century Church and English Society*. Cambridge: Cambridge University Press.

Leach, W.E. (1820) *Malacostraca Podophthalma Britanniae*, no. 17. London: J. Sowerby.

—— (1852) *Molluscorum Britanniae Synopsis: A Synopsis of the Mollusca of Great Britain*. London: Van Voorst.

Leedham-Green, E. (1996) *A Concise History of the University of Cambridge*. Cambridge: Cambridge University Press.

Le Keux, J. (1845) *Memorials of Cambridge: a series of views . . .* London: D. Bogue.

Le Rougetel, H. (1990) *The Chelsea Gardener: Philip Miller, 1691–1771*. London: Natural History Museum Publications.

Lewis, Jan (1992) *Walter Hood Fitch*. London: H.M.S.O.

Lindley, J. (1830) *Introduction to the Natural System of Botany*. London: Longman, Rees, Orme, Brown & Green.

—   (1832) *Introduction to Botany*. London: Longman, Rees, Orme, Brown, Green & Longman.

Lindley, J. & Hutton, W. (1831–7) *The Fossil Fauna of Great Britain*. London: J. Ridgway.

Litchfield, H.E. (ed.) (1904) *Emma Darwin: a Century of Family Letters*. 2 vols. Cambridge: Cambridge University Press.

Lloyd, F.E. (1942) *The Carnivorous Plants*. Waltham, Mass.: Chronica Botanica Co.

Locke, J. (1690) *Essay Concerning Human Understanding*. London: Basset.

Lowe, R.T. (1857) *A Manual Flora of Madeira and the Adjacent Islands . . .* Vol. 1. London: Van Voorst.

Lyell, C. (1830–3) *Principles of Geology*. 3 vols. London: John Murray.

—   (1997) *Principles of Geology*. (ed. with introduction by J. Secord) London: Penguin Classics.

Lyell, K.M. (1906) *Life of Sir Charles J.F. Bunbury*. London: John Murray.

McAlpine, D. (1989) *The Botanical Atlas: a guide to the Practical Study of Plants*, with Introduction by S.M. Walters. London: Bracken Books.

McKie, D. & Mills, J. (1998) The Gardens and Close of Jesus College. *Cambridge: the Magazine of the Cambridge Society*, **42**, 48–53.

Mallalieu, H.L. (1986) *The Dictionary of British Watercolour Artists up to 1920*. 2nd edn. Woodbridge: Antique Collectors' Club.

Markham, R.A.D. (1990) *A Rhino in the High St: Ipswich Museum – the Early Years*. Ipswich: Ipswich Borough Council.

Marr, J.E. & Shipley, A.E. (eds.) (1904) *Handbook to the Natural History of Cambridgeshire*. Cambridge: Cambridge University Press.

Martyn, J. (1729) *The First Lecture of a Course of Botany*. London: R. Reily.

Martyn, T. (1764) *Heads of a Course of Lectures on Botany, read at Cambridge . . .* London.

—   (1807) *The Language of Botany . . .* 3rd edn. London: J. White.

Mearns, B. & Mearns, R. (1992) *Audubon to Xántus: the lives of those commemorated in North American Bird Names*. London: Academic Press.

Mitch, D.F. (1992) *The Rise of Popular Literacy in Victorian England*. Philadelphia, Pa.: University of Pennsylvania Press.

Mitchell, G.F. & Ryan, M. (1986; 2nd edn 1987) *Reading the Irish landscape*. Dublin: Town House.

Morley, C. & ". . . Conchological Members" of Suffolk Naturalists' Society (1938) The Mollusca of Suffolk. *Transactions, Suffolk Naturalists' Society*, **4**, 2–22.

Morsley, C. (1961) A Victorian Parson – Professor. *Country Life*, 11 May 1961, 11–12.

Morton, A.G. (1981) *History of Botanical Science*. New York: Academic Press.

— (1986) *John Hope, 1725–86, Scottish Botanist*. Edinburgh: Edinburgh Botanic Garden (Sibbald) Trust.

Murray, A. (1848) *Hardy perennials in the New Botanic Garden*. Cambridge: Cambridge University Press.

— (1850) *A Catalogue of the Hardy Plants in the Botanic Garden, Cambridge*. Cambridge: Cambridge University Press.

Nash, R. (1990) *Scandal in Madeira: the Story of Richard Thomas Lowe*. Sussex: The Book Guild.

Natusch, S. & Swainson, G. (1988) *William Swainson of Fern Grove* . . . Wellington, N.Z.: New Zealand Founders Society.

Naval Chronicle, 1–40. 1799–1818. London.

Nockles, P.B. (1994) *The Oxford Movement in Context: Anglican High Churchmanship, 1760–1857*. Cambridge: Cambridge University Press.

Oliver, D. (1864) *Lessons in Elementary Botany*. London: Clay, Son & Taylor.

Otter, W. (1824) *Life and Remains of Edward Daniel Clarke*. London: J.F. Dove.

Paley, W. (1818) *A View of the Evidences of Christianity*. Derby: H. Morley.

Pearsall, W.H. (1950; 1971 edn.) *Mountains and Moorlands* (New Naturalist, 11) London: Collins.

Perring, F.H. et al. (1964) *Flora of Cambridgeshire*. Cambridge: Cambridge University Press.

Pettit-Stevens, H.W. *see* Stevens, H.W. Pettit.

Plarr, V.G. (1930) *Plarr's Lives of the Fellows of the Royal College of Surgeons*. Rev. by D'A. Power. 2 vols. London: Royal College of Surgeons.

Pollard, E. (1974) Distribution maps of *Helix pomatia*. *Journal of Conchology*, **28**, 239–42.

— (1975) Aspects of the ecology of *Helix pomatia*. *Journal of Animal Ecology*, **44**, 305–29.

Porter, D.M. (1980) The Vascular plants of Joseph Dalton Hooker's: an enumeration of the plants of the Galapagos Archipelago . . . *Botanical Journal of the Linnean Society*, **81**, 79–134.

Porter, T.M. (1986) *The Rise of Statistical Thinking, 1820–1900*. Cambridge: Cambridge University Press.

Preston, C.D. (1993) The distribution of the oxlip *Primula elatior* (L.) Hill in Cambridgeshire. *Nature in Cambridgeshire*. **35**, 29–60.

Proctor, M., Yeo, P. & Lack, A. (1996) *The Natural History of Pollination.* (New Naturalist, 83) London: HarperCollins.

Ransford, C. (1846) *Biographical sketch of the late Robert Graham, M.D., F.R.S.E.* Edinburgh: Harveian Society.

Raven, C.E. (1950) *John Ray, Naturalist: His Life & Works.* 2nd edn. Cambridge: Cambridge University Press.

Raverat, G. (1952; 1960 edn) *Period Piece: a Cambridge childhood.* London: Faber & Faber.

Report of H.M. Commissioners . . . see Graham.

Roberts, M. (1998) Geology and genesis unearthed. *The Churchman,* **112**, 225–55.

Ross, S. (1962) Scientist: the story of a word. *Annals of Science,* **18**, 65–85.

Royal Commission on Historical Monuments (1972) *An Inventory of Historical Monuments in the County of Cambridge, Vol. 2: North-East Cambridgeshire.* London: H.M. Stationery Office.

Russell, A. (1980) *The Clerical Profession.* London: S.P.C.K.

— (1986) *The Country Parish.* London: S.P.C.K.

— (1993) *The Country Parson.* London: S.P.C.K.

Russell-Gebbett, J. (1977) *Henslow of Hitcham.* Lavenham: Terence Dalton.

Salaman, R.N. (1949) *History and Social Influence of the Potato.* Cambridge: Cambridge University Press.

Schultes, J.A. *see* Hooker, W.J. (1830).

Smith, J.E. (1795) *Syllabus of a Course of Lectures on Botany.* London: printed for the author.

— (1807) *An Introduction to Physiological & Systematical Botany.* London: Longman, Hurst, Rees & Orme.

— (1818) *Considerations respecting Cambridge, more particularly relating to its Botanical Professorship.* London: Longman, Hurst, Rees, Orme & Brown.

— (1824–36) *English Flora.* 5 vols London: Longman, Hurst, Rees, Orme, Brown & Green.

Smith, J.J. (ed.) (1840) *The Cambridge Portfolio.* London: J.W. Parker.

Smythe, C. (1939) A Vanishing Inscription. *Cambridge Review,* 2 June 1939, 445–7.

Spater, G. (1982) *William Cobbett: the Poor Man's Friend.* 2 vols. Cambridge: Cambridge University Press.

Speakman, C. (1982) *Adam Sedgwick, Geologist and Dalesman, 1785–1873.* London & Cambridge: Broad Oak Press.

Bibliography

Sprengel, C.K. (1793) *Das entdeckte Geheimnis der Natur im Bau in der Befruchtung der Blumen.* Berlin: Friedrich Vieweg dem aeltern.

Stafleu, F.A. & Cowan, R.S. (1976–) *Taxonomic literature.* Utrecht: Bohn, Scheltema & Holkema.

Stair-Douglas, Mrs (1881) *The Life and Selections from the Correspondence of William Whewell, D.D.* London: Kegan Paul.

Stevens, H.W. Pettit (1899) *Downing College.* London: F.E. Robinson.

Stevens, P.F. (1994) *The Development of Biological Systematics.* New York: Columbia University Press.

Trevelyan, R. (1978) *A Pre-Raphaelite Circle.* London: Chatto & Windus.

Van Riper, A.B. (1993) *Men among Mammoths: Victorian Science and the Discovery of Human Pre-history.* Chicago: University of Chicago Press.

Venn, J.A. (1940–54) *Alumni Cantabrigiensis.* Pt. II, **1–6**: 1752–1900. Cambridge: Cambridge University Press.

Victoria History of the County of Cambridgeshire and the Isle of Ely. Vols. 1–9. London: Oxford University Press.

Vidler, A.R. (1961; repr. 1990) *The Church in an Age of Revolution.* London: Penguin.

Virgin, P. (1989) *The Church in an Age of Negligence.* Cambridge: James Clarke & Co.

Walker, M. (1988) *Sir James Edward Smith, 1759–1828.* London: Linnean Society of London.

Wallich, N. (1830–32) *Plantae Asiaticae Rariores.* 3 vols. London, Paris, Strasburgh: Treuttel & Würtz.

Walters, S.M. (1981; repr. 1996) *The Shaping of Cambridge Botany.* Cambridge: Cambridge University Press.

Walters, S.M. *see also* Briggs, D. & Walters, S.M.

Ward, K. (1996) *God, Chance and Necessity.* Oxford: One World Publications.

Welch, D. (1961) Water forget-me-nots in Cambridgeshire. *Nature in Cambridgeshire,* **4**, 18–27.

Whewell, W. (1837) *The History of the Inductive Sciences.* 3 vols. London: Parker; Cambridge: Deighton.

Wiggins, I.L. & Porter, D.M. (1971) *Flora of the Galapagos Islands.* Stanford, Ca.: Stanford University Press.

Winstanley, D.A. (1940) *Early Victorian Cambridge.* Cambridge: Cambridge University Press.

Wolffe, J. (ed.) (1995) *Evangelical Faith & Public Zeal: Evangelicals and Society In Britain, 1780–1980.* London: S.P.C.K.

328

Woodward, H.B. (1907) *History of the Geological Society of London*. London: Geological Society.

Woodward, S.P. & Barrett, L. (1858) On the genus Synapta. *Proceedings of the Zoological Society of London*, **26**, 360–67, plate xiv.

Wright, J.M.F. (1827) *Alma Mater, or Seven Years at the University of Cambridge*. 2 vols. London: Black, Young & Young.

Wymer, J.J. & West, R.G. (2000) A Memorial to John Frere. *Quaternary Newsletter (Quaternary Research Association)*, **90**, 27–29.

Youmans, E.A. (1872) *An Essay on the Culture of the Observing Powers of Children Especially in Connection with the Study of Botany*. Edited with notes . . . by Joseph Payne.

— (1880) *The First Book of Botany, Designed to Cultivate the Observing Powers of Children*. New [London] edition. London: Kegan Paul.

III: General Reading

Chadwick, O. (1975) *The Secularisation of the European Mind in the 19th Century*. Cambridge: Cambridge University Press.

Darby, H.C. (ed.) (1938) *A Scientific Survey of the Cambridge District: prepared for Cambridge Meeting of British Association for the Advancement of Science*. London: B.A.A.S.

Ford, A. (1988) *John James Audubon: a Biography*. New York: Abbeville Press.

Jardine, N., Secord, J.A. & Spary, E.C. (eds.) (1996) *Cultures of Natural History*. Cambridge: Cambridge University Press.

Keynes, R.D. (1979) *The Beagle Record: Selections from the Original Pictorial Records and Written Accounts of the Voyage of H.M.S. Beagle*. Cambridge: Cambridge University Press.

Layton, D. (1973) *Science for the people: the Origins of the School Science Curriculum in England*. London: George Allen & Unwin.

Mabberly, D.J. (1985) *Jupiter Botanicus: Robert Brown of the British Museum*. Braunschweig: Cramer; London: British Museum (Natural History).

McLeod, H. (1984) *Religion and the Working Class in Nineteenth-century Britain*. London: MacMillan.

Morrell, J. & Thackray, A. (1981) *Gentlemen of Science: Early Years of the British Association for the Advancement of Science*. Oxford: Clarendon Press.

O'Brien, M. (1995) *'Perish the Privileged Orders': a Socialist History of the Chartist Movement*. London: Redwords.

Raven, C.E. (1953) *Natural Religion and Christian Theology, Vol. 1: Science and Religion; Vol. 2: Experience & Interpretation.* (Gifford Lectures, 1951 and 1952) Cambridge: Cambridge University Press.

Rehbock, P.F. (1983) *The Philosophical Naturalists.* Wisconsin: University Press.

Steers, J.A. (ed.) (1965) *The Cambridge Region: prepared for the Meeting of the British Association . . . 1965.* Cambridge: B.A.A.S.

Taylor, A. (1998) *Archaeology of Cambridgeshire, Vol. 2: South East Cambridgeshire & the Fen Edge.* March: Cambridgeshire County Council.

Thomas, K. (1983) *Man and the Natural World.* London: Allen Lane.

Thompson, E.P. (1991) *The Making of the English Working Class.* London: Penguin.

Thomson, D. (1950) *England in the Nineteenth Century.* (Pelican History of England, 8) London: Penguin.

Wardle, D. (1970) *English Popular Education, 1780–1970.* Cambridge: Cambridge University Press.

Yeo, R. (1993) *Defining Science: William Whewell, Material Knowledge, and Public Debate in Early Victorian Britain.* Cambridge: Cambridge University Press.

Index

In addition to the main text, Appendices 4 and 5 have been indexed, and substantial references have also been added from the end-notes. Black-and-white Figures are indicated by italic page numbers, and coloured Plates are given in bold.